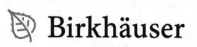

# Applications of Unmanned Aerial Vehicles in Geosciences

Edited by
Tomasz Niedzielski

Previously published in *Pure and Applied Geophysics* (PAGEOPH),
Volume 175, No. 9, 2018

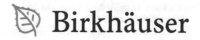

*Editor*
Tomasz Niedzielski
Department of Geoinformatics
and Cartography
University of Wrocław
Wrocław, Poland

ISSN 2504-3625
ISBN 978-3-030-03170-1

Library of Congress Control Number: 2018960748

Cover illustration: The photo has been taken and prepared by Dr. Waldemar Spallek

This book is published under the imprint Birkhäuser, www.birkhauser-science.com by the registered company Springer Nature Switzerland AG
The registered company address is: Gewerbestrasse 11, 6330 Cham, Switzerland

# Contents

Applications of Unmanned Aerial Vehicles in Geosciences: Introduction.................. 1
*Tomasz Niedzielski*

Aeromagnetic Surveying with a Rotary-Wing Unmanned Aircraft System:
A Case Study from a Zinc Deposit in Nash Creek, New Brunswick, Canada .......... 5
*Michael Cunningham, Claire Samson, Alan Wood and Ian Cook*

Detecting Surface Changes from an Underground Explosion in Granite Using
Unmanned Aerial System Photogrammetry ..................................... 19
*Emily S. Schultz-Fellenz, Ryan T. Coppersmith, Aviva J. Sussman, Erika M. Swanson and
James A. Cooley*

Applicability of Unmanned Aerial Vehicles in Research on Aeolian Processes ............. 39
*Algimantas Česnulevičius, Artūras Bautrėnas, Linas Bevainis, Donatas Ovodas and Kęstutis Papšys*

UAV and SfM in Detailed Geomorphological Mapping of Granite Tors: An Example of
Starościńskie Skały (Sudetes, SW Poland) ..................................... 53
*Marek Kasprzak, Kacper Jancewicz and Aleksandra Michniewicz*

Geovisualisation of Relief in a Virtual Reality System on the Basis of Low-Level
Aerial Imagery ......................................................... 69
*Łukasz Halik and Maciej Smaczyński*

Detection and Mapping of the Geomorphic Effects of Flooding Using UAV
Photogrammetry ....................................................... 83
*Jakub Langhammer and Tereza Vacková*

Assessment of Riverine Morphology and Habitat Regime Using Unmanned Aerial
Vehicles in a Mediterranean Environment ..................................... 107
*Elias Dimitriou and Eleni Stavroulaki*

Application of Low-Cost Fixed-Wing UAV for Inland Lakes Shoreline Investigation ........ 123
*Tomasz Templin, Dariusz Popielarczyk and Rafał Kosecki*

Automated Snow Extent Mapping Based on Orthophoto Images from Unmanned
Aerial Vehicles ........................................................ 145
*Tomasz Niedzielski, Waldemar Spallek and Matylda Witek-Kasprzak*

Multitemporal Accuracy and Precision Assessment of Unmanned Aerial System
Photogrammetry for Slope-Scale Snow Depth Maps in Alpine Terrain ............. 163
*Marc S. Adams, Yves Bühler and Reinhard Fromm*

UAS as a Support for Atmospheric Aerosols Research: Case Study...................... 185
*Michał T. Chiliński, Krzysztof M. Markowicz and Marek Kubicki*

**Can Clouds Improve the Performance of Automated Human Detection in Aerial Images?** .................................................................. **203**

Tomasz Niedzielski and Mirosława Jurecka

**Technical Report: Unmanned Helicopter Solution for Survey-Grade Lidar and Hyperspectral Mapping** ....................................................... **217**

Ján Kaňuk, Michal Gallay, Christoph Eck, Carlo Zgraggen and Eduard Dvorný

**Technical Report: The Development and Experience with UAV Research Applications in Former Czechoslovakia (1960s–1990s)** ..................... **235**

Jaromír Kolejka and Ladislav Plánka

Pure Appl. Geophys. 175 (2018), 3141–3144
© 2018 The Author(s)
https://doi.org/10.1007/s00024-018-1992-9

**Pure and Applied Geophysics**

# Applications of Unmanned Aerial Vehicles in Geosciences: Introduction

TOMASZ NIEDZIELSKI[1]

In the last decade, unmanned aerial vehicles (UAVs)—informally known as drones—became standard tools for acquiring spatial data to support various geoscientific analyses. Although the most common applications of drones are associated with acquiring aerial images and processing them using photogrammetric methods to make maps, the miniaturization of geophysical instruments has recently opened new opportunities to install them on-board drones (Hatch 2017). Modern UAVs have longer endurance than their predecessors and therefore heavier payload, like geophysical sensors, can be hung under drones. Geophysicists make use of drone-based mapping (e.g., Bemis et al. 2014) as well as employ various UAV-mounted geophysical sensors (Hatch 2017). Nowadays, drones enable the measurements of the Earth's subsurface and allow geoscientists to observe its surface, the hydrosphere or the troposphere.

The most common geoscientific application of UAVs uses aerial imagery to reconstruct the Earth's elevations through producing digital surface models and to reconstruct land cover through generating orthophotos. The two products are produced by the Structure-from-Motion (SfM) algorithm (Westoby et al. 2012) which generates sparse and dense point clouds on a basis of visible-light or near-infrared aerial images. The increase in the quality of georeferencing, achieved mainly through the use of ground control points and real-time kinematic technology, led to the reproducibility and repeatability of multi-temporal spatial data (Clapuyt et al. 2016). As a consequence, the UAV-based multi-temporal digital surface models and orthophotos became suitable for detecting changes in the Earth's surface. These UAV-based maps can be used as base maps, as noticed by an environmental geophysicist (Hatch 2016).

Not only digital cameras but also more advanced geophysical sensors are also mounted on-board drones. Underground survey can be carried out, for instance, by UAV-borne ground penetrating radar (Chandra and Tanzi 2018) or drone-mounted magnetometers (Versteeg et al. 2007). Underwater survey may be conducted using UAVs equipped with the light detection and ranging (LIDAR) sensor, with laser pulses in green domain of the spectrum (Mandlburger et al. 2016). The Earth's surface may also be monitored by standard LIDAR (Lin et al. 2011) or multispectral cameras (Ahmed et al. 2017). Observations of the troposphere are conducted using dedicated meteorological sensors mounted on-board drones (Spiess et al. 2007) or utilizing standard UAV equipment such as the Pitot tubes and the Global Navigation Satellite System receivers (Niedzielski et al. 2017).

The variety of UAV applications—covering the investigations into the Earth's subsurface, its surface, the hydrosphere and the troposphere—were discussed during the 23rd Cartographic School "Applications of unmanned aerial vehicles in geosciences" which took place in Świeradów-Zdrój (Poland) on 8–10 June 2016. In the aftermath of the conference and the associated workshop, its participants and other researchers contributed to this topical issue of Pure and Applied Geophysics.

Cunningham et al. (2018) carried out a demonstration UAV survey of zinc deposits in Nash Creek, New Brunswick, Canada. They approached the problem from a geophysical perspective and utilized a cesium vapor magnetometer mounted on-board a

[1] Department of Geoinformatics and Cartography, Faculty of Earth Sciences and Environmental Management, University of Wrocław, pl. Uniwersytecki 1, 50-137 Wrocław, Poland. E-mail: tomasz.niedzielski@uwr.edu.pl

UAV. The paper contributes to our understanding of potentials and limitations of UAV-based magnetic survey, especially with respect to well-established ground survey and manned aircraft survey. Yet another UAV-supported analysis of underground activity was presented by Schultz-Fellenz et al. (2018). The authors investigated the influence of controlled underground chemical explosion in granite on uplift, subsidence, surface fractures and morphological change. The SfM-based pre- and post-shot digital surface models allowed the authors to quantify surface changes of the physical terrain. The authors claimed that the SfM-based methodology "provides valuable data to link with other geological and geophysical techniques".

Change detection based on UAV digital surface models can be a very difficult problem when the physical terrain is highly dynamic and unstable. Such difficulties occur while investigating aeolian environments such as, for instance, sand dunes. Česnulevičius et al. (2018) attempted to utilize the SfM method for dune mapping and inferring its dynamics. Based on the analysis of data collected in the Curonian Spit, they formulated a few recommendations about the usability of UAVs for monitoring aeolian environments. The problem of surface stability does not occur on rock outcrops as presented by Kasprzak et al. (2018). They carried out a feasibility study about the use of the SfM method for mapping granite tors, on the example of Starościńskie Skały in the Sudetes in southwestern Poland. They concluded that the approach enables the identification of rock micro-topography, including complex joint systems, weathering pits, rills and karrens. The SfM method can therefore complement indirect geophysical measurements of rock joint geometry. Visualization of Earth's topography is an important element of the inference on its changes. Halik and Smaczyński (2018) integrated the virtual reality technology with UAV-based digital surface models, and visualized the natural aggregate mine in western Poland. For many years, virtual reality has been perceived as an added value in exploration geophysics (Midttun et al. 2000), and thus its integration with the UAV-based digital surface models may be used to support geophysical surveying of the physical terrain.

The UAV-supported determination of the dynamics of fluvial forms, which impact hydrological processes, may enhance or validate results obtained using hydrogeophysical methods. Langhammer and Vacková (2018) investigated fluvial landforms that are formed or modified by flooding. The authors studied the specific snow-melt flood episode that occurred in December 2015 in the Šumava Mountains in Czechia. Riverine morphology was also the topic of the paper by Dimitriou and Stavroulaki (2018) who used SfM to determine spatial and temporal patterns of erosion and deposition in the Spercheios river basin in Greece. Another water-related problem was tackled by Templin et al. (2018) who conducted a detailed fieldwork in the Suskie Lake in northern Poland, and determined its shoreline using a UAV survey.

Also, UAVs contributed to geophysics of snow. Niedzielski et al. (2018) utilized the UAV-based orthophotos from study sites located in the Izerskie Mountains in southwestern Poland to estimate snow extent. The k-means classification was applied to discriminate between snow-free and snow-covered terrain. Adams et al. (2018) addressed the issue of accuracy and precision of snow depth reconstructions using SfM, with a particular emphasis put on the slope scale. They carried out the field experiment in the Tuxer Alps in Austria.

Meteorological applications of UAVs go beyond spatial survey of terrain and include the production of vertical profiles of the atmosphere. Chiliński et al. (2018) carried out a case study of atmospheric aerosols at two stations (Świder and Warsaw) located in Poland. They utilized a UAV with micro-aethalometer and radiosonde on-board, and offered a concept of sounding atmospheric aerosols using UAVs. Atmospheric phenomena, such as for instance clouds, usually influence remote sensing data. Niedzielski and Jurecka (2018) carried out a field experiment in the Izerskie Mountains in southwestern Poland, the aim of which was to automatically detect persons in photographs taken by a UAV operating in the bottom of low-altitude clouds. They argued that there exists a certain level of intermediate cloud-driven image blurriness for which the performance of the search algorithm increases.

The majority of the described studies made use of micro UAVs, carrying limited payload. Rarely, geoscientists utilize mini UAVs, with maximum take-off weight up to a few tens of kilograms, with a considerable payload capacity. One of such examples was presented in the technical report by Kaňuk et al. (2018) who used an unmanned helicopter with on-board LIDAR and camera or on-board hyperspectral pushbroom scanner, the sensors which are also dedicated to carry out geophysical measurements. This topical issue ends up with a historical view of drones. Kolejka and Plánka (2018) offered the technical report on the development of UAVs in former Czechoslovakia, beginning their story in mid-1960s. They published photographs of first remotely controlled aircrafts in that country and published historical aerial data collected by visible-light and multispectral cameras mounted on UAVs.

This topical issue highlights recent advances in UAV applications in geosciences, including the use of various on-board cameras and sensors as well as data processing methods. Since the volume is a selection of case studies carried out with the use of UAVs, it may be interesting to scientists and, predominantly, to students and practitioners who will be provided with the materials to follow real-world examples of the UAV applications in Earth sciences. The case studies cover a wide range of problems solved with the support offered by drones, including the monitoring of surface and subsurface of terrain as well as the investigations into phenomena acting in the hydrosphere and the troposphere.

### Acknowledgements

The Guest Editor would like to thank Dr Renata Dmowska, the Editor-in-Chief for topical issues and book reviews of Pure and Applied Geophysics, for offering valuable support during the editorial work. Special thanks should also go to all reviewers who read the manuscripts and offered valuable remarks. Last but not least, Dr Ismail Gultepe and Prof. Andrzej Icha are acknowledged for handling Guest Editor's manuscripts.

### REFERENCES

Adams, M. S., Bühler, Y., & Fromm, R. (2018). Multitemporal accuracy and precision assessment of unmanned aerial system photogrammetry for slope-scale snow depth maps in alpine terrain. *Pure and Applied Geophysics*. https://doi.org/10.1007/s00024-017-1748-y.

Ahmed, O. S., Shemrock, A., Chabot, D., Dillon, C., Williams, G., Wasson, R., et al. (2017). Hierarchical land cover and vegetation classification using multispectral data acquired from an unmanned aerial vehicle. *International Journal of Remote Sensing, 38,* 2037–2052.

Bemis, S. P., Micklethwaite, S., Turner, D., James, M. R., Akciz, S., Thiele, S. T., et al. (2014). Ground-based and UAV-based photogrammetry: A multi-scale, high-resolution mapping tool for structural geology and paleoseismology. *Journal of Structural Geology, 69A,* 163–178.

Česnulevičius, A., Bautrėnas, A., Bevainis, L., Ovodas, D., & Papšys, K. (2018). Applicability of unmanned aerial vehicles in research on aeolian processes. *Pure and Applied Geophysics*. https://doi.org/10.1007/s00024-018-1785-1.

Chandra, M., & Tanzi, T. J. (2018). Drone-borne GPR design: Propagation issues. *Comptes Rendus Physique, 19,* 72–84.

Chiliński, M. T., Markowicz, K. M., & Kubicki, M. (2018). UAS as a support for atmospheric aerosols research: Case study. *Pure and Applied Geophysics*. https://doi.org/10.1007/s00024-018-1767-3.

Clapuyt, F., Vanacker, V., & Van Oost, K. (2016). Reproducibility of UAV-based earth topography reconstructions based on structure-from-motion algorithms. *Geomorphology, 260,* 4–15.

Cunningham, M., Samson, C., Wood, A., & Cook, I. (2018). Aeromagnetic surveying with a rotary-wing unmanned aircraft system: A case study from a zinc deposit in Nash Creek, New Brunswick, Canada. *Pure and Applied Geophysics*. https://doi.org/10.1007/s00024-017-1736-2.

Dimitriou, E., & Stavroulaki, E. (2018). Assessment of riverine morphology and habitat regime using unmanned aerial vehicles in a Mediterranean environment. *Pure and Applied Geophysics*. https://doi.org/10.1007/s00024-018-1929-3.

Halik, Ł., & Smaczyński, M. (2018). Geovisualisation of relief in a virtual reality system on the basis of low-level aerial imagery. *Pure and Applied Geophysics*. https://doi.org/10.1007/s00024-017-1755-z.

Hatch, M. (2016). Environmental geophysics: Using drones to create base maps. *Preview, 185,* 31–32.

Hatch, M. (2017). Environmental geophysics: Developments in miniaturisation technology. *Preview, 189,* 32–33.

Kaňuk, J., Gallay, M., Eck, C., Zgraggen, C., & Dvorný, E. (2018). Technical report: Unmanned helicopter solution for survey-grade lidar and hyperspectral mapping. *Pure and Applied Geophysics.* https://doi.org/10.1007/s00024-018-1873-2.

Kasprzak, M., Jancewicz, K., & Michniewicz, A. (2018). UAV and SfM in detailed geomorphological mapping of granite tors: An example of Starościńskie Skały (Sudetes, SW Poland). *Pure and Applied Geophysics.* https://doi.org/10.1007/s00024-017-1730-8.

Kolejka, J., & Plánka, L. (2018). Technical report: The development and experience with UAV research applications in former Czechoslovakia (1960s–1990s). *Pure and Applied Geophysics.* https://doi.org/10.1007/s00024-018-1807-z.

Langhammer, J., & Vacková, T. (2018). Detection and mapping of the geomorphic effects of flooding using UAV photogrammetry. *Pure and Applied Geophysics.* https://doi.org/10.1007/s00024-018-1874-1.

Lin, Y., Hyyppa, J., & Jaakkola, A. (2011). Mini-UAV-borne LIDAR for fine-scale mapping. *IEEE Geoscience and Remote Sensing Letters, 8,* 426–430.

Mandlburger, G., Pfennigbauer, M., Wieser, M., Riegl, U., & Pfeifer, N. (2016). Evaluation of a novel UAV-borne topobathymetric laser profiler. *The International Archives of Photogrammetry, Remote Sensing and Spatial Information Sciences, 41,* 933.

Midttun, M., Helland, R., & Finnstrom, E. (2000). Virtual reality—adding value to exploration and production. *The Leading Edge, 19,* 538–544.

Niedzielski, T., & Jurecka, M. (2018). Can clouds improve the performance of automated human detection in aerial images? *Pure and Applied Geophysics.* https://doi.org/10.1007/s00024-018-1931-9.

Niedzielski, T., Skjøth, C., Werner, M., Spallek, W., Witek, M., Sawiński, T., et al. (2017). Are estimates of wind characteristics based on measurements with Pitot tubes and GNSS receivers mounted on consumer-grade unmanned aerial vehicles applicable in meteorological studies? *Environmental Monitoring and Assessment, 189,* 431.

Niedzielski, T., Spallek, W., & Witek-Kasprzak, M. (2018). Automated snow extent mapping based on orthophoto images from unmanned aerial vehicles. *Pure and Applied Geophysics.* https://doi.org/10.1007/s00024-018-1843-8.

Schultz-Fellenz, E. S., Coppersmith, R. T., Sussman, A. J., Swanson, E. M., & Cooley, J. A. (2018). Detecting surface changes from an underground explosion in granite using unmanned aerial system photogrammetry. *Pure and Applied Geophysics.* https://doi.org/10.1007/s00024-017-1649-0.

Spiess, T., Bange, J., Buschmann, M., & Vörsmann, P. (2007). First application of the meteorological Mini-UAV 'M2AV'. *Meteorologische Zeitschrift, 16,* 159–169.

Templin, T., Popielarczyk, D., & Kosecki, R. (2018). Application of low-cost fixed-wing UAV for inland lakes shoreline investigation. *Pure and Applied Geophysics.* https://doi.org/10.1007/s00024-017-1707-7.

Versteeg, R., McKay, M., Anderson, M., Johnson, R., Selfridge, B., & Bennett, J. (2007). *Feasibility study for an autonomous UAV-magnetometer system (No. INL/EXT-07-13386).* Idaho Falls: Idaho National Lab.

Westoby, M. J., Brasington, J., Glasser, N. F., Hambrey, M. J., & Reynolds, J. M. (2012). 'Structure-from-Motion' photogrammetry: A low-cost, effective tool for geoscience applications. *Geomorphology, 179,* 300–314

(Published online  September 12, 2018)

Pure Appl. Geophys. 175 (2018), 3145–3158
© 2017 Springer International Publishing AG, part of Springer Nature
https://doi.org/10.1007/s00024-017-1736-2

Pure and Applied Geophysics

CrossMark

# Aeromagnetic Surveying with a Rotary-Wing Unmanned Aircraft System: A Case Study from a Zinc Deposit in Nash Creek, New Brunswick, Canada

MICHAEL CUNNINGHAM,[1] CLAIRE SAMSON,[1] ALAN WOOD,[2] and IAN COOK[2]

*Abstract*—Unmanned aircraft systems (UASs) have been under rapid development for applications in the mineral exploration industry, mainly for aeromagnetic surveying. They provide improved detection of smaller, deeper and weaker magnetic targets. A traditional system flying an altitude of 100 m above ground level (AGL) can detect a spherical ore body with a radius of $\sim$ 16 m and a magnetic susceptibility of $10^{-4}$ buried at a depth of 40 m. A UAS flying at an altitude of 50 or 2 m AGL would require the radius to be 11 or 5 m, respectively. A demonstration survey was performed using the SkyLance rotary-wing UAS instrumented with a cesium vapour magnetometer in Nash Creek, New Brunswick, Canada. The UAS flew over a zinc deposit featuring three magnetic anomalies. It acquired repeatable data that compared well with upward continuation maps of ground magnetic data. Dykes or faults that are dipping eastward at 25° and are approximately 1.5 m wide fit the observed response of the three anomalies captured on the UAS magnetic data.

**Key words:** Unmanned aircraft systems, Magnetics, Airborne surveying, Mineral exploration.

## 1. Introduction

Over the past decade there has been rapid development of unmanned aircraft systems (UASs) as well as of their payloads (Pajares 2015). UASs are becoming a desirable alternative or a complementary approach for remote sensing since they allow for high versatility and flexibility in comparison to traditional airborne and satellite systems. UAS surveys require smaller team sizes than their traditional counterparts and remove many risks for the pilots, such as low altitude flying and operator fatigue (Kroll 2013).

Furthermore, UASs are also capable of flying at very low altitudes (some as low as 2 m AGL) and at much slower flight speeds, giving them the ability to acquire higher resolution data, spatially and temporally.

The geophysical magnetic method is extensively used to map several different mineral deposits, such as: volcanic massive sulphides; mafic and ultra-mafics; porphyry copper and molybdenum; magmatic bodies; and skarns (Dentith and Mudge 2014). Currently, magnetic surveying with UASs is gaining momentum in the mineral exploration industry. There is a demand for cost-effective, high-resolution, small-size surveys to be flown at lower altitudes than by traditional methods. UASs represent an opportunity to conduct such surveys. Presently, there are various UASs under development for aeromagnetic surveying. Initially many of these systems were fixed-wing UASs such as the GeoRanger (Fugro/CGG—now retired), which was flown for several missions (Anderson and Pita 2005); the GeoSurv II (Sander Geophysics and Carleton University) (Caron et al. 2011, 2014; Samson et al. 2010); and the AeroVision (Abitibi Geophysics and GEM Systems) (Dion-Ortega 2015). Recently, rotary-wing UASs have become increasingly popular, including: the Scout B1-100 UAV helicopter (Aeroscout) (Eck and Imbach 2011); the UAV-MAG (Pioneer Aerial Surveys Ltd.) (Burns 2017; Parvar 2016); and the AirBIRD (GEM Systems) (Gordon 2016). Stratus Aeronautics has also developed two UASs for aeromagnetic surveying; the fixed-wing Venturer UAS and the SkyLance rotary-wing UAS (Wood et al. 2016; Cunningham 2016).

This paper presents theoretical and field survey results that aim to assess the capabilities of UASs

[1] Department of Earth Sciences, Carleton University, Ottawa, ON, Canada. E-mail: Michael.Cunningham@carleton.ca
[2] Stratus Aeronautics, #123, 3191 Thunderbird Crescent, Burnaby, BC V5A 3G1, Canada.

compared to traditional magnetic surveying systems and methods, specifically for mineral exploration purposes. Firstly, forward modelling is used to determine the detection capabilities of UASs in general. Secondly, comparisons between a ground magnetic survey and an aeromagnetic survey performed by the SkyLance UAS over a zinc deposit in Nash Creek, New Brunswick, Canada, are presented.

## 2. Theory

Three magnetic survey platforms were considered for the theoretical analysis (Table 1). The Cessna Grand Caravan was selected to represent a traditional fixed-wing survey platform. The Venturer UAS (Wood et al. 2016; Cunningham 2016) and the SkyLance UAS (Fig. 1) were selected to represent a fixed-wing UAS and a rotary-wing UAS, respectively.

The objective of the theoretical analysis is to determine the minimum detection limits and the spatial resolution of each system. The detection limits encompass the minimum detectable size and magnetic susceptibility of a body, and the maximum depth below ground level that the body can still be detectable by each system. The spatial resolution is the minimum separation between two magnetic bodies for which each body can be detected individually within the resultant signal (Elkins and Hammer 1938). Two bodies that have a large enough

Figure 1
The SkyLance rotary-wing UAS mounted with cesium vapour magnetometer at the end of a boom

separation will create two distinct peaks in the magnetic data. As the bodies are brought closer, the signals from each body will progressively blend together. For this analysis, the following parameters were considered: (1) the design parameters of each system; (2) mathematical models of spherical magnetic ore bodies; (3) a realistic noise level; and (4) realistic spatial sampling intervals.

### 2.1. Magnetic Model

The magnetic spherical anomaly was selected for this analysis due to the simplicity of the derivation and calculations compared to that of other geological features (i.e. thin sheets, cylinders). Derived from

Table 1

*Specification of the three aeromagnetic survey systems*

| System | Cessna Grand Caravan | Venturer UAS | SkyLance UAS |
|---|---|---|---|
| Length | 12.67 m | 2.74 | 1.0 m |
| Wingspan | 19.03 m | 4.95 | 1.0 m |
| Crew size | 2 + people | 2 people | 2 people |
| Mass | Approx. 3995 kg | Approx. 55 kg (with fuel) | Approx. 20 kg |
| Mass of payload | Variable | 8.28 kg | 5 kg |
| Endurance | 9–10 h at 200 km/h | 9–10 h at 100 km/h | 30 min at 35 km/h |
| Max. flight speed | 343 km/h | 120 km/h | 37 km/h |
| Sensors | Four geometrics cesium vapour magnetometers | Two geometrics G-823A cesium vapour magnetometer | Geometrics G-823A cesium vapour magnetometer |
| Flight speed (km/h) | 200 | 100 | 32 |
| Spatial sampling interval (m) | 5.6 | 2.8 | 0.9 |

Spatial sampling intervals have been calculated for a 10-Hz sampling frequency

Telford et al. (1976), a three-dimensional magnetic model was built. The magnetic anomaly induced by the Earth's magnetic field is described in three Cartesian components: $B_x$, $B_y$, and $B_z$:

$$B_x = \frac{4\pi R^3 kD}{3(x^2 + y^2 + z^2)^{5/2}} \left[ H_0 \sin(\beta)\left(2x^2 - y^2 - z^2\right) \right.$$
$$\left. - H_0 \cos(\beta)3xy + Z_0 3xz \right], \tag{1}$$

$$B_y = \frac{4\pi R^3 kD}{3(x^2 + y^2 + z^2)^{5/2}} \left[ H_0 \sin(\beta)3xy - H_0 \cos(\beta) \right.$$
$$\left. \left(-x^2 + 2y^2 - z^2\right) + Z_0 3xz \right], \tag{2}$$

and

$$B_z = \frac{4\pi R^3 kD}{3(x^2 + y^2 + z^2)^{5/2}} \left[ H_0 \sin(\beta)3xz - H_0 \cos(\beta)3yz \right.$$
$$\left. + Z_0\left(-x^2 - y^2 + 2z^2\right) \right]. \tag{3}$$

Variables $H_0$ and $Z_0$ are the horizontal and vertical components of the external Earth's magnetic field, respectively, $\beta$ is the declination of the Earth's magnetic field, $k$ is the magnetic susceptibility of the body, $R$ is the body radius, and $x$, $y$, and $z$ are the spatial coordinates of the location in space where the magnetic field is being calculated. The $x$ and $y$ coordinates represent the horizontal position of the location of interest, and the $z$ coordinate represents vertical position. In this case, the vertical position is the sum of the flight altitude of the survey system under consideration and the limiting depth of the spherical body. $D$ is the demagnetization effect (Eq. 4), which results in a reduction of the magnetic field depending on the shape of the magnetic body. $N$ varies depending on the shape of the body under consideration. No demagnetization occurs for a rod-like body with magnetization along the long-axis and with a small cross-section compared to its length. A maximum demagnetization occurs for a thin sheet that is magnetized in a direction normal to its surface. For a spherical body, $N = 4\pi/3$ and

$$D = \frac{1}{1 + Nk} \text{ where } (0 \leq N \leq 4\pi). \tag{4}$$

The resultant total magnetic field is shown in Eq. 5,

$$B_{\text{model}} = \sqrt{B_x^2 + B_y^2 + B_z^2}. \tag{5}$$

Aeromagnetic noise was characterized using data from a test flight of the Venturer UAS performed on October 2, 2013 (Wood et al. 2016) to provide a realistic scenario. The fourth difference levels of the Venturer were within industry standards of $\pm$ 0.05 nT (Coyle et al. 2014). Through the use of long wavelength filters, regional and geological effects were filtered out so that only noise from wing flexing remained. The filtered data follow a normal distribution centred at 0 nT and with a standard deviation of 0.059 nT. For analyses to follow, random noise following this distribution was added to synthetic data. The spatial sampling intervals for the three survey systems of interest are also provided in Table 1. Small anomalies will go undetected if the spatial sampling interval is too coarse.

The Chi-squared ($\chi^2$) goodness-of-fit (GOF) test was employed to determine whether an anomaly could be successfully detected. This statistical testing does not rely on the selection of arbitrary criteria such as requiring a fixed number of points above the noise level for a successful detection. Anomalous data points do not need to be detected consecutively, which makes the approach robust even in the presence of noise. Furthermore, statistical testing provides a quantitative metric indicating how likely the anomaly will be successfully detected.

The observed signal is a combination of the total magnetic intensity (TMI) from a spherical body plus randomly generated noise. The expected signal is simply the noise-free TMI from the spherical body. The number of degrees of freedom is equal to the number of data points in the signal. The required significance level has been set to 0.01. This value was selected to reduce the possibility of a type 1 statistical error; the rejection of a result that should be accepted (Peck and Devore 2012).

## 2.2. Minimum Detection Limits

Repeated tests (100 per set of parameters) were performed to determine the likelihood that the observed signal was detected. Magnetic susceptibility varied between $10^{-5}$ and $10^4$, the radius between 0 and 30 m, and the limiting depth between 0 and 100 m. A cumulative distribution was computed and a detection success rate was imposed.

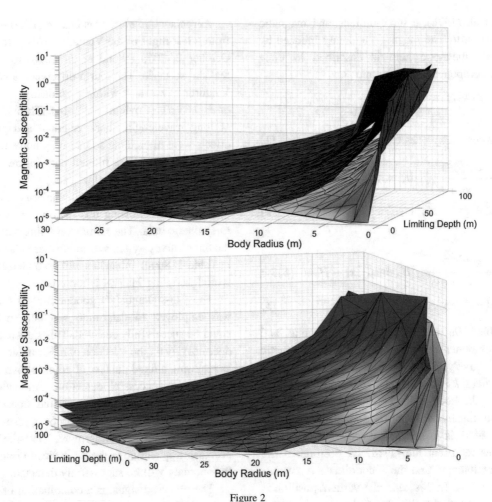

Figure 2
Surface plots of the minimum detection limits for a spherical body with a 90% success rate for each survey system with respect to the magnetic susceptibility, body radius, and limiting depth. Blue—100 m flight altitude; red—50 m flight altitude; and green—2 m flight altitude

Figure 2 shows surfaces representing the minimum values for each body parameter for which the corresponding magnetic anomaly will be detectable with a success rate of 90%. For large body radii and/or small limiting depths, the magnetic susceptibility can be as low as approximately $10^{-5}$ and the body will be considered detectable. As the radius decreases and/or the limiting depth increases, magnetic susceptibility must increase for a successful detection.

Profile plots are shown in Fig. 3 where the effects of varying two body parameters, while the third parameter is kept constant, for both 50 and 90% success rates. In the top plot, the source body is set to have a limiting depth of 20 m and the relationship

between the body radius and magnetic susceptibility is shown. In order to continue to successfully detect the body as its radius decreases, the magnetic susceptibility must increase rapidly. The middle plot compares the limiting depth and magnetic susceptibility for a body with a radius set to 10 m. As limiting depth increases, the magnetic susceptibility also needs to increase for a successful detection. Lastly, the bottom plot presents the relationship between the limiting depth and radius of the spherical body when the magnetic susceptibility is held constant at $10^{-4}$. A linear relationship between the limiting depth and radius is observed. As the limiting depth increases, the radius of the body must also increase for the body to be detectable.

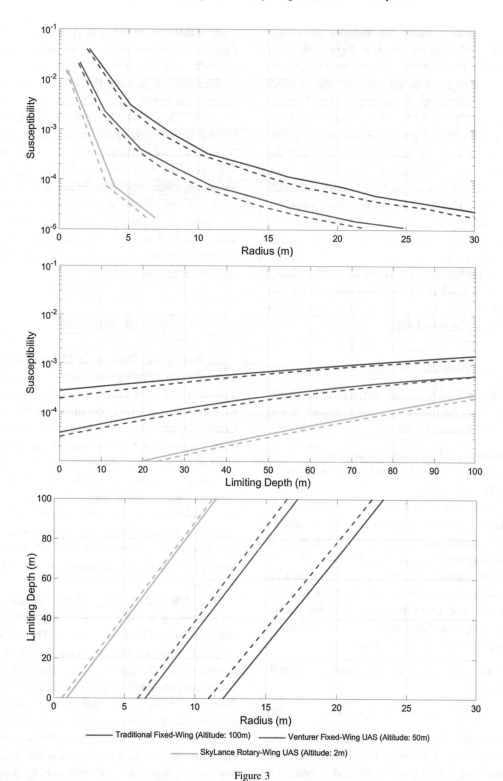

Figure 3
Minimum detection limits for a spherical magnetic body. The top plot is for a limiting depth of 20 m and the middle plot is for body radius of 10 m. The detectable region is above the curves. The bottom plot is for a magnetic susceptibility of $10^{-4}$. The detectable region is right of the curves. The solid and dashed lines represent 90 and 50% success rates, respectively

Figure 3 clearly shows the superior theoretical detectability afforded by the low flight altitude of UASs. Such altitudes are practically applicable only for UASs as opposed to piloted aircrafts. A UAS flying at an altitude of 50 m will detect a weakly magnetic $(10^{-4})$ ore body with a radius of 10 m buried at a limiting depth of approximately 30 m. A UAS flying at an altitude of 2 m will detect the same body at a limiting depth of approximately 80 m. The body will not be detected by the traditional fixed-wing aircraft. Figure 3 also reveals that improvements in detectability are almost the same when lowering the flight altitude from to 100 to 50 m, and from 50 to 2 m, due to the fact the TMI falls off as a function of inverse distance cubed. In practice, flying a middle-size UAS at an altitude slightly higher or lower than its nominal flight altitude will have little overall impact on detectability.

### 2.3. Spatial Resolution

In order to mathematically determine the spatial resolution limit, the following conditions must be met (Elkins and Hammer 1938):

1. Resolution is only considered for profiles that are produced from two identical bodies, in this case, two spherical bodies.
2. Each body can be represented by a symmetric function, $(s)$, where $s$ is a horizontal coordinate along the profile under consideration. The origin at $s = 0$ corresponds to the axis of symmetry.
3. $x$ is the horizontal coordinate along the profile. At the point $x = a$ (and $x = -a$), where is the location of the centre of one of the bodies along the profile, $(x)$ is analytical. The origin, $x = 0$, is located halfway between the two bodies.

The TMI is the sum of the contribution from each sphere over all $x$:

$$\Psi(x) = \Phi(x + a) + \Phi(x - a). \tag{6}$$

To compute the spatial resolution limit, the second derivative of the $\Phi(x)$ function needs to be determined and set to equal zero and solving for $a$ when $x = a$. This provides half of the resolution limit because each body will be offset from the origin by an amount $a$. The total separation between each

body is therefore equal to 2. At a separation smaller than 2 it will not be possible to distinguish the magnetic signature from the two bodies.

The relation between the spatial resolution limit, limiting depth, and radius of two neighbouring bodies for each flight altitude for the three survey systems of interest is presented in Fig. 4. The magnetic susceptibility does not affect the resolution limit so it is not displayed. Figure 4 shows that as flight altitude decreases, spatial resolution increases. For two bodies with a depth of 40 m and radius of 10 m, the minimum detectable separation between the two bodies is 67.5 m for an altitude of 100 m AGL, 45.0 m for an altitude of 50 m AGL, and 22.5 m for an altitude of 2 m AGL.

### 3. Study Site

In August and October, 2015, co-located ground magnetic and aeromagnetic surveys were performed just outside of Nash Creek, New Brunswick, Canada, over known magnetic anomalies (Fig. 5). The Nash Creek property, whose mineral rights are owned by Slam Exploration Ltd., is located off the southern coast of Chaleur Bay, New Brunswick, and south of the Gaspé Peninsula, Québec. The property lies between two northeast trending orogenic belts; the Aroostook-Percé (north) highlands and the Miramichi highlands (south) (Bongajum 2011). It is situated approximately 50 km northwest of the Bathurst Mining Camp which is located in the Miramichi highlands.

The Bathurst Mining Camp features several occurrences of volcanic massive sulphides deposited within Cambro-Ordovician rocks. The Nash Creek property, however, sits on younger rocks of Lower Devonian age (Bongajum 2011). The rocks in Nash Creek are composed of volcanic breccias, siltstones, limestones, mafic flows, rhyolites, and tuffs where the main source of mineralization is within the Dalhousie group. Brown (2007) suggests that the mineralization occurred within a failed rift system filled with shallow water. Volcanic and sedimentary rocks were deposited in a half-graben where a fault boundary allowed for the flow of hydrothermal fluid causing sulphide mineral accumulation containing zinc, lead,

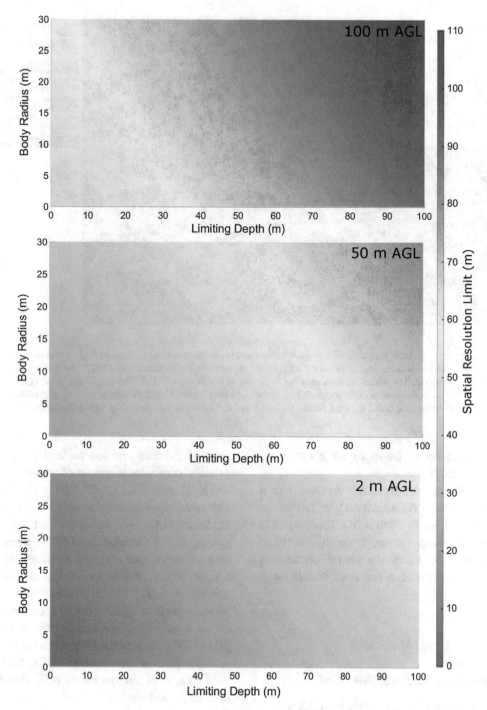

Figure 4
Comparison of the spatial resolution for different flight altitudes with respect to the limiting depth and radius of two neighbouring bodies

and silver. The property lies on the western edge of the Jacquet River Syncline and is bounded by the normal Black Point-Arleau Brook Fault to the west, with three subparallel faults defining the small Nash Creek graben (Fig. 5) (Walker 2010). Extensive exploration, primarily drilling, has been performed on

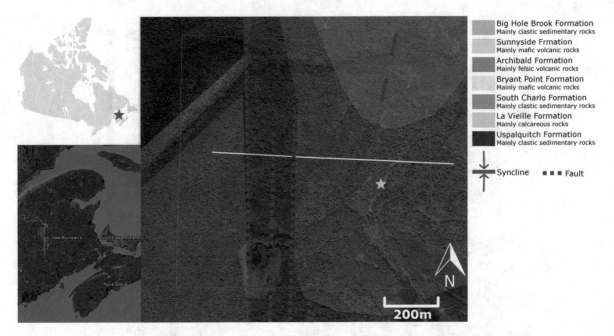

**Figure 5**

Nash Creek, NB survey area. Left: general location maps where red star marks the survey site. Right: geological map modified after Walker (2010). The red box outlines the ground magnetic survey area (the survey area is bound by latitudes of 47.880° and 47.889°, and by longitudes of − 66.098° and − 66.112°. The white line represents the flight line of the SkyLance UAS. The purple star is the take-off and landing location; and the blue star is the location of the magnetic base station. The westernmost blue dotted line is the trace of the Black Point-Arleau Brook fault, and the other blue dotted lines are related secondary faults. The red dashed line is the axis of the Jacquet River Syncline

the Nash Creek property for approximately 60 years (Ugalde et al. 2007).

The Nash Creek area is densely forested with a few access roads for local residential use. Trees were observed to reach up to 20 m AGL. Topography is hilly, with elevations varying from 80 to 117 m above sea level (ASL) over a lateral distance of between 500 and 1000 m. A few areas feature small cliffs.

## 4. Methods

### 4.1. Description of the Rotary-Wing UAS

The SkyLance UAS is a rotary-wing UAS (Fig. 1 and Table 1) currently being developed by Stratus Aeronautics to perform high-resolution aeromagnetic surveys for mineral exploration. The system has been built to minimize the magnetic interference from the UAS frame on the magnetometers. Four lithium polymer batteries power the rotors as well as other onboard avionics and the payload electronics. The avionics components consist of an autopilot system and a differential RTK GPS. The payload consists of a fluxgate magnetometer mounted at the front of the SkyLance UAS, used for aircraft attitude and navigation, and a Geometrics G-823A cesium vapour magnetometer, used to record aeromagnetic data. The cesium vapour magnetometer has a sampling frequency of 5 Hz. The differential RTK GPS unit provides positioning accuracy of approximately 5 cm in the horizontal plane and 3.6 cm in the vertical plane. Corrections to the GPS unit were streamed via a radio link to the onboard GPS antenna. The SkyLance UAS also measures ASL altitude with a barometric altimeter.

### 4.2. UAS Flights

The SkyLance UAS performed three test flights on October 17, 2015, along the line shown in Fig. 5. Time constraints and deteriorating weather conditions limited the number of possible flights. On that day, the weather started out sunny in the morning and as the day progressed the weather became overcast with temperatures ranging between 0 and 7 °C, respectively. Wind speeds were low (approximately 8–10 km/h); however, there were the occasional wind gusts. Space Weather Canada reported quiet solar activity for the sub-auroral zone where the survey was located; minimal magnetic noise would be due from solar activity. To ensure the SkyLance flew safely over all trees, and to ensure visual line of sight operations, the survey was flown at an average flight altitude of 80 m AGL (170 m ASL).

Prior to the first flight, the system was allowed to idle while recording, initially with the rotors off for approximately 2 min and then with the rotors on and the system hovering for another 2 min (Fig. 6). The time at which the power to the rotors is initiated is shown in blue. There is a drop in TMI of approximately 0.5 nT when the rotors are turned on. This drop is believed to be related to the orientation of rotors and to the direction of magnetic field they produce, which results in a slight reduction of the Earth's magnetic field. Other than this small drop in TMI, no effects on the TMI are observed due to the use of the rotors. After the rotors are turned on, there is a 0.3 nT increase in TMI over a 150-s period. This may be caused by a combination of effects such as: increased rotor rotation speeds; increased power

output to the rotors; static magnetization of onboard components; and/or motions of the UAS (i.e. ascending or vibrations).

Each of the three SkyLance UAS flights covered a further distance than the previous along the flight line shown in Fig. 5, providing repeated measurements. Flight altitude was stable; the standard deviation of the altitude is only ± 1 m along the flight lines. The UAS attitude was also consistent, with approximately ± 5° variations in pitch, and ± 2° in roll, which are primarily associated with wind gusts. The average pitch is negative due to the flight mechanics; a helicopter needs to point its nose down to fly forwards.

The fourth difference was calculated for each of the flights using raw data (Fig. 7). The maximum variation is approximately ± 0.4 nT, with one standard deviation equal to ± 0.04 nT.

Due to wind gusts, the UAS would occasionally oscillate causing a transient high-frequency noise in the aeromagnetic data. This effect is shown in Figs. 8 and 9. This high-frequency noise was removed for each flight via filtering the 0.6 Hz oscillation. As seen in Fig. 10, after diurnal corrections and frequency filtering, the TMI data show that high degree of repeatability was achieved along the sections of the line that were covered by each flight.

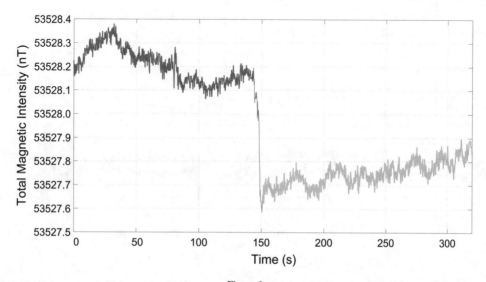

Figure 6
SkyLance UAS magnetic data during system power-up prior to surveying. Data are corrected for diurnal effects. Red—magnetic data prior to the rotors being turned on; blue—rotor powering on; green—rotors on, system hovering

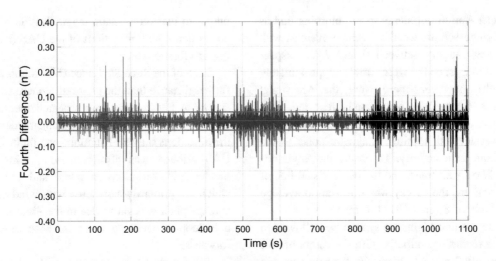

Figure 7
SkyLance UAS fourth difference data. Blue—flight 1; red—flight 2; black—flight 3. The one standard deviation envelope (± 0.04 nT) is indicated by green horizontal lines

### 4.3. Ground Survey

The ground magnetic survey was executed over a 3-day period, August 11–13, 2015. As seen on the satellite image (Fig. 5), the property is heavily forested. There are many areas that contain dense brush and tree canopy as well as steep slopes, making it difficult to traverse and causing poor GPS signal quality. GPS signal was lost at various points; in order to remedy this issue, the recorded magnetic data position was interpolated along the walking paths by assuming a constant walking velocity.

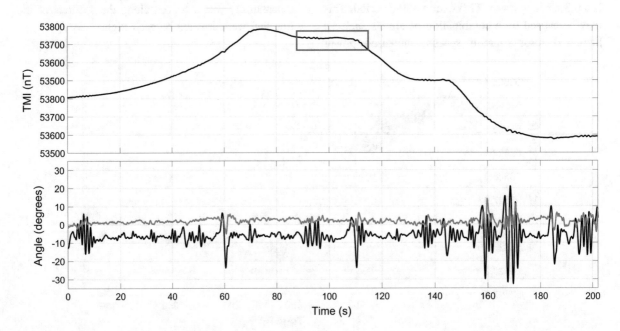

Figure 8
Relating transient high-frequency noise in aeromagnetic data (top plot) with pitch (black) and roll (grey) oscillations (bottom plot). Time is from beginning of flight line. Magnetic data is from flight 3

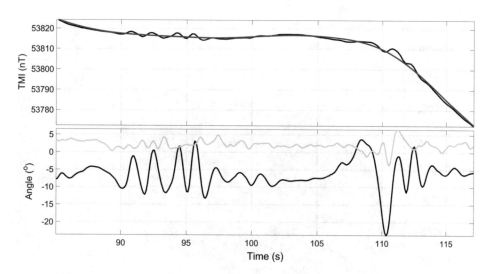

Figure 9
Zoomed in portion of Fig. 8 between 90s and 105s. Top plot—TMI with oscillations (black) and TMI with oscillations removed (blue). Bottom plot—roll (grey) and pitch (black)

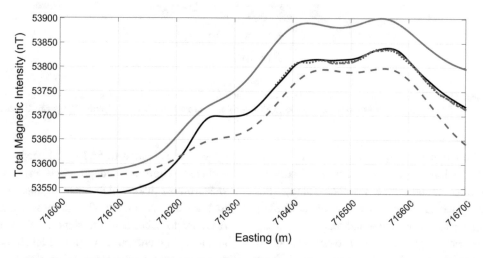

Figure 10
SkyLance UAS magnetic data in blue—flight 1; red—flight 2; and black—flight 3. Upward continued to 80 m AGL of the ground magnetic survey in green. Response of the three magnetic targets modelled as thin sheets at 80 m AGL in magenta

## 5. Results and Discussion

The ground magnetic map, computed using minimum curvature interpolation, is shown in Fig. 11. Borehole casings from previous exploration drilling are scattered around the eastern survey area, which produce a strong magnetic signature. Any recording that fell within 15 m of a borehole was removed. The map shows three north-trending oblong anomalies. Which align with the Black Point-Arleau Brook fault

and the related secondary faults located immediately to the east (Fig. 5).

Upward continuation of the ground magnetic data to 80 m AGL is presented in Fig. 12 with data from the third UAS flight overlaying it. A profile plot along the flight line shows that the trends of the two datasets match very well (Fig. 10). There is, however, a DC shift of approximately 20–50 nT in TMI between the upward continued ground data and the aeromagnetic data. This is possibly due to: (1) an effect from

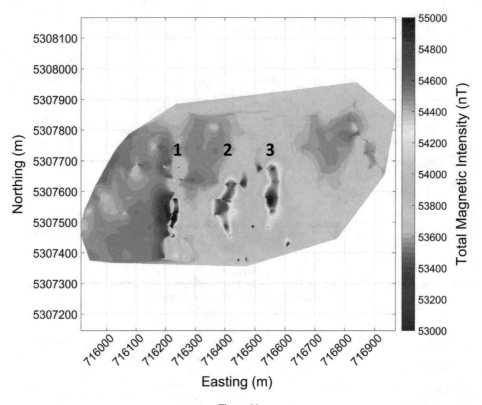

Figure 11
Ground magnetic map of the survey area exhibiting three prominent anomalies labelled 1, 2, and 3. The colour bar is limited to 53,000 and 55,000 nT instead of the maximum data range of 50,306 to 58,494 nT to improve visualization. Coordinate system: UTM. Zone: 19T

the upward continuation algorithm used; (2) differences in the TMI due to the time of year the surveys were performed (August and October, 2015)—the International Geomagnetic Reference Field (IGRF) has a – 15 nT shift from the ground magnetic survey date to the aeromagnetic survey date; (3) static magnetic sources on the UAS; (4) instrument drift on the Overhauser magnetometer used to acquire the ground magnetic survey data; and/or (5) DC shift in magnetic readings between the two different magnetometers used.

The magnetic anomalies coincide with diabase dykes observed in the field (Veglio, E. (2017). Personal communication. November 1, 2017) and therefore were modelled as thin sheets. Figure 10 shows the magnetic response, calculated by forward modelling, of three magnetic dykes with a thickness of 1.5 m and a dip of at 20° to the east, separated by approximately 150 m. The magnetic susceptibility used for modelling was $2 \times 10^{-1}$. The two eastern

dykes are estimated to be located deeper in the subsurface (limiting depth $\sim$ 40 m) than the westernmost dyke which outcrops and was observed in the field. A 53,565 nT shift was applied to the response to bring it to the same level as the ground and aeromagnetic data. At ground level, the response matches the ground magnetic data closely. As seen in Fig. 10, the response also matches the trends of the ground and aeromagnetic data well for an altitude of 80 m AGL.

## 6. Conclusions

Compared with traditional airborne survey systems, one of the main advantages of UASs is their lower flight altitudes. Flying closer to the ground allows recording a fainter signal; deeper, smaller, or weakly magnetic bodies are being detected. The estimates for minimum detection limits presented

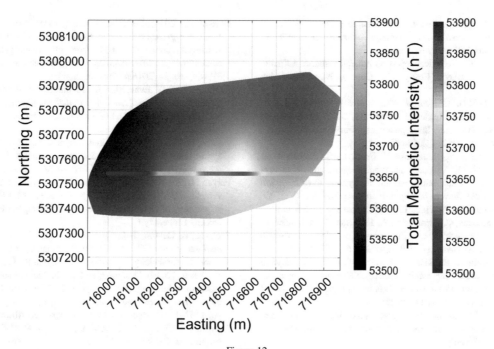

Figure 12
Aeromagnetic survey data from flight 3 (in colour) overlaying the upward continued to 80 m AGL ground magnetic data in black and white.
Coordinate system: UTM. Zone: 19T

here were determined using test flight data from the Venturer fixed-wing UAS. Detection capability is likely to increase in the future when the design of UAVs will improve, leading to further reduction in noise levels. Another advantage of UASs is their capability of flying at slower speeds than manned aircraft which allows them to better delineate closely spaced magnetic targets.

Before UAV surveys can become a viable alternative to ground surveys or manned aircraft surveys, their potential must be assessed in a variety of terrains and field conditions. The Nash Creek UAS survey is a contribution towards that overarching goal. Flying at an altitude of 80 m AGL, the Sky-Lance rotary-wing UAS magnetic data successfully captured three anomalies which had been previously identified on the ground magnetic data. Modelling concluded that the two easternmost anomalies were from two separate bodies; however, this combination of targets and flight parameters was approaching the spatial resolution limit of the UAS. Stability during flight is an area for improvement, with the objective of attenuating the high-frequency pitch oscillations

occurring during wind gusts, which would allow the UAS to fly closer to the ground, leading to an overall increased detection capability.

*Acknowledgements*

We thank Dr. Bernd Milkereit from the Department of Earth Sciences at the University of Toronto for facilitating access to the Nash Creek property as well as lending us the use of his ground magnetic survey equipment. We also thank Andrew Hay from Carleton University for assisting with the ground magnetic survey data collection. This project was partially funded through an ENGAGE grant from the Natural Sciences and Engineering Research Council of Canada (NSERC) to C. Samson.

REFERENCES

Anderson, D., & Pita, A. (2005). *Geophysical surveying with GeoRangerTM UAV* (pp. 67–78). Reston: American Institute of Aeronautics and Astronautics Inc.

Bongajum, E. (2011). *Investigating seismic wave scattering in heterogeneous environments and implications for seismic imaging*. Toronto: University of Toronto.

Brown, D. (2007). Technical report on mineral resource estimate, Nash Creek project, Restigouche County, New Brunswick Canada

Burns, M. (2017). *UAV-MAG—The leader in unmanned geophysics surveying*. Retrieved August 03, 2017, from Pioneer Aerial Surveys Ltd. http://pioneeraerialsurveys.com/Pioneer_Aerial_Surveys_Promo_2017.pdf.

Caron, R., Samson, C., Straznicky, P., Ferguson, S., Archer, R., & Sander, L. (2011). Magnetic and magneto-gradiometric surveying using a simulated unmanned aircraft system. In *81st Annual meeting of the Society of Exploration Geophysicists* (Vol. 30, pp. 861–865). San Antonio, TX: SEG Expanded Abstracts.

Caron, R., Samson, C., Straznicky, P., Ferguson, S., & Sander, L. (2014). Aeromagnetic surveying using a simulated unmanned aircraft system. *Geophysical Prospecting, 62,* 352–363.

Coyle, M., Dumont, R., Kiss, F., & Miles, W. (2014). *Geological Survey of Canada aeromagnetic surveys: Design, quality assurance, and data dissemination*. Ottawa: Geological Survey of Canada. https://doi.org/10.4095/295088.

Cunningham, M. (2016). Aeromagnetic surveying with unmanned aircraft systems. M.Sc. Thesis. Department of Earth Sciences, Carleton University

Dentith, M., & Mudge, S. T. (2014). *Geophysics for the mineral exploration geoscientist*. Cambridge (UK): University Press.

Dion-Ortega, A. (2015, June). *Abitibi Géophysique lance le tout premier drone magnétométrique*. Montreal, QC, Canada. Retrieved November 09, 2015, from https://www.lesaffaires.com/dossier/exploration-miniere/abitibi-geophysique-lance-le-tout-premier-drone-magnetometrique/579253

Eck, C., & Imbach, B. (2011). Aerial magnetic sensing with an UAV helicopter ISPRS. *International Archives of the Photogrammetry, Remote Sensing and Spatial Information Sciences,* XXXVIII-1/C22, 81–85.

Elkins, T., & Hammer, S. (1938). The resolution of combined effects, with applications to gravitational and magnetic data. *Geophysics*. https://doi.org/10.1190/1.1439512.

Gordon, R., (2016). *The development of magnetometer UAV platforms for geoscience*. GEM Systems, Toronto, ON, Canada. Retrieved August 03, 2017, from http://www.gemsys.ca/wp-content/uploads/2016/09/GEM_UAV_for-Geoscience_EAGE_SPAIN_sep_24_2016.pdf?lbisphpreq=1

Kroll, A. (2013). Evaluation of an unmanned aircraft for geophysical survey. In *23rd international geophysical conference and expedition* (pp. 1–4). Melbourne, Australia: ASEG Extended Abstracts.

Pajares, G. (2015). Overview and current status of remote sensing applications based on unmanned aerial vehicles (UAVs). *Photogrammetric Engineering and Remote Sensing, 81,* 281–329.

Parvar, K. (2016). Development and evaluation of unmanned aerial vehicle (UAV) magnetometry systems. M.A.Sc. Thesis. Department of Geological Sciences and Geological Engineering, Queens University.

Peck, R., & Devore, J. L. (2012). *Statistics—The exploration & analysis of data* (7th ed.). Boston: Cengage Learning.

Samson, C., Straznicky, P., Laliberte, J., Caron, R., Ferguson, S., & Archer, R. (2010). Designing and building an unmanned aircraft system for aeromagnetic surveying. In *80th Annual meeting of the Society of Exploration Geophysicists* (Vol. 29, pp. 1167–1171). Denver, CO: SEG Expanded Abstracts.

Telford, W., Geldart, L., Sheriff, R., & Keys, D. (1976). *Applied geophysics*. Cambridge: Cambridge University Press.

Ugalde, H., L'Heureux, E., & Milkereit, B. (2007). An integrated geophysical study for orebody delineation, Nash Creek, New Brunswick. In *Proceedings of exploration 07: Fifth decennial international conference on mineral exploration* (pp. 1055–1058).

Walker, J. A. (2010). *Stratigraphy and lithogeochemistry of Early Devonian volcano-sedimentary rocks hosting the Nash Creek Zn-Pb-Ag Deposit, northern New Brunswick* (pp. 52–97). Geological Investigations in New Brunswick for 2009. New Brunswick Department of Natural Resources; Lands Minerals and Petroleum Division, Mineral Resource Report 2010-1

Wood, A., Cook, I., Doyle, B., Cunningham, M., & Samson, C. (2016). Experimental aeromagnetic survey using an unmanned air system. *The Leading Edge, 35,* 270–273. https://doi.org/10.1190/tle35030270.1.

(Received March 27, 2017, revised November 20, 2017, accepted November 27, 2017, Published online December 5, 2017)

Pure Appl. Geophys. 175 (2018), 3159–3177
© 2017 The Author(s)
This article is an open access publication
https://doi.org/10.1007/s00024-017-1649-0

Pure and Applied Geophysics

# Detecting Surface Changes from an Underground Explosion in Granite Using Unmanned Aerial System Photogrammetry

Emily S. Schultz-Fellenz,[1] Ryan T. Coppersmith,[2] Aviva J. Sussman,[1] Erika M. Swanson,[1] and James A. Cooley[3]

*Abstract*—Efficient detection and high-fidelity quantification of surface changes resulting from underground activities are important national and global security efforts. In this investigation, a team performed field-based topographic characterization by gathering high-quality photographs at very low altitudes from an unmanned aerial system (UAS)-borne camera platform. The data collection occurred shortly before and after a controlled underground chemical explosion as part of the United States Department of Energy's Source Physics Experiments (SPE-5) series. The high-resolution overlapping photographs were used to create 3D photogrammetric models of the site, which then served to map changes in the landscape down to 1-cm-scale. Separate models were created for two areas, herein referred to as the test table grid region and the nearfield grid region. The test table grid includes the region within ~40 m from surface ground zero, with photographs collected at a flight altitude of 8.5 m above ground level (AGL). The near-field grid area covered a broader area, 90–130 m from surface ground zero, and collected at a flight altitude of 22 m AGL. The photographs, processed using Agisoft Photoscan® in conjunction with 125 surveyed ground control point targets, yielded a 6-mm pixel-size digital elevation model (DEM) for the test table grid region. This provided the ≤3 cm resolution in the topographic data to map in fine detail a suite of features related to the underground explosion: uplift, subsidence, surface fractures, and morphological change detection. The near-field grid region data collection resulted in a 2-cm pixel-size DEM, enabling mapping of a broader range of features related to the explosion, including: uplift and subsidence, rock fall, and slope sloughing. This study represents one of the first works to constrain, both temporally and spatially, explosion-related surface damage using a UAS photogrammetric platform; these data will help to advance the science of underground explosion detection.

## 1. Introduction

Underground explosions can produce changes that may be observed at the surface and recorded by sensors at a range of different standoff distances. The Source Physics Experiment (SPE), a series of controlled underground chemical explosive experiments conducted under the auspices of the US Department of Energy's National Nuclear Security Administration (NNSA), aims to advance the physical understanding of seismic wave propagation through geologic media (Snelson et al. 2013). An ancillary objective within SPE is to quantify and analyze permanent ground surface damage resulting from the small underground conventional chemical explosive experiments. The ground surface damage information presented here helps validate other real-time field geologic and geophysical measurements (including surface fracture mapping, downhole accelerometer instrumentation, and seismic wave analyses recorded from near-field and farfield sensors; e.g., Snelson et al. 2013; Patton 2015; Larmat et al. 2017), and provides critical input into predictive models (e.g., Larmat et al. 2015).

The formation of subtle surface features from underground activities, including explosions, depends upon the geologic environment, depth of burial, and yield. While databases exist describing surface features associated with underground nuclear explosions (e.g., Garcia 1997; Grasso 2001; Cong et al. 2007), details of the mapped features were constrained by the data capture method and the target of the mapping program. Previous studies have described morphological changes, such as fracturing and shifting of underground infrastructure, resulting from

[1] Earth and Environmental Sciences Division, Los Alamos National Laboratory, Los Alamos, NM 87545, USA. E-mail: eschultz@lanl.gov
[2] Coppersmith Consulting, Inc., 2121 N. California Blvd Ste. 290, Walnut Creek, CA 94596, USA.
[3] Gresham Smith and Partners, 511 Union St, 1100 Nashville City Center, Nashville, TN 37219, USA.

underground conventional explosions (e.g., Cattermole and Hansen 1962; Phang et al. 1983; Yu et al. 2014; Chowdhury and Wilt 2015), but have not described detailed changes at the ground surface or explored changes at the cm- to sub-cm scale. Similarly, while airborne and satellite remotely sensed methods have been employed to detect morphological changes and dynamic geological processes such as landslides, flooding, and coastal change (e.g., Gutierrez et al. 2001; Glenn et al. 2006; LePrince et al. 2008; Niethammer et al. 2010; Scaioni et al. 2014; Pelletier and Orem 2014; Warrick et al. 2016), the meter-scale data resolution range may overlook important, but smaller features. The acquisition, analysis, and interpretation of cm-scale change detection data from ground-based and ground-proximal methods provide important constraints for, and validation of, the possible range of scales for relevant features, thereby allowing us to better constrain the mechanisms that govern the damage patterns expressed at the ground surface.

Recent studies have advanced the use of photogrammetry into the field of geologic mapping and landform change detection (e.g., James and Varley 2012; Fonstad et al. 2013; Hugenholtz et al. 2013; Tuffen et al. 2013; Micheletti et al. 2015; Piras et al. 2017). To apply photogrammetry to morphological change detection studies in national and global security applications, such as the detection of smaller and/or deeper events, data collections need to be nimble, rapid, non-invasive, and non-destructive while providing high-resolution data. When signatures are subtle, data captured from greater standoff distances may not possess the resolution needed to detect changes. Furthermore, the methods required to capture ground-based lidar and traditional mapping data may destroy signatures of significance. Low-altitude (<100 m above ground level) unmanned aerial systems (UAS) with photogrammetric assets meet these needs and position this technology to grow further into geologic signature and change detection for national and global security application spaces. Here we present data and methods from UAS photogrammetry collections, executed with a dense grid of ground control points, before and after an underground conventional explosive experiment. We demonstrate that small-yield underground explosions

in granite elicit a detectable surface expression of damage. We validate the proof-of-concept for applicability of a UAS-based Structure-from-Motion (SfM) workflow for fine-scale surface change-detection studies and test the operation's ability to produce results comparable to terrestrial lidar. We compare the observed changes with the regional geologic setting and discuss the benefits of integrating imagery and topographic data in the interpretation of signatures from underground activities.

## 2. Setting

The field area is located within the Nevada National Security Site (NNSS), formerly the Nevada Test Site (Fig. 1). The data collection site is a 250 m × 200 m region broadly centered on the SPE emplacement borehole (Fig. 2) and is underlain by the Climax Stock. The Climax Stock is a Cretaceous-aged, fine- to medium-grained coarsely porphyritic quartz monzonite, comprised largely of quartz, potassium feldspar, plagioclase, and minor biotite (Orkild et al. 1983). Syn- and post-emplacement hydrothermal alteration of the Climax Stock produced moderate argillic alteration of plagioclase throughout the intrusion. The Boundary Fault, a basin-bounding normal fault active in the Quaternary, separates the monzonite stock from nearby Yucca Flat. This fault strikes north- to northeast and has accommodated between ~250 m (Orkild et al. 1983) and 600 m (Maldonado 1977) of throw. Dominant joint sets are N64W, 90; N32W, 22 NE; and N35E, 90 (Houser and Poole 1960).

The field site (Fig. 2) consists of a flat-graded region serving as the test table, a bermed sump southeast of the test table (known as the "muck pit") to trap and hold drilling fluids and runoff, an ~7 m high quartz monzonite rock outcrop west of the southwesterly corner of the test table, and a broad region of native terrain outside of the margins of the test table that includes undulating topography, a small east–west oriented drainage, and scrubland vegetation. This study captured data in association with the fifth SPE test in the series (hereafter referred to as SPE-5). SPE-5 involved the detonation of chemical high explosives within a cylindrical canister

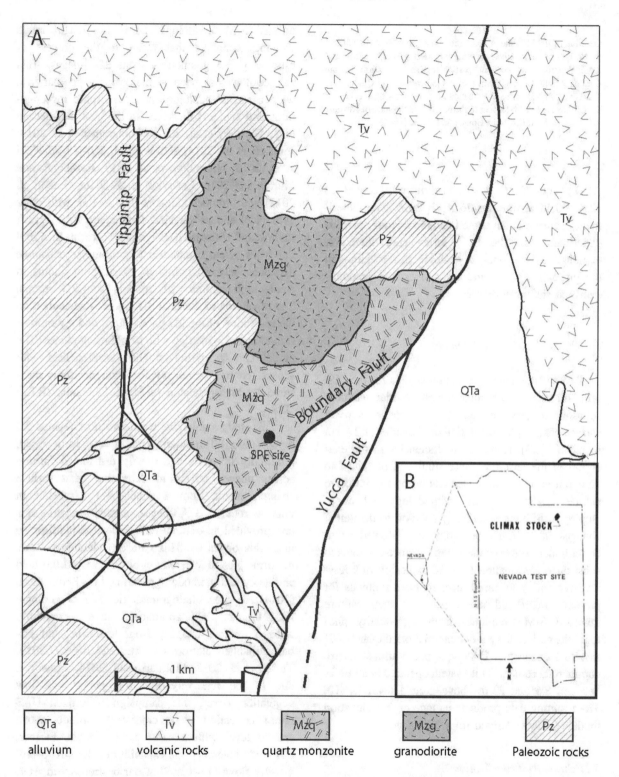

| QTa | Tv | Mzq | Mzg | Pz |
|---|---|---|---|---|
| alluvium | volcanic rocks | quartz monzonite | granodiorite | Paleozoic rocks |

emplaced in a 91.4-cm diameter borehole with a centroid charge depth of 76.5 m. Above the subsurface explosive canister, the emplacement hole was backfilled with stemming materials, including gravel and poured grout. The underground experiment, a casting of PBXN-114 with a 5035-kg trinitrotoluene

◄Figure 1
Location of the SPE site. **a** Simplified geologic map of the Climax Stock quartz monzonite, showing exposed geologic units and regional structures; note in particular the orientation of the Boundary fault. Geologic unit prefixes indicate age; *Q* Quaternary, *T* Tertiary, *M* Mesozoic, *P* Paleozoic. **b** *Inset* showing location of the Climax stock with respect to the NNSS and the state of Nevada. Modified from Houser and Poole (1961)

(TNT) equivalent, was successfully executed at 13:49.00 local time on 26 April 2016. The experiment occurred in the same borehole as four previous experiments in the SPE series and, therefore, exploited a pre-damaged geologic media, although the pre-existing geologic damage existed predominantly at shallower depths.

## 3. Methods

Past surface change detection studies for the SPE program used conventional terrestrial lidar scanning solutions to produce datasets with spatial vertical surface change ($\Delta z$) resolution on the order of 2.8 cm locally (Fig. 3). However, inefficiencies in the execution of the terrestrial lidar studies at the site and potential ground disturbance caused by the intrusive nature of the collection methodology led us to employ a different technology solution to document changes in surface topography and allowed us to better understand how the ground surface responds to subsurface deformation from an explosion in a less-invasive and more rapid manner. Field methods for pre-experiment and post-experiment data capture involved SfM techniques from high-quality photographs collected by a commercial off-the-shelf DJI Inspire 1 quadcopter UAS equipped with a standard, commercial camera. This system operated at altitudes between 8.5 and 22 m above ground level (AGL). The resulting data products include digital elevation models (DEMs) derived from 2D imagery.

### 3.1. Structure-From-Motion

With a paramount need to acquire low-cost, time-efficient, high-resolution topographic data, this study utilized SfM technology. SfM is a useful range viewing technique that refers to the process of estimating 3D geometries from 2D images. This methodology has become increasingly popular in a wide range of geologic and geomorphic investigations and standoff distances in the past half-decade (e.g., Westoby et al. 2012; Niethammer et al. 2010; Johnson et al. 2014; Bemis et al. 2014; Micheletti et al. 2015; Nouwakpo et al. 2015; Reshetyuk and Martensson 2016; Carrivick et al. 2016; Piras et al. 2017) and provides valuable data to link with other geological and geophysical techniques. The Agisoft Photoscan® software package is a powerful SfM tool that uses photogrammetric algorithms to calculate distances between points in overlapping photos to create high-resolution 3D models from 2D images. The resolution of the final 3D model is dependent on the resolution of the 2D photographs and the ratio of frontal and side overlap with neighboring photos. The model can be optimized and georeferenced with ground control points (GCPs) with positions measured by high-precision real-time kinematic (RTK) GPS.

This study's objective was to acquire detailed 3D topographic information from a site in a rapid, minimally invasive, and evenly distributed fashion across uneven terrain to determine what surface changes result from a small-yield explosion in granitic rock. A UAS-borne photogrammetric system provided an optimal data collection platform to meet this objective. This campaign largely follows the principles of Westoby et al. (2012) and the field and analytical methodology outlined by Bemis et al. (2014) for SfM photogrammetric data of geologic field targets. SfM approaches have successfully utilized oblique imagery from terrestrial and airborne manned platforms (James and Robson 2012; Tuffen et al. 2013; Molinari et al. 2014). However, due to the relatively low topography and low vegetation cover, this campaign prioritized UAS photo collection at a consistent altitude above ground level with a nadir camera angle. Further, this study leveraged the capability of the DJI Inspire 1 to fly slowly ($\sim 1$ m/s) at our desired 8.5 m AGL low ceiling to capture the highest resolution images possible.

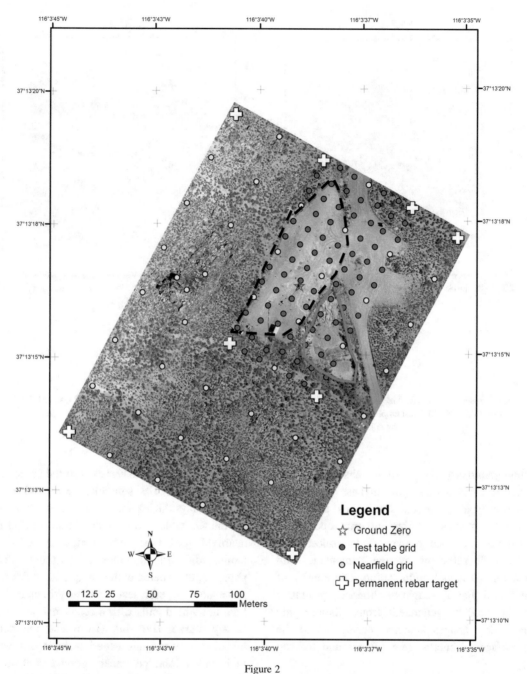

Figure 2

Study area with locations of test table grid (*blue dots*), near-field grid (*yellow dots*), and permanent (*white crosses*) ground control point targets. SPE-5 emplacement hole at surface ground zero shown as *red star*. Boundaries of test table (*black dashed line*) and muck pit (*solid orange line*) shown

### 3.2. GCP Grid Design and Installation

While recent studies have explored direct georeferencing of UAS photogrammetry data (e.g., Jozkow and Toth 2014; Carbonneau and Dietrich 2016), this study required that final data products be accurately georeferenced for high-resolution change detection. This required a careful and sufficiently dense grid of surveyed ground control points (GCPs) to translate

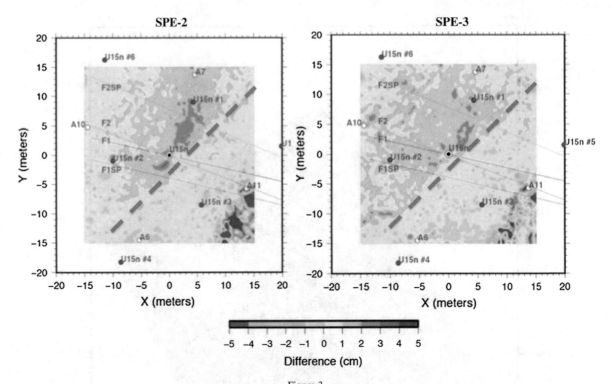

Figure 3
Surface vertical change detection (Δz) maps from analyses of pre- and post-test terrestrial lidar data for previous SPE tests (SPE-2, *left*; SPE-3, *right*). Coordinate 0,0 is the SPE test emplacement hole (labeled U15n). *Green dashed line* represents a geologic feature along which surface Δz appears to align. Modified from Schultz-Fellenz et al. (2013)

the photogrammetry data into an absolute spatial reference frame and obtain the highest quality and accuracy geospatial output. The GCPs served as passive pilot guidance during airborne operations, provided the reference framework to collect photographs with sufficient overlap percentage, and allowed for 3D model optimization. The density of the GCP grid (Fig. 2) sought to eliminate systematic errors in models generated from photographic datasets using images captured orthogonal to the target surface of interest (e.g., James and Robson 2014).

Within the field site, a Topcon HiPer V base and rover unit utilized real-time kinematic (RTK) differential GPS procedures to stakeout approximate locations for 126 temporary and permanent GCPs. The baseline horizontal accuracy of the HiPer V GPS is 10 mm + 1 ppm, and vertical accuracy is 15 mm + 1 ppm. GPS site control was established using the Online Positioning User Service (OPUS) available from the National Geodetic Survey. Overall

control of the SPE site was established based on three independent OPUS solutions performed on three separate days. GCPs were installed at 10 m spacing for the test table grid area, and 30 m spacing for the nearfield grid area. The temporary GCPs were laminated, brightly-colored 21.6 cm × 27.9 cm paper targets labeled with a unique identifier (a letter and a number), each secured to the ground by 7.6 cm-long household nails with a center nail acting as the survey marker (Fig. 4a). Given that the field site intended to host future experiments, it was prescient to install several permanent ground control monuments well away from the region of anticipated damage to ensure that future ground control can reference undisturbed monuments (Fig. 4b). These permanent targets consisted of 1 m square black-and-white striped plastic fiducial sheets, secured to the ground with a combination of rebar, 10.16 cm nails with 5.08 cm washers, and rocks. The team surveyed the centerpoint of each GCP for absolute spatial control, using the same Topcon RTK differential

Figure 4
Examples of the temporary (*top*) and permanent (*bottom*) GCP markers in the study area

GPS, to acquire high-fidelity position information for approximately 30 s (30 epochs) per temporary target and 3 min (180 epochs) per permanent target, with redundant checks on all permanent targets. Targets were surveyed prior to airborne operations during the pre-experiment collection and resurveyed following airborne operations for the post-experiment collection.

## 3.3. UAS Platform and Airborne Operations

This campaign utilized a DJI Inspire I dual-control quadcopter UAS, operated with master (flight) and slave (camera) controls. For the test table grid region, the UAS operated at an 8.5-m AGL flight ceiling. For the near-field grid region, the UAS operated at a 22-m AGL flight ceiling. From these positions, the UAS collected predominantly nadir still photography using its standard onboard X3 FC350 12 megapixel camera. The field collection area also served as the launch, landing, and recharging point

for the UAS. The nominal flight time for a single battery for the UAS was approximately 17 min. Due to external constraints on deployment time, the team required no fewer than eight spare batteries, with continuous recharging capability, to ensure multiple hours of operation and data collection.

After completion of ground checks and sensor calibration, the team flew the UAS in a planned flight pattern to obtain appropriate photo overlap. For optimal data quality, the team required >65% side photo overlap and >80% frontal photo overlap across the entirety of the field area. This was accomplished using the GCPs as reference and a flight operations workflow that captured the same GCP in multiple frames (Fig. 5). Further, calibration of the flight altitudes to the GCP spacing achieved the desired photo framing. Using the slave controller for image capture operations, the team could view and capture each GCP in no fewer than nine separate positions within individual images (e.g., within single frames, each GCP was captured in the top left/center/right, center left/center/right, and the bottom left/center/right), thus yielding the required image overlap. The high-resolution photographs collected at high overlap and a low flight ceiling created a seamless and spatially robust point cloud dataset. The establishment of flight boundaries set 10–20 m beyond the study area ensured full coverage and avoided the zone of boundary artifacts that can occur during photogrammetric processing. Due to the variation of vegetation thickness and topographic in the study area, the collection coordinated the timing of certain flight paths with favorable lighting conditions whenever possible: data collection over highly vegetated and high relief locations occurred closer to mid-day to reduce shadowing, whereas the flatter, less vegetated locations were flown during the lower light angle hours. Since the SPE-5 test date had been set for 26 April 2016 and the test was densely instrumented with other scientific and technical diagnostics, a fixed data collection schedule was developed by the SPE campaign leadership. For UAS photogrammetry, the pre-test data collection occurred across eight daylight flight hours on 19 and 20 April 2016 in clear, cloudless conditions. The explosion was executed on schedule on 26 April 2016. The post-test data collection took place across six daylight

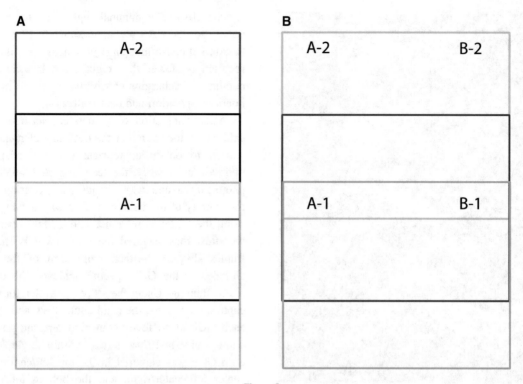

Figure 5
Schematic showing the positioning of GCPs in image frames to secure the required amount of photo overlap for 3D model generation. Three photos were taken at each GCP target. *Left panel* shows photo frame (*square*) centered on GCP target A1 in the *top*, *center*, and *bottom* of the camera frame. *Right panel* shows photo centered between the targets providing sufficient side overlap

flight hours on 27 and 28 April 2016 under evenly cloudy skies.

After the first day of each data collection, all captured images underwent review to confirm sufficient photo overlap and the focus of the imagery. Areas with insufficient overlap or poor quality photos were reflown on the second day.

### 3.4. Geospatial Data Processing and DEM Development

Aerial imagery data, in conjunction with surveyed GCPs and Agisoft Photoscan®, served as the basis for developing the 3D models. After removing poorly focused, blurry, or lower-quality photos, between 1060 and 1620 photographs per model were imported into Agisoft Photoscan® for processing and model generation (Table 1).

The software first aligns the photographs based on finding 'like' points in the overlapping portions of the images. Then, manual assignment of survey points within the photographs was performed for every location where the GCPs are visible.

The more accurate locations of the surveyed GCP targets optimize the locations of the aligned photos, both with respect to all other photos in the model and in an absolute spatial reference frame. After location optimization, the dense 3D point cloud is generated—the most computationally demanding part of the analysis. Figure 6 summarizes the steps of this process.

### 4. Results

Results of the photo data collection and subsequent DEM models are summarized in Table 1 and discussed below. The models relied heavily on the survey control that provided 1 cm + 1 ppm horizontal and 1.5 cm + 1 ppm vertical resolution at

Table 1

*Parameters and model information for the four DEMs generated in this study*

| Model | Camera locations and image overlap | | | | | Model optimization | | | Final point cloud and DEM results | |
|---|---|---|---|---|---|---|---|---|---|---|
| | Number of images | Aligned photos | Flying altitude (m) | Coverage area (km²) | Ground resolution (mm/pix) | GCP targets | Total error— GCP camera optimization (m) | Reprojection error based on GCPs (pix) | Point count | DEM resolution (cm/pix) |
| Preshot test table grid | 1241 | 1240 | 8.24 | 0.0181 | 3.07 | 92 | 0.0066 | 1.95 | 480,768,917 | 0.0614 |
| Postshot test table grid | 1504 | 1504 | 9.23 | 0.0141 | 3.44 | 65 | 0.0069 | 2.1 | 416,011,453 | 0.0687 |
| Preshot near-field grid | 1060 | 1060 | 23.4 | 0.0521 | 8.74 | 53 | 0.0147 | 2.04 | 281,877,253 | 1.75 |
| Postshot near-field grid | 1623 | 1623 | 21.5 | 0.0548 | 8.1 | 45 | 0.0154 | 2.02 | 349,781,020 | 1.62 |

GCP locations. The results of the survey portion of the study included the high-resolution position capture of 126 GCP targets for the pre-test survey. After photo collection for the pre-test effort, 38 GCP targets moved from non-test related occurrences (e.g., weather). Therefore, the post-shot survey yielded 88 GCP locations.

### 4.1. Photo Overlap and Alignment Analysis

Table 1 also presents a summary of the number of photos analyzed and aligned in each model. The field campaign yielded excellent photo coverage, with photo overlap typically exceeding nine photos per point throughout each model (Fig. 6). These values indicate that the flight workflow shown in Fig. 5 was executed correctly with each GCP target captured in at least nine photos. This resulted in very high ground resolution, based on image quality rather than spatial control, and also generated orthoimagery as a data product.

### 4.2. GCP Input and Dense Point Cloud Construction

Introduction of absolute positions of GCPs to the models occurred after the photo alignment stage, through manual placement of GCP markers on the center nail as seen in the GCP target indicated in Fig. 4. This action took place approximately 800 times for each model, given the number of GCP targets and the coverage of at least 9 photos per target. (At the time of publication, no automated process to perform this step is known.) As described earlier, even though fewer GCP targets were available for use in the post-test model, sufficient special GCP coverage existed in the remaining 88 preserved post-test GCP network to ensure a low re-projection error (Table 1). After GCP placement, the model was optimized based on these locations. The total error based on this optimization is provided in Table 1. Sub-centimeter ($\sim$6.5 mm) error was achieved for the test table grid locations, whereas $\sim$1.7 cm total error was achieved for near-field grid locations.

During dense point cloud generation, the "High" resolution and "Aggressive" filtering settings in Agisoft Photoscan® were used. The total points in the four models presented in Table 1 show consistent coverage for the test table grid and near-field grid locations across the pre- and post-shot models. The point clouds registered no data gaps due to the densely overlapping photo coverage and extension of flight paths 10–20 m beyond the desired data capture region.

The GCP network was also comparatively examined from pre-test to post-test, to assess whether the underground explosion resulted in any lateral translation ($\Delta x$ and/or $\Delta y$) of portions of the area.

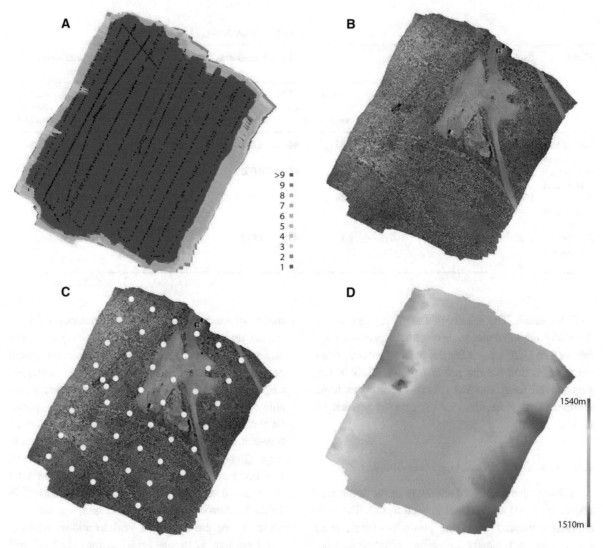

Figure 6
Simplified steps of the Agisoft Photoscan® workflow including: **a** photo overlap analysis; color legend indicates number of photos available at a given location, **b** point cloud creation, **c** georectification from GCPs, and **d** DEM construction

### 4.3. Hillshade DEMs

Following georectification, the dense point cloud was processed into a DEM. As indicated in Table 1, the resolution of the test table grid area DEMs was approximately 6.5 mm and nearfield grid area were approximately 1.7 cm. Figure 7 presents hillshade maps derived from the DEMs for the test table grid region pre- and post-shot models. These models generated details sufficient to map fine-scale features on and near the test table (Fig. 8). Some of the

features visible in the dataset include small (1–2 cm diameter) cobbles, GCP paper targets, cables, rills, tire tracks, sandbags, and footprints. Additionally, more spatially extensive features such as zones of uplift and subsidence at various locations around the test table, large (40–50 cm diameter) boulders, rock fractures, and surface grading or filling are detectable. The DEMs resulting from the test table grid region contain minimal noise, since the construction of the DEM relies largely on the dense GCP network.

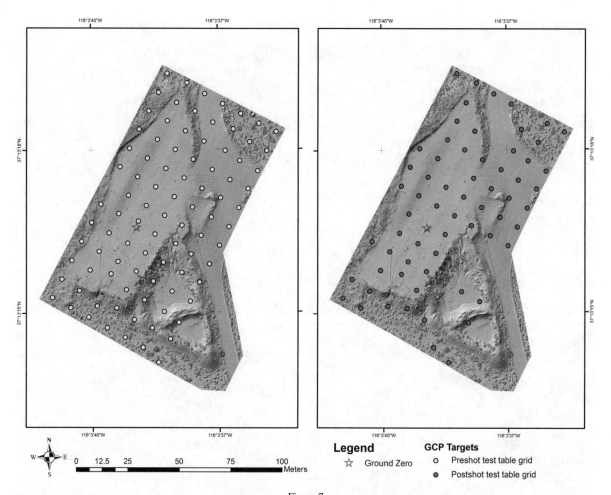

Figure 7
Hillshade maps derived from 6.5 mm DEMs from the test table grid region. The *left panel* is the pre-test model and the *right panel* is the post-test model. Ninety-two GCP targets (*yellow dots*) were surveyed in the pre-test model, whereas 65 GCP targets (*blue dots*) were surveyed in the post-test model

Areas of the near-field grid DEM off the graded test table, however, do show noise, likely resulting from vegetation in these areas. This noise is ameliorated by the resurvey of GCPs in the near-field grid post-experiment, indicating no vertical change in the position of GCPs.

A major feature in the near-field grid region is a granite outcrop (Fig. 9). Individual boulders, fractures, and GCP targets were mapped in this outcrop for change analysis. A raster differencing of the pre-test from the post-test DEMs provides both quantification and visualization of the spatial extent of vertical surface deformation ($\Delta z$) resulting from the SPE-5 test (Fig. 10).

### 4.4. Data Resolution

Within the near-field grid area, a total of 1060 overlapping photos were collected at an average 22 m AGL flight ceiling. Processing these photos in Agisoft led to the generation of a dense point cloud (5410 points/m$^2$) and a 1.7-cm hillshade DEM (Fig. 8). Data of this resolution permit the accurate mapping of discrete surface fractures (including strike and dip analyses) and $\Delta z$ and volumetric difference analyses (when used as a change-detection tool).

Data collection in the test table grid region yielded 1241 overlapping photos collected at an average 9 m AGL flight ceiling. The Agisoft

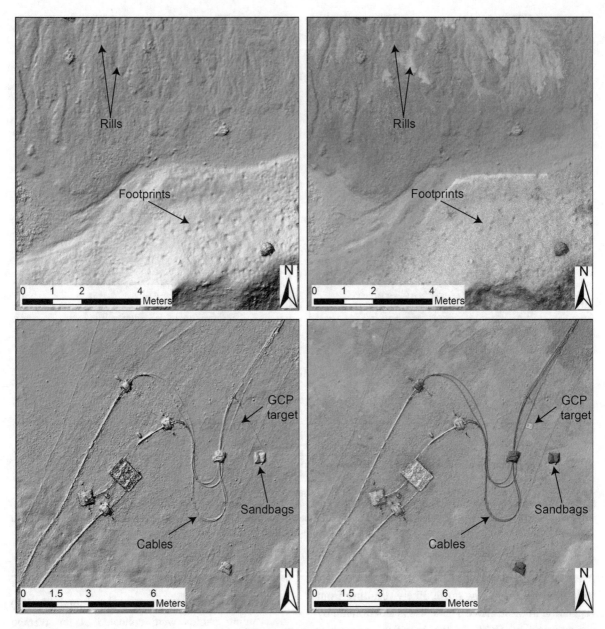

Figure 8
Examples of high-resolution hillshade DEMs (*left panels*) plus draped orthoimagery (*right panels*; with 50% DEM transparency) from the test table grid post-shot model. *Upper images* footprints in the lower half of the image on the slope of the muck pit, and localized rills with centimeter-scale incision. *Lower images* surface topographic relief detected on 2 cm wide cables, sandbags, and laminated paper GCP targets

processing of this photo suite generated an ultra-dense point cloud (26,560 points/m$^2$) and a 6.1-mm grid size DEM. Data of this resolution permits discretion of very subtle features, including footprints, cables, laminated sheets of paper, tire tracks, and surface fractures of 1 cm relief (Fig. 8).

## 5. Discussion

Four composite orthoimages were produced for the field area—one each pre-experiment and post-experiment, for both the test table grid and the near-field grid. These orthoimages serve as a fully archival

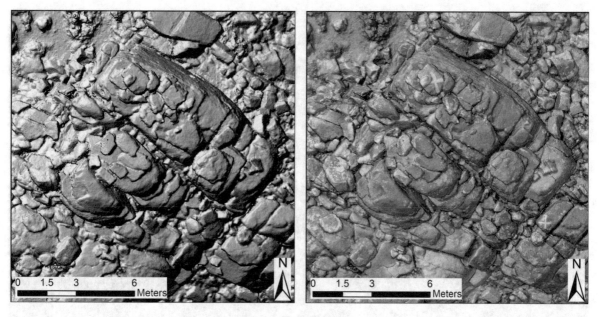

Figure 9
Hillshade map of the 7 m-quartz monzonite outcrop west of the test table derived from 1.7 cm DEM from the near-field grid region. The *right panel* includes an orthoimage draped over the DEM with 50% transparency. Individual cobbles and boulders, fractures, and GCP points (*white crosses*) are readily observable

dataset of the visible conditions of the field area before and after SPE-5. Figure 11 shows pre- and post-experiment comparative imagery within 5 m of the emplacement hole.

Within the post-experiment orthoimagery, surface fractures near surface ground zero are visible in a variety of orientations, with the greatest quantity of visible fractures appearing to be within 8 m of the emplacement hole. Also visible are regions of color change on the ground surface not present in the pre-test orthoimagery. These discolored regions are surface deposits caused by geysering of perched groundwater via two instrumentation boreholes on the test table, as well as the evacuation of grout and gravel stemming material from the emplacement hole in the minutes following SPE-5 zero-time (Fig. 11).

This study compared DEMs, not point clouds, to assess site changes resulting from the underground experiment. The $\Delta z$ DEMs show that up to 18 cm uplift was recorded within a 2-m radius of surface ground zero (Figs. 12, 13a). Within 5 m of surface ground zero, a maximum of 7 cm of uplift was detected. The addition of material to the ground surface as a result of the evacuation of stemming material from the emplacement hole is visible (Fig. 12). Experiment-induced uplift of at least 2 cm is present over 70% of the graded test table. At the northern and southwesternmost corners of the graded test table, up to 2 cm of subsidence was recorded over a broad region. The steep margins of the muck pit exhibited raveling and erosion in certain locations, and redeposition of that material in downslope locations (Fig. 12).

The surfaces changes presented in this study can be compared to those measured by ground-based lidar for earlier experiments SPE-2 and SPE-3 (see Fig. 3). These experiments were shallower and smaller, but had the same scaled depth of burial, and thus would be expected to have broadly similar surface uplifts, at least within the same order of magnitude. The lidar-derived surface uplifts showed a maximum of 3 cm uplift on the graded test table, with contours showing elongation in a northeast-southwest orientation (Fig. 3). Outside the gravel-covered area within 2 m of the shot, the photogrammetry-derived surface uplifts after SPE-5 show a maximum of 4 cm of uplift, with contours also elongated in a northeast–southwest direction

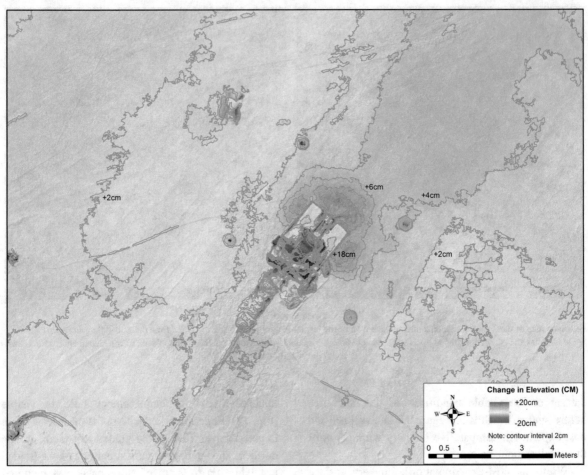

Figure 10
Contour map showing Δz near-surface ground zero in the test table grid region

(Fig. 10). The similarity in magnitude and orientation of uplift from two different techniques provides confidence in these trends. Aligning broadly with the orientation of the Boundary Fault and with a major joint set identified by Orkild et al. (1983), our data suggest that the surface damage manifested from SPE-5 exploits a pre-existing structural grain in the Climax Stock host rock. Additional evidence is seen in the measured horizontal translation of GCPs following SPE-5 (Fig. 13b), which shows that 31 of 88 GCPs surveyed post-experiment record as much as 2.64 cm of lateral translation from their pre-experiment locations. Of those 31 GCPs with Δx and/or Δy, 29 represent motion away from surface ground zero, and over half express vectors of motion towards the east and south.

Compared to the 2.8-cm resolution terrestrial lidar data presented in Fig. 3 (Sussman et al. 2012; Schultz-Fellenz et al. 2013), the flexible and versatile UAS surface change characterization process provides DEM data of sub-centimeter resolution and captured data without requiring personnel to access the site and potentially destroy signatures. In addition to drastically increasing time efficiency in the field, the UAS photogrammetry method also permitted greater efficiency in computational processing time. Whereas the generation of DEMs from terrestrial lidar in prior SPE campaigns had taken weeks (due largely to the meshing of multiple scan setups), generation of DEMs from photogrammetric data took a matter of hours. The UAS method also offers more complete coverage of the study area. The image

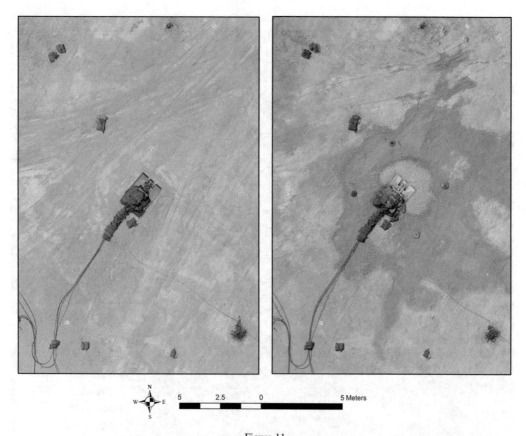

Figure 11

Orthophotos from surface ground zero in the test table grid region. *Left panel* from the pre-test dataset, *right panel* from the post test dataset. The region of darker material on the ground surface was a result of stemming material evacuation from the emplacement hole. A small fan of lighter-colored gravel is visible immediately adjacent to the northeast side of the emplacement hole

capture density and multiple flight passes over features during different times of day and varying sun angles minimized data gaps in shadowed regions.

UAS photogrammetry provides a direct technique to quantify surface spall of underground explosions at fine-scale resolution. The data resolution acquired in this investigation is critical for constraining collection requirements in applying this technology to monitoring and verification of underground explosions. This investigation shows that UAS-borne photogrammetry has the potential to further explore rapidly detected anomalous seismic signals in challenging terrain with centimeter- to sub-centimeter scale resolution.

Spallation results as the tensile stresses of the downgoing shock wave reflect off the free (ground) surface. Spall, a visible and measurable signature of geomaterial damage, is both a record of surface damage and a cue to additional damage in the subsurface (Patton et al. 2005, 2012; Sussman et al. 2016). The subsurface environment (i.e., material and fracture properties) imparts subtle yet significant impacts on the propagation, speed, and potential phase shifts of seismic waves resulting from the underground explosion (cf. Patton et al. 2012). Furthermore, quantification of surface spall, including total mass and spatial extent, can clarify fundamental modes of seismic surface waves and, therefore, improve confidence in moment tensor results from explosively induced seismic events (Patton 1991).

## 6. Conclusions

The use of low-altitude UAS-borne photogrammetry, along with a careful ground control point grid,

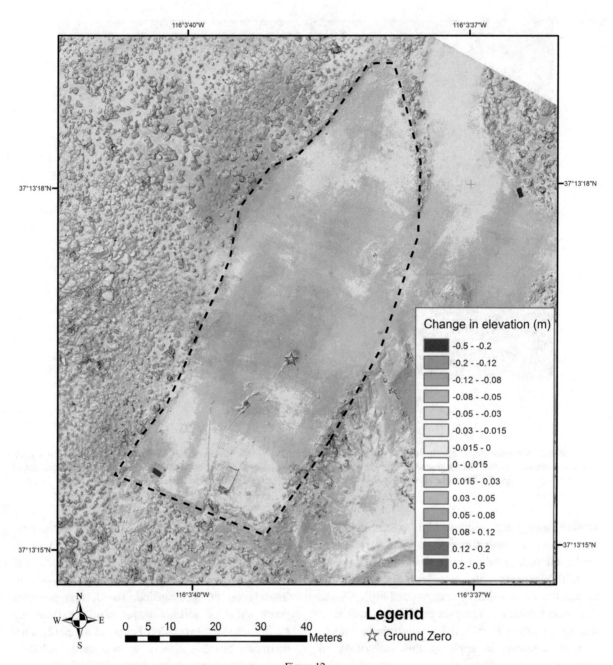

Figure 12
Surface $\Delta z$ across the test table grid region

produced DEMs with as dense as 6.1 mm grid size with little to no systemic error. The resolution of these datasets is sufficient to perform DEM difference calculations to detail discrete regions of uplift, subsidence, surface fractures, erosion, and deposition related to a controlled underground chemical explosion in granite. The rapid data collection occurred over less than eight flight hours per collection and fully covered the survey area. The method of collection ensured precision survey control on data while preserving subtle features.

Comparisons to terrestrial lidar over a section of the same region from previous tests in the SPE series show that while both techniques demonstrated the

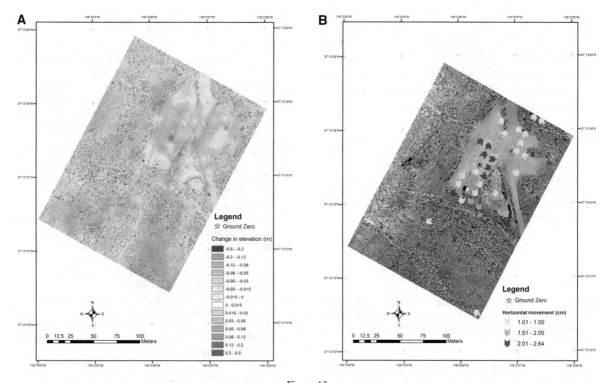

Figure 13
**a** Surface $\Delta z$ across the near-field grid region. **b** Horizontal translation ($\Delta x$ and/or $\Delta y$) as detected by repeat survey of GCPs. *Color* of *arrow* indicates range of movement; *arrow* vector points in direction of movement

capability to detect surface changes from the underground explosion, the UAS-based technique was faster, offered superior data resolution, had fewer coverage gaps, provided a less-invasive collection methodology, and acquired orthoimagery in addition to elevation information.

UAS photogrammetry can evaluate surface topographic changes created from small-yield underground explosions and has amplified impact when conducted in a manner that permits control from pre- and post-test conditions. The measured vertical surface topographic change resulting from the small-yield underground explosion carried out for the SPE-5 shot was on the order of 1–5 cm and locally as much as several tens of cm. The maximum amount of topographic change was concentrated at the ground surface at, and immediately north of, the emplacement hole. When used both before and after execution of underground chemical explosions, this technique can be used to determine the baseline topography and quantify the explosion-induced changes to the surface. Our work suggests that the geologic environment plays a significant role in how, how much, and where the surface responds to underground explosions.

*Acknowledgements*

The Source Physics Experiments (SPE) would not have been possible without the support of many people from several organizations. The authors thank NNSA, the Office of Defense Nuclear Nonproliferation Research and Development, and the SPE working group, a multi-institutional, interdisciplinary group of scientists and engineers. This work relied on the skill, dedication, and focus of our UAS pilot, Michael Grimler of Los Alamos National Laboratory's Security Division, for safe and successful airborne operations. Katherine Norskog and Steven Clement of Los Alamos National Laboratory, Leon Berzins and Beth Dzenitis of Lawrence Livermore National Laboratory, TJ Williams of Sandia National Laboratories, and Jesse Bonner and Robert Ziehm of

National Security Technologies, LLC provided essential field and logistics support. Los Alamos National Laboratory performs work for the US Department of Energy under contract DE-AC52-06NA25396. This document is unclassified and has been approved for unlimited release (LA-UR-16-29246).

## REFERENCES

Bemis, S. P., Mickelthwaite, S., Turner, D., James, M. R., Akciz, S., Thiele, S. T., et al. (2014). Ground-based and UAV-based photogrammetry: a multi-scale, high-resolution mapping tool for structural geology and paleoseismology. *Journal of Structural Geology, 69,* 163–178.

Carbonneau, P. E., & Dietrich, J. T. (2016). Cost-effective non-metric photogrammetry from consumer-grade sUAS: implications for direct georeferencing of structure from motion photogrammetry. *Proceedings of Landforms and Earth Surfaces.* doi:10.1002/esp.4012.

Carrivick, J. L., Smith, M. W., & Quincy, D. J. (2016). *Structure from motion in the geosciences.* New York: Wiley Blackwell.

Cattermole, J. M., & Hansen, W. R. (1962). Geologic effects of the high-explosive tests in the USGS tunnel area, Nevada Test Site, US Geological Survey Professional Paper 382-B.

Fonstad, M. A., Dietrich, J. T., Courville, B. C., Jensen, J. L., & Carbonneau, P. E. (2013). Topographic structure from motion: a new development in photogrammetric measurement. *Earth Surface Processes and Landforms, 38,* 421–430.

Garcia, M. N. (1997). Field and photogrammetric methods for mapping nuclear induced surface effects at the Nevada Test Site, Nye County. Nevada: US Geological Survey Open-File Report 97-695.

Glenn, N. F., Streuker, D. R., Chadwick, D. J., Thackray, G. D., & Dorsch, S. J. (2006). Analysis of LiDAR-derived topographic information for characterizing and differentiating landslide morphology and activity. *Geomorphology, 73,* 131–148.

Grasso, D. N. (2001). GIS surface effects archive of underground nuclear detonations conducted at Yucca Flat and Pahute Mesa, Nevada Test Site, Nevada, US Geological Survey Open-File Report 01-272.

Gutierrez, R., Gibeaut, J. C., Smyth, R. C., Hepner, T. L., Andrews, J. R., Weed, C., Guteliuz, W., & Mastin, M. (2001). Precise airborne lidar surveying for coastal research and geohazards applications. In *Internationall Archives of Photogrammetry and Remote Sensing* (Vol. XXXIV-3/W4), Annapolis.

X. Cong, Gutjahr, K. H., Schlittenhardt, J., & Soergel, U. (2007). Measurement of surface displacement caused by underground nuclear explosions by differential SAR interferometry. In *International Society of Photogrammetry and Remote Sensing* (Vol. XXXVI-1/W51), Hannover.

Houser, F. N., & Poole, F. G. (1960). Age relations of the Climax composite stock, Nevada Test Site, Nye County, Nevada; USGS Misc. Investigations Map I-328, scale 1:4800.

Hugenholtz, C. H., Whitehead, K., Brown, O. W., Barchyn, T. E., Moorman, B. J., Le Clair, A., et al. (2013). Geomorphological mapping with a small unmanned aircraft system (sUAS): feature detection and accuracy assessment of a photogrammetrically-derived digital terrain model. *Geomorphology, 194,* 16–24.

James, M. R., & Robson, S. (2012). Straightforward reconstruction of 3D surfaces and topography with a camera: accuracy and geoscience application. *Journal of Geophysical Research, 117,* F03017. doi:10.1029/2011JF002289.

James, M. R., & Robson, S. (2014). Mitigating systematic error in topographic models derived from UAV and ground-based image networks. *Earth Surf Processes and Landforms, 39,* 1413–1420.

James, M. R., & Varley, N. (2012). Identification of structural controls in an active lava dome with high-resolution DEMs: Volcan de Colima, Mexico. *Geophysical Research Letters, 39,* L22303. doi:10.1029/2012GL054245.

Johnson, K., Nissen, E., Saripalli, S., Arrowsmith, J. R., McGarey, P., Scharer, K., et al. (2014). Rapid mapping of ultrafine fault topography with structure from motion. *Geosphere, 10*(5), 18.

Jozkow, G., & Toth, C. (2014). Georeferencing experiments with UAS imagery. *ISPRS Annals of the Photogrammetry, Remote Sensing, and Spatial Information Sciences, II-1,* 25–29. doi:10.5194/isprsannals-II-1-25-2014.

Larmat, C., Rougier, E., & Patton, H. (2017). Apparent explosion moments from Rg waves recorded on SPE. *Bulletin of the Seismological Society of America, 107,* 43–50.

Larmat, C. S., Steedman, D. W., Rougier, E, Delorey, A., & Bradley, C. R. (2015). Coupling hydrodynamic and wave propagation modeling for waveform modeling of SPE. In *EOS Trans AGU,* abstract#S53B-2799, 2015 AGU Fall Meeting, San Francisco.

Leprince, S., Berthier, E., Ayoub, F., Delacourt, C., & Avouac, J.-P. (2008). Monitoring Earth system dynamics with optical imagery. *EOS, 89,* 1–2.

Maldonado, F. (1977). Summary of the geology and physical properties of the Climax stock, Nevada Test Site: USGS Open File Report 77-356.

Micheletti, N., Chandler, J. H., & Lane, S. N. (2015). Investigation the geomorphological potential of freely available and accessible structure-from-motion photogrammetry using a smartphone. *Earth Surface Processes and Landforms, 40,* 473–486.

Molinari, M., Medda, S., & Villani, S. (2014). Vertical measurements in oblique aerial imagery. *ISPRS International Journal of Geo-Information, 3,* 914–928.

Niethammer, U., Rothmund, S., James, M. R., Travelletti, J., & Joswig, M. (2010). UAV-based remote sensing of landslides. *International Archives of Photogrammetry, Remote Sensing, and Spatial Information Sciences, XXXVIII*(Part 5), 496–501.

Nouwakpo, S. K., Weltz, M. A., & McGwire, K. (2015). Assessing the performance of structure-from-motion photogrammetry and terrestrial LiDAR for reconstructing soil surface microtopography of naturally vegetated plots. *Earth Surface Processes and Landforms, 41,* 308–322. doi:10.1002/esp.3787.

Orkild, P. P., Townsend, D. R., & Baldwin, M. J. (1983) Chapter A; Geologic investigations. In *Geologic and Geophysical*

*Investigations of Climax Stock Intrusive*, Nevada, USGS Open File Report 83-377.

Patton, H. J. (1991) Seismic moment estimation and the scaling of the long-period explosion source spectrum. In S. R. Taylor, H. J. Patton & P. G. (Eds.) *Richards explosion source phenomenology*. American Geophysical Union, Washington, D.C. doi:10.1029/GM065p0171.

Patton, H. J. (2015). New insights into the explosion source from SPE. In *EOS Trans. AGU*, abstract S51F-05, 2015 AGU Fall Meeting, San Francisco.

Patton, H. J., Bonner, J. L., & Gupta, I. N. (2005). Rg excitation by underground explosions: insights from source modeling the 1997 Kazakhstan depth-of-burial experiment. *Geophysical Journal International, 163,* 1006–1024.

Patton, H. J., Larmat, C., Rougier, E., Rowe, C. A., & Yang, X. (2012). *Seismic studies of Source Physics Experiments conducted at the Nevada National Security Site using close-in (<2 km) waveforms, 2012 Monitoring Research Review*. Albuquerque: Ground-based Nuclear Explosion Monitoring Technologies.

Pelletier, J. D., & Orem, C. A. (2014). How do sediment yields from post-wildfire debris-laden flows depend on terrain slope, soil burn severity class, and drainage basin area? Insights from airborne-LiDAR change detection. *Earth Surface Processes and Landforms, 39,* 1822–1832.

Phang, M. K., Simpson, T. A., & Brown, R. C. (1983) Investigation of blast-induced underground vibrations from surface mining, US Department of Interior Office of Surface Mining Report.

Piras, M., Taddia, G., Forno, M. G., Gattaglio, M., Aicardi, I., Dabove, P., et al. (2017). Detailed geologic mapping in mountain areas using an unmanned aerial vehicle: application to the Rodoretto Valley, NW Italian Alps. *Geomatics, Natural Hazards and Risks, 8,* 137–149.

Reshetyuk, Y., & Martensson, S.-G. (2016). Generation of highly accurate digital elevation models with unmanned aerial vehicles. *The Photogrammetric Record, 31,* 143–165. doi:10.1111/phor.12143.

Scaioni, M., Longoni, L., Mellilo, V., & Papini, V. (2014). Remote sensing for landslide investigations: an overview of recent achievements and perspectives. *Remote Sensing, 6,* 1–53.

Schultz-Fellenz, ES, AJ Sussman, RE Kelley, and DI Cooper, 2013, Post-shot surface damage detected with lidar at the Source Physics Experiment Site. In *EOS Trans. AGU*, abstract S31E-04, 2013 AGU Fall Meeting, San Francisco.

Snelson, C. M., Abbott, R. E., Broome, S. T., Mellors, R. J., Patton, H. J., Sussman, A. J., et al. (2013). Chemical explosion experiments to improve nuclear test monitoring. *Eos, 94,* 237–239.

Sussman, A. J., Schultz-Fellenz, E. S., Broome, S. T., Townsend, M. J., Abbott, R. E., Snelson, C. M., Cogbill, A., Conklin, G., Mitra, G., & Sabbeth, L. (2012). Characterization of the source physics experiment site. In *EOS Trans. AGU*, abstract S21C-03, 2012 AGU Fall Meeting, San Francisco.

Sussman, A. J., Swanson, E., Wilson, J., Townsend, M., & Prothro, L. (2016). Multi-scale fracture damage associated with underground chemical explosions. In *EOS Trans. AGU*, abstract MR41A-2686, 2016 AGU Fall Meeting, San Francisco.

Tuffen, H., James, M. R., Castro, J. M., & Schipper, C. I. (2013). Exceptional mobility of an advancing rhyolitic obsidian flow at Cordon Caulle volcano in Chile. *Nature Communications*. doi:10.1038/ncomms3709.

Warrick, J. A., Ritchie, A. C., Adelman, G., Adelman, K., & Limber, P. W. (2016). New techniquest to measure cliff change from historical oblique aerial photographs and structure-from-motion photogrammetry. *Coastal Research, 33,* 39–55.

Westoby, M. J., Brasington, J., Glasser, N. F., Hambrey, M. J., & Reynolds, J. M. (2012). 'Structure from Motion' photogrammetry: a low-cost, effective tool for geoscience applications. *Geomorphology, 179,* 300–314.

A. H. Chowdhury, & Wilt, T. E. (2015). Characterizing explosive effects on underground structures, US Nuclear Regulatory Commission report NUREG/CR-7201.

Yu, H., Yuan, Y., Yu, G., & Liu, X. (2014). Evaluation of influence of vibrations generated by blasting construction on an existing tunnel in soft soil. *Tunneling and Underground Space Technology, 43,* 59–66.

(Received  January 12, 2017, revised  July 27, 2017, accepted  August 8, 2017, Published online  August 19, 2017)

Pure Appl. Geophys. 175 (2018), 3179–3191
© 2018 Springer International Publishing AG, part of Springer Nature
corrected publication August 2018
https://doi.org/10.1007/s00024-018-1785-1

# Applicability of Unmanned Aerial Vehicles in Research on Aeolian Processes

ALGIMANTAS ČESNULEVIČIUS,[1] ⓘ ARTŪRAS BAUTRĖNAS,[1] LINAS BEVAINIS,[1] DONATAS OVODAS,[1] and KĘSTUTIS PAPŠYS[1]

*Abstract*—Surface dynamics and instabilities are characteristic of aeolian formation. The method of surface comparison is regarded as the most appropriate one for evaluation of the intensity of aeolian processes and the amount of transported sand. The data for surface comparison can be collected by topographic survey measurements and using unmanned aerial vehicles. Time cost for relief microform fixation and measurement executing topographic survey are very high. The method of unmanned aircraft aerial photographs fixation also encounters difficulties because there are no stable clear objects and contours that enable to link aerial photographs, to determine the boundaries of captured territory and to ensure the accuracy of surface measurements. Creation of stationary anchor points is irrational due to intense sand accumulation and deflation in different climate seasons. In September 2015 and in April 2016 the combined methodology was applied for evaluation of intensity of aeolian processes in the Curonian Spit. Temporary signs (marks) were installed on the surface, coordinates of the marks were fixed using GPS and then flight of unmanned aircraft was conducted. The fixed coordinates of marks ensure the accuracy of measuring aerial imagery and the ability to calculate the possible corrections. This method was used to track and measure very small (micro-rank) relief forms (5–10 cm height and 10–20 cm length). Using this method morphometric indicators of micro-terraces caused by sand dunes pressure to gytia layer were measured in a non-contact way. An additional advantage of the method is the ability to accurately link the repeated measurements. The comparison of 3D terrain models showed sand deflation and accumulation areas and quantitative changes in the terrain very clearly.

**Key words:** Climate indicators, unmanned aerial vehicles, aeolian processes, imagery measurement accuracy.

**Electronic supplementary material** The online version of this article (https://doi.org/10.1007/s00024-018-1785-1) contains supplementary material, which is available to authorized users.

[1] Department of Cartography and Geoinformatics, Vilnius University, M.K.Čiurlionis St. 21/27, 03101 Vilnius, Lithuania. E-mail: algimantas.cesnulevicius@gf.vu.lt

## 1. Introduction

The course and intensiveness of aeolian processes are determined by several important factors, such as surface deposition and degree of weathering thereof, meteorological conditions and vegetation. Aeolian processes are highly dynamic and their dynamicity is determined by sudden weather changes. Changeable weather strongly determines the choice of research methodology and its actual application when conducting research on aeolian processes. The classical methods (granulometric investigation of the deposition, fixation of the drifting sand mass with the help of sand catchers, fixed weather registration methods) allow collecting fundamental data that describe the process. On the other hand, the local nature of related observations does not allow giving an exact assessment of the spread and changes characteristic of aeolian processes in larger areas. The obtained exact local data need to be interpolated (in reliable cases) or extrapolated (in much less reliable cases).

Distance air methods are applicable in the process of spatial evaluation of the intensiveness of aeolian processes. However, the classical aerial photography or space photography requires major investment. Over the last 10 years the arrival of cheap unmanned aerial vehicles boasting a rather stable flight has resulted in a wide application of contactless research methods that are exact, reliable, relatively cheap and expeditious.

The unmanned aerial vehicles (UAV) images are widely used to capture static objects when many pictures of a static subject, taken at different angles, are combined through the "structure-from-motion" (SfM) workflow to create a digital surface model (DSM) and a composite image with images acquired using a typical digital camera. This method is also suitable for surface dynamic processes, but there emerges a problem of identifying the exact position of the surface points. This problem has been addressed in numerous publications (Clapuyt et al. 2015; Fonstad et al. 2013; Harwin and Lucieer 2012; James and Robson 2012, 2014; James et al. 2017; Mancini et al. 2013; Niedzielski et al. 2016; Smith et al. 2016; Westoby et al. 2012).

In 2015–2016 aerial photography enabled by unmanned aerial vehicles was used to research the drifting sand dunes on the Curonian Spit. The aim of the research was to develop a sequence of methodological actions allowing to obtain reliable and exact data on sand accumulations, surface microforms and related dynamics.

## 2. Object of Research

The Curonian Spit is a macroform of littoral origin located on the southeaster coast of the Baltic Sea (Fig. 1). The formation of the Curonian Spit started at the end of the Pleistocene. The melting of the last glaciation ice cover witnessed various development stages of the Baltic Sea (the Baltic Ice Lake, Littorina Sea, Mia, Yoldia Sea) that caused changes in the sea level (Bitinas et al. 2001, 2002, 2005; Damušytė 2009; Kabailienė et al. 2009; Kondratienė and Damušytė 2009; Molodkov et al. 2010).

Littoral erosion severely affected the western shores of the Semba Peninsula. The prevailing waves and sea currents carried the weathering deposition flood to the north. The Curonian Spit together with the sea coast and the Curonian Lagoon make up a complex geological–geomorphological unit. In addition to aeolian processes, this area boasts unique micro-tectonic processes characteristic of the dunes and related to various deformations of the gytia formed in the freshwater basin (Fig. 2). Unique geological–geomorphological processes resulted in the formation of a freshwater gytia lake which surfaces on the shore of the Curonian Lagoon. Geological–geomorphological structures of the kind are also found in other places on the Baltic Sea coast (Lampe et al. 2011; Sergeev et al. 2017). Aeolian sand accumulations weigh on the gytia layer and make it sink. This causes the formation of micro-terraces on the eastern slope of the dune–ridge the height and width of which change after every autumn–winter season.

The formation of the spit ended some 4000 years ago. Due to natural afforestation, intensive geomorphological processes occurred only on the sea coast (littoral accumulation) and in separate non-afforested areas (aeolian accumulation). After the Curonian Spit was populated, its deforestation started, which finally led to more intensive aeolian deflation–accumulation processes. The anthropogenic impact was especially strong in the thirteenth to fourteenth century when the spit became a communication route between Konigsberg and Klaipėda (Fig. 3).

During the wars of the seventeenth to eighteenth century the spit was subjected to intensive deforestation and suffered from fires which resulted in large completely deforested areas covered in drifting sand. At the end of the eighteenth century the major part of the Curonian Spit was already an open area affected by aeolian deflation and accumulation processes. Afforestation started only in the mid-nineteenth century. Since then the area of drifting sand dunes has continued decreasing. At the beginning of the twenty-first century there were only three remaining spots with drifting sand dunes: Juodkrantė–Pervalka, Nida–Pilkopa and Pilkopa–Rasytė. As of the mid-nineteenth century the area of drifting sand dunes became four times smaller in size (Fig. 1, Table 1). In 2015 there were only four segments of drifting sand dunes on the Curonian Spit. The largest of them is 11.1 km$^2$ in size and it covers a narrow 50–100 m strip of the coastal beach and the protective dune–ridge.

## 3. Research Methodology

The analysis of sand resources and sand balance requires frequent and repeated mapping efforts. Classical geodesic over-ground measurements aimed at researching aeolian dynamics is time and money consuming. Aerial photography aided by unmanned aerial vehicles is a much cheaper and more efficient way to research the microforms of aeolian relief and assess the volume of drifting sand. However, this method has some disadvantages that limit its applicability to a certain extent. First of all, it has to do

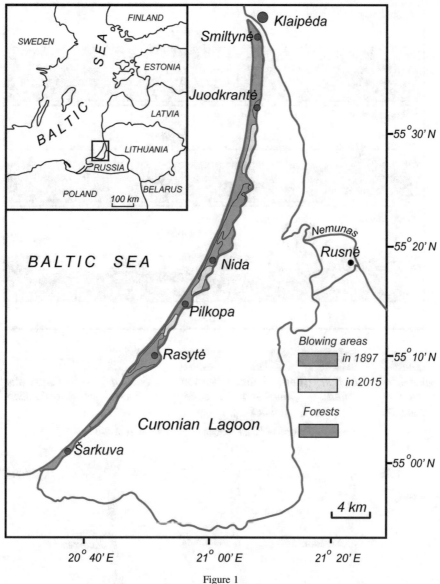

Figure 1
Location of the Baltic Sea and Curonian Spit

with the actual meteorological situation (wind, rain) that has direct impact on the flight of an unmanned aerial vehicle and the quality of aerial photographs. When planning respective surface measurements, weather forecasts for the next couple of days were considered.

Meteorological conditions directly impact the intensiveness of aeolian processes. When capturing any changes to the ground surface, it is important to consider the following two aspects: the beginning and

the end of strong wind periods. The comparison of aerial images of a ground surface taken before the very arrival of a storm and right after it calms down is the most efficient way to analyse the volume of drifting sand and changes in the form of aeolian relief.

Mapping of the dune surface was performed with the help of unmanned aerial vehicle DJI Inspire and performing geodesic over-ground measurements of the surface. The following parameters of spatial

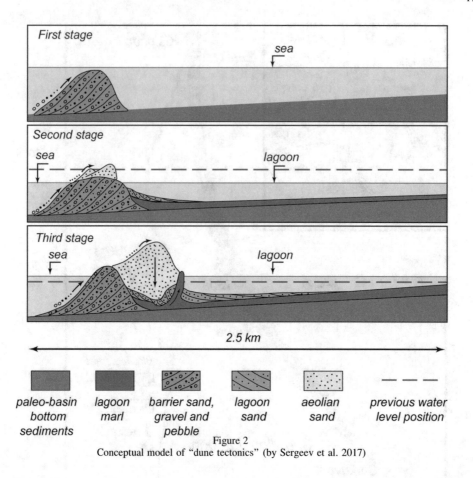

Figure 2
Conceptual model of "dune tectonics" (by Sergeev et al. 2017)

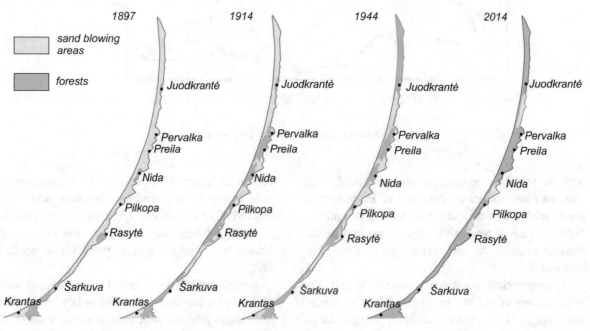

Figure 3
Sand blowing areas and forestry changes in the Curonian Spit since 1897

Table 1

*Sand blowing areas and forestry changes in the Curonian Spit (1849–2015)*

| Year | Forest areas | | Blowing areas | |
|------|------|------|------|------|
| | km$^2$ | % | km$^2$ | % |
| 1849–1860 | 9.8 | 5.8 | 159.2 | 94.2 |
| 1914 | 35.7 | 21.1 | 133.3 | 78.9 |
| 1944 | 48.9 | 28.9 | 120.1 | 71.1 |
| 2015 | 133.3 | 78.9 | 35.6 | 21.1 |

Table 2

*UAV DJI Inspire camera specification*

| | |
|------|------|
| Camera type | X3 |
| Model | FC350 |
| Total pixel | 12.76 M |
| Effective pixel | 12.4 M |
| Maximal image size | 4000 × 3000 |
| ISO range | Photo—100–1600 |
| | Video—100–3200 |
| The electronic shutter speed | 8–1/8000 s |
| A field of view (FOV) | 94° |
| Different supported file formats | Photo: JPEG, DNG. Video: MP4/MOV (MPEG-4 AVC/H.264) |
| Types of electronic media | Micro SD. Maximal capacity 64 GB. Class 10 or UHS-1 |

position of the unmanned aerial vehicle and the photo camera determine the accuracy of photo images (Table 2).

To compare the accuracy of the results, the special produced marks as ground control points (GCP) were prepared to cover the area of the dunes subject to research. A GCP is a characteristic point whose coordinates are known. At least three GCPs are needed to scale, rotate and locate the model, but it is recommended to have between 5 and 10 GCPs, well distributed over the area, depending on the terrain, the difficulty of the area and the height differences. The described workflow was considered in our project.

GCPs were coordinated with the help of Trimble, a precise GPS device. The planned position of the coordinated marks (GCP) was established with the accuracy of 7 mm and the absolute altitude was established with 12 mm accuracy. The GCPs were used to georeference the project. It was inevitable to use the network of marks (GCPs) because on the surface of drifting sand there are no clear contours that would be suitable for the purpose of establishing

fixed points. Moreover, the over-ground fixation of marks (GCPs) helped ensure the accuracy of aerial image analysis. Peripheral segments of aerial images contain inadmissible distortions of the position of points exceeding 5 mm and determined by environmental conditions (wind gusts, lighting). Yet another important reason for the distortions to occur was the fact that the area subjected to research, that is the Nida–Pilkopa stripe, is located close to the Lithuanian–Russian border where the internet connection is not always reliable. This had a direct impact on the ability to control the flight of the unmanned aerial vehicle.

The flight mission was planned using the mobile app Pix4dcapture. This mobile app helped to define flight mission to the map area and to customize mapping parameters such as flight altitude, overlap and so on. After the mission, the data were transferred to Pix4D desktop, where full-resolution images were analysed using advanced tools on Pix4Dmapper Pro.

Pix4Dmapper Pro software was applied to process the images. This software is based on automatically finding thousands of common points between images. When two or more key points on two different images are found to be the same, the software is able to match key points. Then each group of correctly matched key points were generated to one 3D point cloud. The more key points are available, the more accurately 3D point cloud can be computed. A very important rule is to maintain high overlap between the images.

The recommended overlap for most cases is at least 75% frontal overlap (with respect to the flight direction) and at least 60% side overlap (between flying tracks). It is recommended that the camera should be maintained at a constant height over the terrain/object as much as possible to ensure the desired ground sampling distance (GSD). Higher overlap (85% frontal overlap and 70% side overlap) was applied in this particular case and the GSD equalled 2 cm/pixel.

## 4. Climate Indicators

An extensive analysis of climate indicators covering a 30-year period enabled to establish the most suitable periods to measure the aeolian relief on the Curonian Spit with the help of unmanned aerial vehicles. The following three indicators that directly impact the aeolian processes were analysed: wind regime, precipitation and air temperature.

The most intensive aeolian processes occur when the weather is hot, dry and windy. Compared to the rest of Lithuania, the Curonian Spit has a unique climate. A respective analysis of climate indicators was performed based on the data of the meteorological stations located in Klaipėda and Nida.

Air temperature is an extremely important indicator that has impact on aeolian processes. During the cold season aeolian processes come to a standstill or are very slow and take place only when storms arrive. Due to low temperatures, a frost crust forms on the surface of the sand thus protecting it from deflation. The cold season on the Curonian Spit lasts for 2 months on average, that is, January and February. Cold years have a longer cold season that lasts for 3 months with the average daily temperature below zero in December as well.

Figure 4 ▶
Multiannual average wind speed registered at meteorological stations in Nida and Klaipėda: **a** December–February (Nida); **b** March–May (Nida); **c** June–August (Nida); **d** September–November (Nida); **e** multiannual average wind speed (Klaipėda); **f** multiannual maximum wind speed (Klaipėda)

Yet another important indicator that impacts aeolian processes is the amount of precipitation and its distribution in time. There is a strong relationship between aeolian processes and the number as well as the duration of precipitation-free periods. The analysis of precipitation data for the period of recent 30 years provided by the meteorological station in Nida showed that the dry periods are rather evenly distributed throughout the seasons. The average duration of precipitation-free periods in spring (March–May) is 15 days, in summer (June–August)—13 days, and in autumn (September–November)—10 days, respectively. Nonetheless, the duration of dry periods varies from year to year. For example, in the spring of 1992 the dry period lasted for 30 days and covered the whole of May; in the spring of 2000 the dry period lasted for 25 days and covered almost the whole of May; and in the spring of 2009 the dry period lasted for 26 days and covered almost the whole of April. The minimum duration of a precipitation-free period in spring was 6 days and the maximum lasted for 30 days. In the summer the longest dry period lasted for 29 days (July 1994) and the shortest was 5 days long, respectively. The longest precipitation-free period in autumn was registered in 1998. It lasted for 21 days and spread over the months of September and October. The October of 2005 boasted 19 dry days.

Wind is the key climatic factor that determines aeolian processes. The speed, direction and timing of wind vary from season to season (Fig. 4). The analysis of multiannual climatic data provided by the meteorological station in Nida shows that the Curonian Lagoon exerts major impact on wind dynamics. According to the data of the meteorological station in Nida, eastern, southeastern and southern winds with the average speed of 5–6 m/s prevail at all seasons. In the meantime, the data of the coastal meteorological station in Klaipėda show that southwestern, western

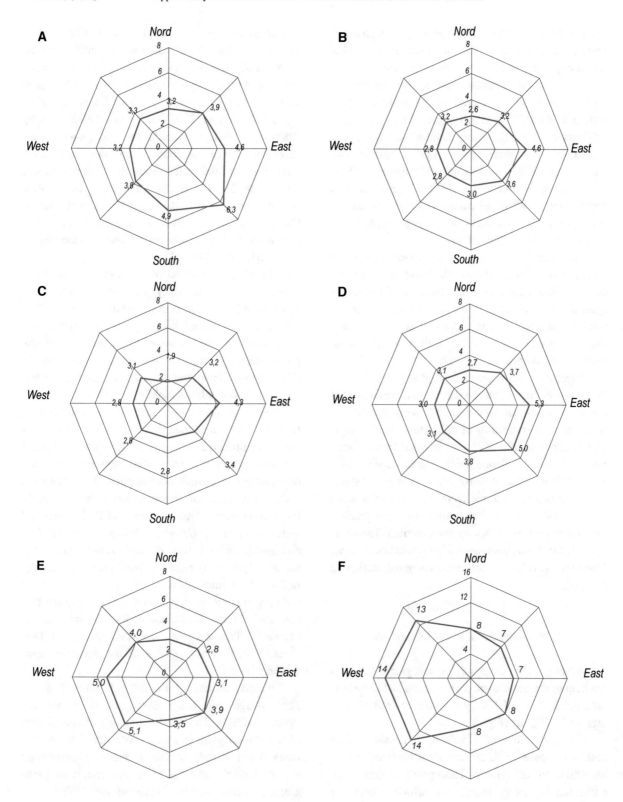

and northern winds with the average speed of 5.0 m/s prevail on the seacoast. The maximum average wind speed registered at the meteorological station in Nida is 15 m/s. In the meantime, the maximum average daily wind speed registered at the meteorological station in Klaipėda is much higher. In the spring, it amounts to 20 m/s (2000), in summer—18 m/s (1981 and 2002), in autumn—26 m/s (1981), and in winter—26 m/s (1982), respectively. Northwestern (51% of all cases), southwestern (38% of all cases) and southeastern (11% of all cases) winds prevail among those with the maximum speed registered, while others make up a mere 1% only.

The identified differences in wind regime are determined by the position of the two meteorological stations. The meteorological station in Klaipėda is open to winds blowing in all directions. The meteorological station in Nida is located at the foot of a dune–ridge that acts as a shelter from southwestern, western and northwestern winds.

Episodic measurements of wind speed and direction allow claiming that the wind regime registered at the meteorological station in Klaipėda provides a more accurate reflection of the actual wind regime that is characteristic of the dune–ridge in the Curonian Spit. The correlation coefficient of wind speed is $r = 0.988$, and that of wind direction is $r = 0.998$.

When planning the flights of the unmanned aerial vehicle, attempts were made to fit into the precipitation-free periods (in spring and autumn). The exact dates of the flights were determined based on weather forecasts provided by the meteorological station in Klaipėda.

## 5. Research Results and Discussion

Mapping of the dune surface was performed with the help of the unmanned aerial vehicle DJI Inspire-1. As a result, 209 aerial images were taken covering an area of 82,350 $m^2$ (Fig. 5).

The application of UAV for dynamic surface research is faced with an important issue of precise identification of points. This problem has been addressed by many researchers whose works are mentioned in the introduction to the article. The blowing aeolian surfaces are also dynamic, so additional over-ground methods have been used to ensure precision identification of the surface points.

With the aim to assess the accuracy of obtained measurements, a network of basic marks, including 11 marks as GCPs, were coordinated and the covering of an area of 23,400 $m^2$ was created (Fig. 5). Additional photogrammetric and topographic measurements were performed on the said area. Photographs of the surface were taken from a relative altitude of 50 m. Extreme differences in absolute altitude varied from 0.16 m (water level on the Curonian Lagoon) to 31.5 m (dune crest). The coordinates of the GCPs were established with the help of GPS device Trimble and the coordinate system LKS-94. The planned position of GCPs ($X$, $Y$) was established with the accuracy of 0.007 m and their absolute altitudes ($Z$) were established with 0.012 m accuracy. The actual accuracy of the results was determined by the technical characteristics of the photo camera attached to the unmanned aerial vehicle DJI Inspire-1, including the methods used to control the vehicle and related flight parameters.

In order to assess the impact of the methods used to control the unmanned aerial vehicle on the accuracy of obtained measurements, three different methods to control the vehicle and to merge the obtained photo images were applied. The respective flight parameters of the unmanned aerial vehicle (GPS coordinates, flight altitude and flight direction) were recorded on the photo images in EXIF (exchangeable image format) and XMP (extensible metadata platform) formats. The Centre for Cartography at Vilnius University developed special software (author Artūras Bautrėnas) to perform further analysis and assessment of the said information (Table 3). The software was titled Dron_Exif_Dat. Figure 6 depicts a fragment of the software management toolbar.

The coordinates of the central point in every photo image were established using the coordinate system WGS84. The obtained values of coordinates of the central point in every photo image were then recalculated based on the rectangular coordinate system LKS-94 and linked to the values of topographic measurements (Zakarevičius 1996). The position of the central point in every photo image is presented in Fig. 7 (red dots).

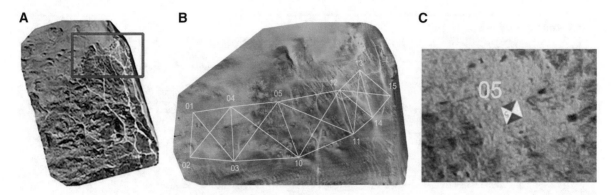

Figure 5
Mapped dune surface: **a** total mapped territory. **b** Verification territory (GCPs) to assess the accuracy of taken measurements. **c** Coordinated mark—GCP

Table 3

*Options used to capture photo images of the surface and deviation of the coordinates of photo images from geodesic measurements*

| Flight control and image merging methods | Photo images (units) | Deviation of the coordinates of photo images from geodesic measurements | | |
|---|---|---|---|---|
| | | $X$ (m) | $Y$ (m) | $Z$ (m) |
| Manual control of the unmanned vehicle and photography. Orientation and merging of photo images is based on the coordinates of the central point in every photo image | 38 | 1.533 | 1.834 | 2.072 |
| Automatic control of the unmanned vehicle and photography. Orientation and merging of photo images is based on the coordinates of the central point in every photo image | 31 | 0.352 | 0.416 | 0.293 |
| Automatic control of the unmanned vehicle and photography. Orientation and merging of photo images is based on the basic marks network | 46 | 0.020 | 0.022 | 0.036 |

In case of manual control of the unmanned vehicle and photography, the flight route was rather chaotic. Moreover, the overlapping of some photo images was reduced to the minimum (southwestern section of the route) which resulted in low accuracy of image merging. In case of automatic control of the unmanned aerial vehicle and photography, the merging of photo images was done based on the coordinates of the central point in every aerial image and the basic marks network. The number of photographs taken was smaller, but the accuracy of image merging was much higher. In the first case the overlapping of photo images was 60%, and in the second case—80%, respectively.

Yet another important factor influencing the accuracy of image merging was the position of the camera at the time of taking pictures. In case of the unmanned aerial vehicle DJI Inspire-1, the camera is fixed onto a special gimbal allowing compensating for the vibration of the flying vehicle and the skew angle, but the position of the camera is not completely stable. The relative altitude of the camera varied from 49.4 to 50.5 m and the yawn varied within the limits of ± 0.85°. It must be noted that the flight pitch variation within the limits of ± 15° had practically no impact on the photo camera, because the gimbal helped to stabilise it. Therefore, only in some cases the camera pitch was under 0.1° (Fig. 8).

Even though the timing of measurements was selected based on the extensive analysis of multiannual climatic indicators, the major fluctuations of the photo camera resulted from wind gusts. At the time of

47

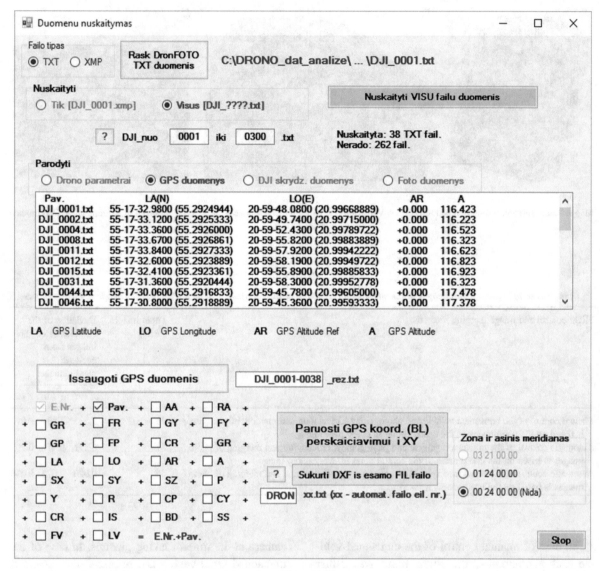

Figure 6
Fragment of Dron_Exif_Dat software management toolbar

taking the measurements wind speed at the altitude of 2 m did not exceed 5–6 m/s. Meanwhile, at the altitude of 50 m wind speed was much higher and difficult to forecast. Having in mind the instability of the flight performed by the unmanned aerial vehicle, an additional full software package Pix4DMapper Pro was used for the purpose of merging photo images. The software allows analysing and transforming high-resolution photo images captured with the help of an unmanned aerial vehicle and creating 2D and 3D models in addition to simplifying the process of

uploading, analysing and storing photo images in a variety of popular data formats. Software package Pix4Dmapper Pro enables aero-triangulation and camera calibration. It allows creating a point cloud, in addition to linking images to geographic coordinates, generating DSM and geomosaic.

To eliminate potential planning-related inaccuracies in terms of points, including altitude-related inaccuracies, a methodology aimed at comparing the coordinates of marks in photo images with respective results of geodesic measurements was developed

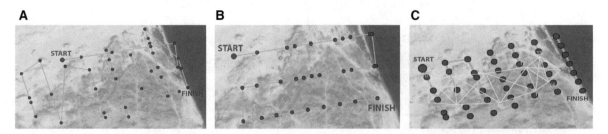

Figure 7

Flight trajectory: **a** manual control of the unmanned vehicle and photography. Orientation and merging of photo images is based on the coordinates of the central point in every photo image. **b** Automatic control of the unmanned vehicle and photography. Orientation and merging of photo images is based on the coordinates of the central point in every photo image. **c** Automatic control of the unmanned vehicle and photography. Orientation and merging of photo images is based on the GCPs

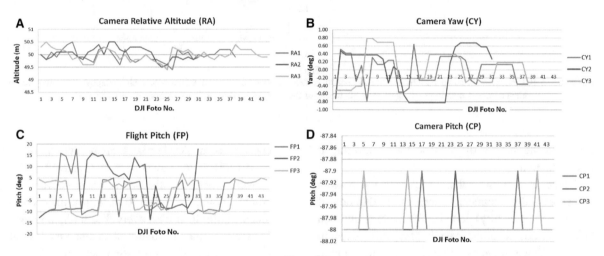

Figure 8
Indexes of flight parameters

(Skeivalas 2001). It is evident that the accuracy of establishing coordinates directly depends on the chosen methods of flight control and merging of photo images. The most accurate mapping is achieved when using a basic network to merge photo images that allows creating respective cross sections of microforms (micro-terraces) and 3D models of the surfaces that are being mapped (Fig. 9).

## 6. Conclusions

1. The intensiveness of aeolian processes occurring on the Curonian Spit is highly dependent on three major factors: wind, air temperature and precipitation volume. Eastern, southeastern and southern

winds with the average speed of 5–6 m/s prevail in the central part of the Curonian Spit at all seasons (according to the data of the meteorological station in Nida). Southwestern, western and northern winds with the average speed of 5.0 m/s prevail in the northern part of the spit (according to the data of the meteorological station in Klaipėda). The maximum average wind speed is up to 15 m/s and the maximum average daily wind speed in spring is 20 m/s, in summer— 18 m/s, in autumn—26 m/s, and in winter—26 m/ s, respectively.

During the cold season aeolian processes come to a standstill or are very slow and occur only when storms arrive. Due to low temperature, a frost crust

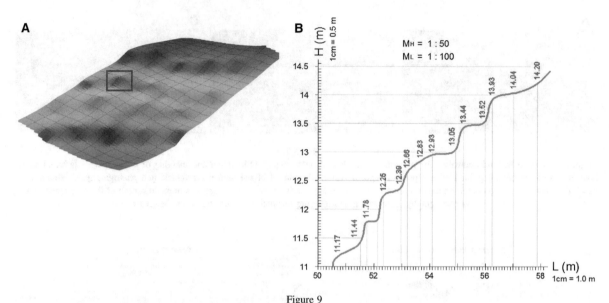

Figure 9

Basic marks (GCPs) network-based options for the automatic control of unmanned aerial vehicles and photography, including image
orientation and merging opportunities. **a** 3D model of the mapped surface. **b** Cross section of micro-terraces

forms on the surface of the sand thus protecting it from deflation.

There is a strong relationship between aeolian processes and the number as well as the duration of precipitation-free (dry) periods. The average duration of dry periods in spring (March–May) is 15 days, in summer (June–August)—13 days, and in autumn (September–November)—10 days, respectively. The longest dry periods were registered in 1992 (30 days), 2000 (25 days) and 2009 (29 days).

2. When unmanned aerial vehicles are used to capture images of a drifting sand surface, identification of the position of photo images and their accuracy becomes a problem. One of the issues is related to the stability of an unmanned aerial vehicle and the camera attached to it during the flight. Yet another issue is related to the fact that there are no clear contours on the surface of drifting sand. Problems related to the relative altitude above the surface that is being mapped, camera pitch and yaw can be eliminated with the help of coordinated marks as GCPs.

3. Maximum accuracy in merging photo images is achieved when using GCPs. It takes a lot of additional time and effort to develop a network of

coordinated marks. For large areas (over 50 ha) it is expedient to develop a coordinated point network where the distance between coordinated points amounts to 100–150 m with a fixed coordinated point in at least every fourth photo image.

4. In case of automatic control of an unmanned aerial vehicle, when the orientation and merging of photo images is based on the coordinates of the central point in every aerial image, it is possible to achieve a significantly higher accuracy in the process of establishing the coordinates. Having planned for a greater overlapping of photo images (up to 80%) and when taking photographs from a lower than 50 m altitude it is possible to achieve sufficient accuracy necessary to assess the morphographic and morphometric indicators of the microforms of aeolian relief.

### Appendix

URL: http://kc.gf.vu.lt/downloads/linasb/Neringa_parnidis.zip.

REFERENCES

Bitinas, A., Damušytė, A., Hütt, G., Jaek, I., & Kabailienė, M. (2001). Application of the OSL dating for stratigraphic correlation of Late Weichselian and Holocene sediments in the Lithuanian Maritime Region. *Quaternary Science Reviews, 20,* 767–772. https://doi.org/10.1016/S0277-3791(00)00011-1.

Bitinas, A., Damušytė, A., Stančikaitė, M., & Aleksa, P. (2002). Geological development of the Nemunas River Delta and adjacent areas, West Lithuania. *Geological Quarterly, 46(4),* 375–389.

Bitinas, A., Žaromskis, R., Gulbinskas, S., Damušytė, A., Žilinskas, G., Jarmalavičius, D. (2005). The results of integrated investigations of the Lithuanian coast of the Baltic Sea: Geology, geomorphology, dynamics and human impact. *Geological Quarterly, 49(4),* 355–362.

Clapuyt, F., Vanacker, V., & Van Oost, K. (2015). Reproducibility of UAV-based earth topography reconstructions based on structure-from-motion algorithms. *Geomorphology, 260,* 4–15. https://doi.org/10.1016/j.geomorph.2015.05.011.

Damušytė, A. (2009). Late glacial and Holocene subfossil mollusc shells on the Lithuanian Baltic coast. *Baltica, 22,* 111–122. http://www.balticajournal.lt/22.

Fonstad, M. A., Dietrich, J. T., Courville, B. C., Jensen, J. L., & Carbonneau, P. E. (2013). Topographic structure from motion: A new development in photogrammetric measurement. Earth surface process. *Landforms, 38,* 421–430. https://doi.org/10.1002/esp.3366.

Harwin, S., & Lucieer, A. (2012). Assessing the accuracy of georeferenced point clouds produced via multi-view stereopsis from unmanned aerial vehicle (UAV) imagery. *Remote Sensing, 4(6),* 1573–1599. https://doi.org/10.3390/rs4061573.

James, M. R., & Robson, S. (2012). Straightforward reconstruction of 3D surfaces and topography with a camera: Accuracy and geoscience application. *Journal of Geophysical Research, 117,* F03017. https://doi.org/10.1029/2011JF002289.

James, M. R., & Robson, S. (2014). Mitigating systematic error in topographic models derived from UAV and ground-based image networks. Earth surface process. *Landforms, 39,* 1413–1420. https://doi.org/10.1002/esp.3609.

James, M. R., Robson, S., & Smith, M. W. (2017). 3-D uncertainty-based topographic change detection with structure-from-motion photogrammetry: Precision maps for ground control and directly georeferenced surveys. Earth surface process. *Landforms, 42,* 1769–1788. https://doi.org/10.1002/esp.4125.

Kabailienė, M., Vaikutienė, G., Damušytė, A., & Rudnickaitė, E. (2009). Post-glacial stratigraphy and paleoenvironment of the northern part of the Curonian Spit, Western Lithuania. *Quaternary International, 207,* 69–79. https://doi.org/10.1016/j.quaint.2008.12.007.

Kondratienė, O., & Damušytė, A. (2009). Pollen biostratigraphy and environmental pattern of Snaigupėlė Interglacial, Late Middle Pleistocene, western Lithuania. *Quaternary International, 207,* 4–13. https://doi.org/10.1016/j.quaint.2008.11.006.

Lampe, R., Naumann, M., Meyer, H., Janke, W., & Ziekur, R. (2011). Holocene evolution of the Southern Baltic Sea coast and interplay of sea-level variation, isostasy, accommodation and sediment supply. In J. Harff, S. Björck, P. Hoth (Eds.), *The Baltic Sea Basin* (pp. 233–251). http://www.springer.com/gp/book/9783642172199

Mancini, F., Dubbini, M., Gattelli, M., Stecchi, F., Fabbri, S., & Gabbianelli, G. (2013). Using unmanned aerial vehicles (UAV) for high-resolution reconstruction of topography: The structure from motion approach on coastal environments. *Remote Sensing, 5,* 6880–6898. https://doi.org/10.3390/rs5126880.

Molodkov, A., Bitinas, A., & Damušytė, A. (2010). IR-OSL studies of till and inter-till deposits from the Lithuanian Maritime Region. *Quaternary Geochronology, 5,* 263–268. https://doi.org/10.1016/j.quageo.2009.04.004.

Niedzielski, T., Witek, M., & Spallek, W. (2016). Observing river stages using unmanned aerial vehicles. *Hydrology and Earth Systems Sciences, 20,* 3193–3205. https://doi.org/10.5194/hess-20-3193-2016.

Sergeev, A. Y., Zhamoida, V. A., Ryabchuk, D. V., Buynevich, I. V., Sivkov, V. V., Dorokhov, D. V., et al. (2017). Genesis, distribution and dynamics of lagoon marl extrusions along the Curonian Spit, southeast Baltic Coast. *Boreas, 46,* 69–82. https://doi.org/10.1111/bor.12177.

Skeivalas, J. (2001). *Metrologic and geodetic measurement data calculation* (p. 220). Vilnius: Technika.

Smith, M. W., Carrivick, J. L., & Quincey, D. J. (2016). Structure from motion photogrammetry in physical geography. *Progress in Physical Geography, 40(2),* 247–275. https://doi.org/10.1177/0309133315615805.

Westoby, M. J., Brasington, J., Glasser, N. F., Hambrey, M. J., & Reynolds, J. M. (2012). "Structure-from-motion" photogrammetry: A low-cost, effective tool for geoscience applications. *Geomorphology, 179,* 300–314. https://doi.org/10.1016/j.geomorph.2012.08.021.

Zakarevičius, A. (1996). *Coordinate system of Lithuanian geodetic networks and their connection* (p. 199). Vilnius: Technika.

(Received January 19, 2017, revised January 17, 2018, accepted January 22, 2018, Published online February 6, 2018)

Pure Appl. Geophys. 175 (2018), 3193–3207
© 2017 The Author(s)
This article is an open access publication
https://doi.org/10.1007/s00024-017-1730-8

**Pure and Applied Geophysics**

# UAV and SfM in Detailed Geomorphological Mapping of Granite Tors: An Example of Starościńskie Skały (Sudetes, SW Poland)

MAREK KASPRZAK,[1] KACPER JANCEWICZ,[1] and ALEKSANDRA MICHNIEWICZ[1]

*Abstract*—The paper presents an example of using photographs taken by unmanned aerial vehicles (UAV) and processed using the *structure from motion* (SfM) procedure in a geomorphological study of rock relief. Subject to analysis is a small rock city in the West Sudetes (SW Poland), known as Starościńskie Skały and developed in coarse granite bedrock. The aims of this paper were, first, to compare UAV/SfM-derived data with the cartographical image based on the traditional geomorphological field-mapping methods and the digital elevation model derived from airborne laser scanning (ALS). Second, to test if the proposed combination of UAV and SfM methods may be helpful in recognizing the detailed structure of granite tors. As a result of conducted UAV flights and digital image post-processing in AgiSoft software, it was possible to obtain datasets (dense point cloud, texture model, orthophotomap, bare-ground-type digital terrain model—DTM) which allowed to visualize in detail the surface of the study area. In consequence, it was possible to distinguish even the very small forms of rock surface microrelief: joints, aplite veins, rills and karren, weathering pits, etc., otherwise difficult to map and measure. The study includes also valorization of particular datasets concerning microtopography and allows to discuss indisputable advantages of using the UAV/SfM-based DTM in geomorphic studies of tors and rock cities, even those located within forest as in the presented case study.

**Key words:** UAV, SfM, geomorphological mapping, granite tors, Sudetes.

## 1. Introduction

Among characteristic elements of granite landscapes are isolated solid rock residuals known as tors. They may also form clusters or occasionally combine into more complicated landform assemblages called rock cities (Migoń et al. 2017). Their existence is connected with spatial variability of bedrock geological features, mainly jointing patterns, whereas joints also control the shape, evolution and patterns of degradation of individual tors (Linton 1955; Jahn 1974; Dumanowski 1968; Migoń 1996, 2006; Twidale and Vidal Romani 2005). These structural conditions have long been surveyed through traditional geological and geomorphological mapping, occasionally aided by interpretation of aerial photographs where the absence of vegetation allowed one to do this. Fieldwork is usually based on ground-level observations, including geological compass measurements. More recently, this type of studies may involve an analysis of high-resolution (1 × 1 m or more) digital terrain models (DTMs). These datasets are usually derived from airborne laser scanning (ALS)—a method which, in recent years, allowed for a dynamic progress in representing and modeling of Earth surface (e.g. Höfle and Rutzinger 2011; Bishop 2013; Migoń et al. 2013), especially within inaccessible areas or those of complex relief.

In certain situations the ALS data can be substituted by photogrammetric methods using unmanned aerial vehicles (UAV) and digital image post-processing technique known as *structure from motion* (SfM). This combination appears to be cheap and efficient and thus, the use of these methods is currently becoming increasingly popular, mainly in case studies focused on small objects or areas (e.g. James and Varley 2012; Hugenholtz et al. 2013; Mancini et al. 2013; Lucieer et al. 2014; Ryan et al. 2015; Clapuyt et al. 2016; Cook 2017; Dąbski et al. 2017; Marteau et al. 2017; Miziński and Niedzielski 2017). This paper presents a case study which aims to determine to what extent the use of UAV and SfM methods can help to recognize the geomorphological structure of granite tors and supplement landform

[1] Institute of Geography and Regional Development, University of Wrocław, pl. Uniwersytecki 1, 50-137 Wrocław, Poland. E-mail: marek.kasprzak@uwr.edu.pl

inventory attempted through traditional methods. The reference object in this study is the Starościńskie Skały rock city in Rudawy Janowickie Mts. (Sudetes, SW Poland, Fig. 1), recently subject to detailed mapping using the combination of traditional field-based geomorphological mapping (Michniewicz et al. 2016). The latter was based on high-resolution, 1 × 1 m LiDAR-based DTM, which was produced in years 2010–2014 for most part of the area of Poland (Wężyk 2014). Aside from the intention to gather new information about the morphology of the particular group of tors, the main objective of this study was to compare the potentially available methods and datasets concerning granite tors in terms of the degree of detail returned. Contrary to some high altitude, treeless areas where tors occur, this analysis concerns objects located within a forest. Consequently, both field mapping and measurements as well as spatial data post-processing were far more difficult and extra problems had to be solved.

## 2. Methods

The main dataset used in this study is the digital model of the tor group, prepared on the basis of aerial photography. The digital images of 4000 × 3000 px resolution were made by DJI Phantom 3 Professional quadrocopter. It is equipped with the digital camera of 1/2.3″ CMOS sensor, 94° field of view, 20 mm (35 mm format equivalent) focal length and f/2.8 focus (Phantom 3 Professional 2015). Flights during which the photos were taken were conducted on 7th March 2017. The time of flights was set before the start of vegetation season and was carried out in the presence of stable wind conditions, sufficient visibility and lack of snow cover. In order to obtain appropriate image coverage of the study area as well as the high level of detail, orthogonal photos were taken during three consecutive flights at the altitude of 80, 50 and 30 m above the starting point (ca. 65, 35 and 15 m above the top of the highest tor). Additionally, oblique photos were taken during the flight around the group of tors in order to provide information about areas invisible directly from above.

Figure 1
Granite rock city of the Starościńskie Skały tors in the Rudawy Janowickie in the Sudetes, SW Poland (photo M. Kasprzak)

The aforementioned fieldwork yielded a set of 239 digital images, which were used as an input data in the process of digital terrain model creation with the use of SfM method. The essence of SfM lies in using the rules of stereoscopy—a series of overlapped images enables creation of a three-dimensional model. Contrary to traditional photogrammetry, information about camera position and angle is unnecessary (however, the camera position was registered by the UAV built-in GPS/GLONASS receiver), as these data are determined automatically during the performance of an algorithm for identification and matching points from particular images. This first stage of processing results in a sparse point cloud, which is subsequently processed into a dense point cloud—a dataset conveying detailed features of the photographed object (shape and color). Further stages of model creation include generation of mesh and image rendering. The whole procedure is described by Westoby et al. (2012) and Eltner et al. (2016).

The software used in this study is Agisoft Photoscan Professional Edition (Agisoft 2016). The SfM procedure resulted in a point cloud of highest possible density (ultra-high). Next, the point cloud was automatically classified to retrieve a subset of ground-level points. Nevertheless, manual removal of points which in fact represented leaves or tree trunks proved necessary to achieve a better representation of rock surfaces. However, in some areas it was not technically possible. During the data pre-processing, the coordinates of ground control points were added in order to set the model more accurately in geographical space and to prevent potential shape representation errors (Barry and Coakley 2013; James and Robson 2014; Jóźków and Toth 2014; Tonkin and Midgley 2016). Five ground control points (GCP) were marked during fieldwork—they were purposefully set in well-exposed locations (not covered by tree crowns), four of them around the tor group and one in its central point. The coordinates of these control points were measured by GPS RTK (Trimble) device, with the decimeter accuracy.

The whole digital processing procedure resulted in the following data: dense point cloud, texture model, digital terrain model (DTM) and orthophotomap. Subsequently, these datasets were used in the analysis of micromorphology of rock surfaces and enabled a comparison with the results of traditional, ground-based geomorphological mapping (Michniewicz et al. 2016) and with the LiDAR-based DTM of $1 \times 1$ m resolution. This reference dataset is a bare ground type model, which was derived from the ALS point cloud of density 4–6 pts/m$^2$; its horizontal mean error is less than 0.5 m and mean vertical error does not exceed 0.15 m (Wężyk 2014). For purpose of this study both DTMs (UAV/SfM-based and LiDAR-based) were visualized. The UAV/SfM with LiDAR-based DEMs comparison was created in the form of DEMs of difference (DoD). Furthermore, three selected cross-profiles were analyzed to indicate differences between both models. In addition, error measures (absolute errors and RMSE of X, Y and Z coordinates) for ground control points were calculated.

### 2.1. Study Area

The study area is the summit part of the Mt. Lwia Góra (718 m a.s.l.) located in the Rudawy Janowickie ridge—a distinctive geomorphic unit of the West Sudetes (Fig. 2) in the Bohemian Massif. Bedrock in this section of the Rudawy Janowickie is composed of Carboniferous granite, with porphyritic structure (Szałamacha 1969; Sobczyk et al. 2015). The Starościńskie Skały tor group has developed in the most elevated part of Mt. Lwia Góra and consists of many isolated tors separated by wide-opened clefts (Fig. 3). This tor group is a rare example of 'rock city' which is an intermediate granitic landform between smaller simple tors and larger dome-shaped bornhardts. This type of rock formation is more common in the sandstone areas and the 'rock city' formed in the granite bedrock is a rare case. The tor group is 180 m long and has a complex structure, both in vertical and horizontal dimensions. In plan the Starościńskie Skały group divides into two sections, different in terms of morphology and size—the wide (85 m) southeastern part and the relatively narrow northwestern part (25 m). Additionally, the former is characterized by hierarchical structure and include a dome-shaped pedestal, at the top of which smaller tors have developed (Michniewicz et al. 2016). The internal morphology of tors is diverse so that rock towers, pulpits, needles and ridges occur. The height of the singular rock forms is from 9 to 11 m and the highest one (Starościńska Skała tor) reaches 30 m. The morphology of the southeastern part resembles granite domes which are known forms in the adjacent Jelenia Góra Basin (Migoń 1993, 2007), however, it is not a freestanding bornhardt rising above a level surface. A specific feature of this tor group is the abundance of microforms which have survived in the

Figure 3 ▶
Geomorphological sketch of the study area (based on mapping done by Michniewicz et al. 2016). Red square shows the extent of model developed in this study (see Fig. 4b and next)

southeastern section. Several forms have been identified like flared slopes, weathering pits and karren (Michniewicz et al. 2016).

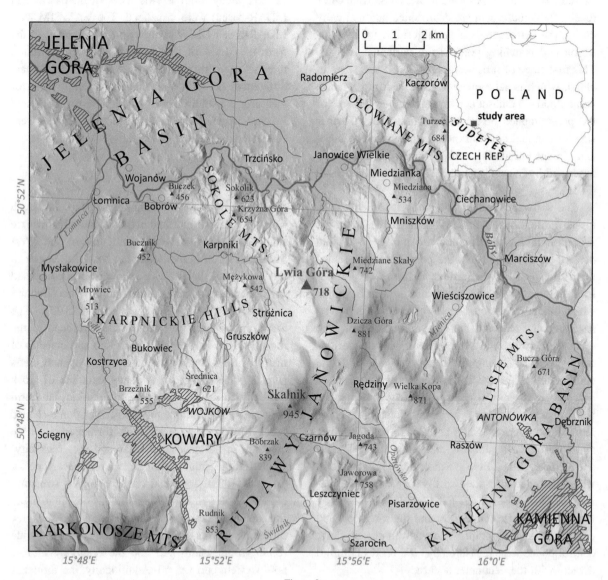

Figure 2
Location map of the Mt. Lwia Góra in the Rudawy Janowickie Mts

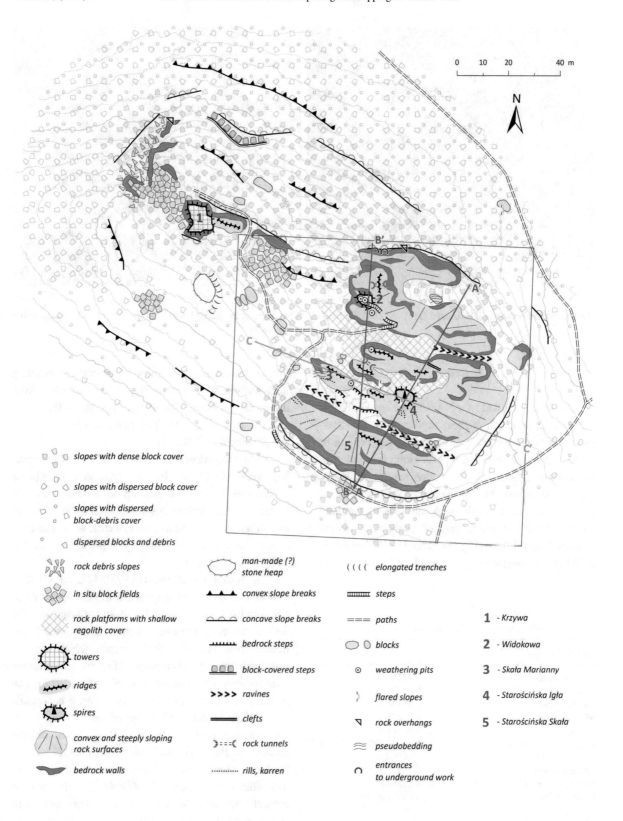

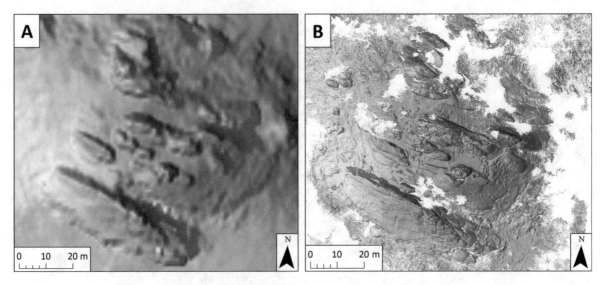

Figure 4
The comparison between LiDAR-based DTM and UAV/SfM-based DTM. Holes in the model result from the lack of data (ground points under forest undergrowth)

## 3. Results

The study resulted in a UAV/SfM-based DTM of the Mt. Lwia Góra surface, representing the Starościńskie Skały tor group. The final raster version of DTM was based on a dense point cloud consisting of 63 million points and the raster resolution is ca. 0.015 × 0.015 m. In consequence, the number of topographic details of the rock surfaces represented by the model is much greater comparing to the LiDAR-based DTM (Fig. 4). The UAV/SfM-based DTM is a bare-ground-type model and areas, where

the unequivocal recognition of points representing land-surface was impossible, were excluded from interpolation. Hence, blank areas appear on the model visualization (Fig. 4b). The basic model properties are presented in Table 1.

A direct comparison between DTMs of the studied area (DoD in Fig. 5) indicates significant differences in imaged tors up to a few (max 10) meters. These differences mainly refer to edges of individual tors as well as concave rock walls and their feet, which are not correctly represented on the LiDAR-based DTM.

The resolution of the UAV/SfM-based DTM is almost two orders of magnitude greater when compared to the LiDAR-based DTM. In consequence, the former not only represents better the height of particular tors, but also their shapes which can be observed on cross-profiles (Fig. 6). Significant improvement of representation of surface details can be observed in case of spires. Moreover, it is possible to visualize in detail all forms of microrelief developed within rock walls as well as small objects located at the rock wall base, such as blocks or boulders. Measurement errors based on ground control points are negligible (Table 2) and indicate correct alignment of tors to reality. The root mean square error (RMSE) of altitude is strongly dependent

Table 1

*Basic properties of the digital terrain model*

UAV/SfM-based DTM

| | |
|---|---|
| Covered area | 0.01242 km$^2$ |
| Raster width | 0.01524 m |
| Raster height | 0.01524 m |
| Min elevation | 671.287 m |
| Max elevation | 718.287 m |
| Projection/datum/units | Poland 1992/ETRS89/m |
| 3d model faces | 12,410,293 |
| 3d model vertices | 6,233,307 |
| Dense point cloud | 63,051,665 |
| Point density | 43.6 points/cm$^2$ |
| Cameras | 239 |
| Aligned cameras | 166 |

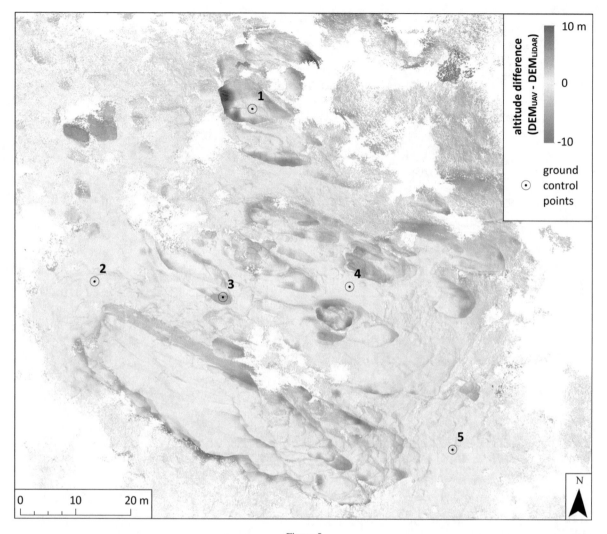

Figure 5
DEMs of difference (DoD): altitude differences between UAV/SfM-based DTM and LiDAR-based DTM

on taking into account in the calculation of GCP no. 3 due its specific location.

An indisputable advantage of UAV/SfM-based elevation data is that they, contrary to the results of field geomorphological mapping, provide detailed information about inaccessible top surfaces of tors (Fig. 7). The model well represents rock surfaces even within clefts or concave bends of rocky surfaces. However, the DTM cannot properly represent surface in areas located underneath dense vegetation cover (coniferous trees, bigger broad-leaved trees or dense bushes). Observations of local microrelief,

based on the UAV/SfM model, are presented below using specific examples (Fig. 8).

### 3.1. Tops of Tors

Morphology of tops of tors depends on the density of joints within a specific granite block. Some tors have relatively oval tops (e.g. Widokowa tor), whereas others are crowned with ridges or singular rounded boulders (Fig. 8a–c). Also, weathering pits were developed practically in all locations where the tops of tors are formed by planar, structurally determined surfaces.

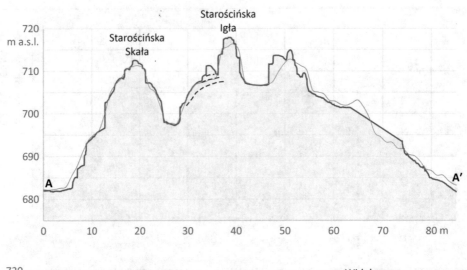

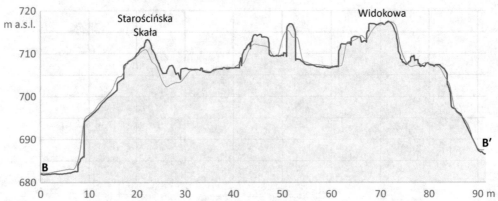

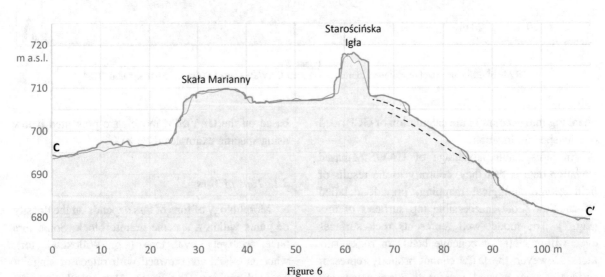

Figure 6

Profiles across tors derived from UAV/SfM-based DTM (colored lines) and LiDAR-based DTM (gray lines). Locations of the crosssections are shown in Fig. 3. Dotted lines show main joints responsible for the shape of the granite dome

Table 2

*Ground control points (GCP) error measures*

| GCP Id | X error [m] | Y error [m] | Z error [m] |
|---|---|---|---|
| 1 | 0.07 | 0.11 | 0.18 |
| 2 | − 0.41 | − 0.02 | 0.49 |
| 3 | 0.05 | 0.08 | 3.47[a] |
| 4 | − 0.12 | 0.10 | − 0.14 |
| 5 | − 0.64 | − 0.22 | 0.31 |
| RMSE | 0.35 | 0.12 | 1.58 |
|  |  |  | 0.31[b] |

X and Y error based on GPS RTK data; Z error based on LiDAR DEM dataset

[a]GCP located on convex rock surface not represented by LiDAR DEM

[b]RMSE calculated with exlusion of GCP no. 3

### 3.2. Joints

There are a few joint systems within the Starościńskie Skały rock city. In general, the spatial pattern of rock residuals within the group of tors is controlled by WNW–ESE trending vertical joints (strike 107°–112°), along which corridors were excavated through preferential weathering. They now separate particular rows of tors. These joints cross with the second system of joints (strike 175°–180°) at an acute angle. The third set is trending N60°E and is particularly visible within the southernmost tor. An important element of joint network is constituted by concentrically arranged, broadly surface-parallel joints (sheeting joints). They are exposed especially within the southeastern part of the tor group, mainly near the edges of rock walls (see also in Fig. 6). Sheeting joints, whose spacing does not exceed 1 m, locally overlap with or grade into horizontal partings known as pseudobedding. The visualization of UAV/SfM-based DTM shows very clearly the contrast between joint density of particular rock walls (Fig. 8d–f). It is possible to distinguish major joints, which separate individual tors, and minor joints which diversify the morphology of rock walls.

### 3.3. Rills and Karrens

On the steeply inclined rock surfaces, the UAV/SfM-based DTM represents in detail linear erosional forms likely developed by periodic water flow. Elongated rills (with lengths 0.4–9.0 m and widths 0.1–1.0 m), some with funnel-like shapes at the points of origin, may be determined by the occurrence of joints so that trickling water exploits a line of structural weakness (Fig. 8g). Likewise, joints may also determine size and density of rills (Fig. 8h). Well-developed rills and karren create a fan-shaped system on the rock wall located east of the Starościńska Igła tor (Fig. 8i). However, their pattern is clearly affected by a diagonal aplite vein.

### 3.4. Weathering Pits

Weathering pits are a common element of microrelief within the Starościńskie Skały tor group. The biggest pits have diameter up to 1 m and are 0.4–0.5 m deep. Apart from pits previously mapped from the ground, the analysis of the UAV/SfM-based DTM enabled us to detect two big (0.5 m diameter) weathering pits which were not documented before (Michniewicz et al. 2016). Both pits are located on tops of tors. High resolution and precision of the DTM enables also to depict shapes of singular pits, including secondary hollows within their floors (Fig. 8j).

### 3.5. Artificial Forms

The UAV/SfM-based DTM shows the steps carved in the granite surface, which lead to the view-point on the Widokowa tor (Fig. 8m, n). The railings and an information board which are located on this tor are not represented on the model, as they are elements of land cover. Consequently, points corresponding to these objects were manually removed during data processing.

### 3.6. Other Objects

The UAV/SfM-based datasets allow also for identification of other, minor forms, which were not mentioned earlier. The three-dimensional visualization of the dense point cloud or textured mesh (as in Fig. 7) reveals the locations of overhangs (flared slopes). Aplite veins and pegmatites can be distinguished as their texture differs from the texture of the

adjacent granite surfaces. The land surface next to the tors is strewn by numerous blocks and boulders (Fig. 8o), which are usually products of rockfall from the rock outcrops or walls. As it was impossible to filter-out all of the elements of land cover, one can

Figure 8 ►
Main features of the Starościńskie Skały microrelief imaged by
UAV/SfM-based DTM. Explanations in the text

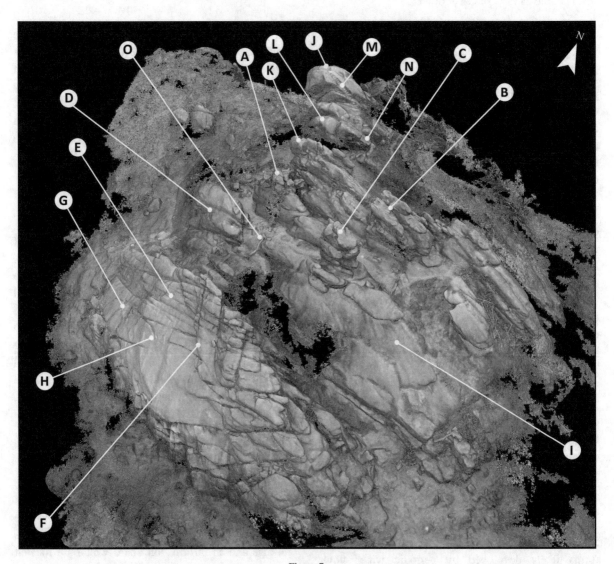

Figure 7
The perspective view of the model of the Starościńskie Skały rock city. Letter tags indicate rock surfaces shown in Fig. 8

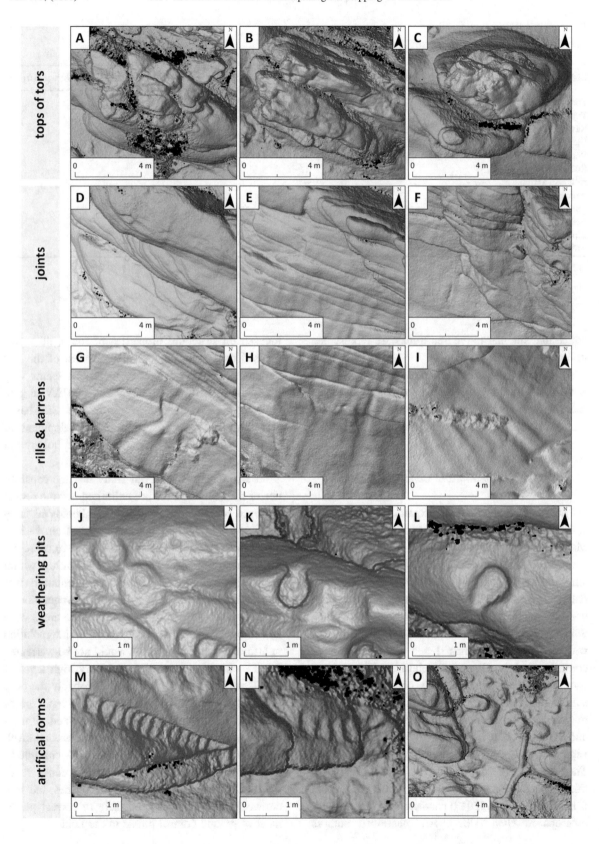

Table 3

*The comparison between quality of data sources in the mapping of tors*

| Characteristic features | Traditional geomorphological mapping | ALS-based DTM (1 × 1 m) | UAV/SfM-based DTM (0.015 × 0.015 m) |
|---|---|---|---|
| Tors walls | + | Without microrelief | + |
| Tops of tors | − | Without microrelief | + |
| Ravines | + | +/− | + |
| Clefts | + | − | + |
| Rock tunnels | + | − | +/− |
| Rock overhangs | + | +/− | + |
| Rills and karren | +/− | − | + |
| Weathering pits | +/− | − | + |
| Joints | +/− | − | + |
| Veins | + | − | + |
| Blocks and debris | + | +/− | + |
| Artificial forms | + | − | + |

'+' possible to distinguish

'+/−' possible to distinguish only in specific conditions or not completely

'−' impossible to distinguish

notice that singular fallen tree trunks are represented in detail on the DTM.

Using the observations reported above as a base, it is possible to offer valorization of particular data sources concerning morphology of singular tors or groups of tors, as presented in Table 3.

## 4. Discussion and Conclusions

This study is the first ever example of using UAV and SfM in analysis of relief within the rock city. Methodically similar case studies focusing on rock surfaces are not numerous. The work is the closest to aims and results obtained by Cruden et al. (2016) who mapped geometry of dykes and dykes network on a wavecut platform at Bingie Bingie point in New South Wales (Australia) using UAV/SfM-based DTM with resolution less than 1 × 1 cm. The broader context of this work was to determine how magma flow was channelized in such networks and how sulfide liquids become trapped in channels to form magmatic sulfide deposits. In similar way Vollgger and Cruden (2016) mapped folds in rocky Cape Liptrap and Cape Peterson in Victoria (Australia). They used high-resolution (7.7 × 7.7 mm) DTM for this purpose. Moreover, basing on the 2.7 × 2.7 cm DTM Chesley et al. (2017) characterized sedimentary outcrops in Utah (USA). All mentioned authors

focused on structural properties of rocks, no works focused on geomorphological properties of the tors have been done so far. Nevertheless, it seems very likely that the increasing popularity of UAV in general and low costs of acquiring the SfM-based elevation data will result in the development of similar studies concerning micromorphology of poorly accessible tor groups.

It also emerges from our study that the capabilities of methods used are not unlimited and the results achieved contain errors and artifacts. Photos taken by the UAV, even though they were taken from various altitudes and angles (James et al. 2017a, b), were not sufficient to represent the whole land-surface within the study area. Proper terrain representation was impossible in areas covered by dense vegetation at the base of tors or in some clefts. These areas were left empty (NoData) as any method of interpolation or artificial image replication would not be warranted due to very high resolution and very high level of details of UAV/SfM-based DTM. Probably the only possibility to digitally represent the land-surface of these areas is to multiply UAV surveys and, primarily, use photos taken manually from the ground level. However, even then it could be very difficult due to terrain complexity, slope steepness and dense vegetation. Covering a land-surface by bushes and tree crowns is also important obstacle to correct placement of ground control points in the field.

Another difficulty which appeared during data processing was the necessity to remove manually points which represented elements of land cover close to rock walls. The experiments with automated filtration of the point cloud did not provide acceptable results. Hence, working out the geometrically correct and artifact-free DTM was the most time-consuming stage of the whole data processing procedure.

There are well-known limitations and systematical errors connected with the SfM procedure, including predisposition to vertical 'doming' of the DTMs surface. They result from the combination of near-parallel imaging directions and inaccurate correction of radial lens distortion (James and Robson 2014). In order to reduce the probability of occurrence of these errors, the authors of this study used GPS-calibrated ground control points. A location of the GCP in the field caused some difficulties, mentioned above. The GCP error measures (Table 2) and an instance of point no. 3 showed that points should not be placed on convex, isolated and relatively small tors.

Notwithstanding the aforementioned limitations of the UAV/SfM-based DTM, the obtained model of group of tors is so detailed that it allows to analyze the microforms developed on the rock surfaces in a similar manner to observations gathered during conventional field mapping. Comparing to the traditional geomorphological mapping, the use of DTM solves the problem concerning mapping of the highest, steepest parts of tors and their tops. It allows to analyze the joint system while minimizing the potential distortion of perspective from the observation point. Thus, it also minimizes the risk of subjective and improper estimation of directions and density of joints in cases when the use of geological compass is impossible. A high level of surface detail representation enables one to distinguish locations of veins, erosional microforms or even small hollows at the bottoms of weathering pits. In sum, it can be stated that the datasets obtained with the use of UAV/ SfM methods (dense point cloud, texture, orthophotomap, DTM) represent rock surfaces with a very high accuracy that allows to determine all minor features characteristic of the rocky relief within the granite area (Figs. 7, 8).

The DTM obtained for the purpose of this case study surpasses the LiDAR-based DTM in terms of the detail involved. It should be emphasized that the airborne LiDAR-based elevation data have recently resulted in a certain breakthrough in geomorphological studies on forested slopes of the Sudetes (Migoń et al. 2013). As the UAV/SfM methods enable considerable improvement of digital surface representation, they may be successfully applied as a supplement to existing elevation models of lower resolution. This combination is especially recommended for areas with numerous rock outcrops, tors or rock cities, e.g. areas with extreme terrain complexity. They are represented in existing cartographical data in a very simplified or clearly inaccurate way (see an example from sandstone landscape given by Migoń and Kasprzak 2015).

## Acknowledgments

This paper is a contribution to the National Science Centre Project: 'The geomorphic significance of tors of the West Sudetes', no. UMO-2016/21/N/ST10/ 03256 (for Aleksandra Michniewicz). The LiDAR data used for this study have been purchased and used with academic license DIO.DFT.D-SI.7211.1619.2015_PL_N, according to the Polish law regulations in the administration of Główny Urząd Geodezji i Kartografii (Head Office of Land Surveying and Cartography). The authors thank Piotr Migoń for helping us to improve the text.

## REFERENCES

Agisoft. (2016). *Agisoft photoscan user manual: Professional* (p. 97). St. Petersburg: Agisoft LLC.

Barry, P., & Coakley, R. (2013). Accuracy of UAV photogrammetry compared with network RTK GPS. *In International Archives of the Photogrammetry, Remote Sensing and Spatial Information Sciences, 2,* 27–31.

Bishop, M. P. (2013). Remote sensing and GIScience in geomorphology: Introduction and overview. In J. Shroder & M. P. Bishop (Eds.), *Treatise on Geomorphology. Remote Sensing and GIScience in Geomorphology* (Vol. 3, pp. 1–24). San Diego: Academic Press.

Chesley, J. T., Leier, A. L., White, S., & Torres, R. (2017). Using unmanned aerial vehicles and structure-from-motion photogrammetry to characterize sedimentary outcrops: An example from the Morrison Formation, Utah, USA. *Sedimentary Geology, 354*(2017), 1–8.

Clapuyt, F., Vanacker, V., & Van Oost, K. (2016). Reproducibility of UAV-based earth topography reconstructions based on Structure-from-Motion algorithms. *Geomorphology, 260,* 4–15.

Cook, K. L. (2017). An evaluation of the effectiveness of low-cost UAVs and structure from motion for geomorphic change detection. *Geomorphology, 278,* 195–208.

Cruden, A., Vollgger, S., Dering, G., & Micklethwaite, S. (2016). High spatial resolution mapping of dykes using unmanned aerial vehicle (UAV) photogrammetry: New insights on emplacement processes. *Acta Geologica Sinica (English Edition), 90*(Supp. 1), 52–53.

Dąbski, M., Zmarz, A., Korczak-Abshire, M., Karsznia, I., & Chwedorzewska, K. J. (2017). UAV-based detection and spatial analyses of periglacial landforms on Demay Point (King George Island, South Shetland Islands, Antarctica). *Geomorphology, 290,* 29–38.

Dumanowski, B. (1968). The influence of petrographical differentiation of granitoids on landforms. *Geographia Polonica, 14,* 93–98.

Eltner, A., Kaiser, A., Castillo, C., Rock, G., Neugirg, F., & Abellán, A. (2016). Image-based surface reconstruction in geomorphometry—merits, limits and developments. *Earth Surface Dynamics, 4,* 359–389.

Höfle, B., & Rutzinger, M. (2011). Topographic airborne LiDAR in geomorphology: a technological perspective. *Zeitschrift für Geomorphologie, 55*(Suppl. 2), 1–29.

Hugenholtz, C. H., Whitehead, K., Brown, O. W., Barchyn, T. E., Moorman, B. J., Le Clair, A., et al. (2013). Geomorphological mapping with a small unmanned aircraft system (sUAS): feature detection and accuracy assessment of a photogrammetrically-derived digital terrain model. *Geomorphology, 194,* 16–24.

Jahn, A. (1974). Granite tors in the Sudeten Mountains. In E. H. Brown & R. S. Waters (Eds.), *Progress in Geomorphology* (Vol. 7, pp. 53–61). London: Institute of British Geographers, Special Publication.

James, M. R., & Robson, S. (2014). Mitigating systematic error in topographic models derived from UAV and ground-based image networks. *Earth Surface Processes and Landforms, 39,* 1413–1420.

James, M. R., Robson, S., d'Oleire-Oltmanns, S., & Niethammer, U. (2017a). Optimising UAV topographic surveys processed with structure-from-motion: Ground control quality, quantity and bundle adjustment. *Geomorphology, 280,* 51–66.

James, M. R., Robson, S., & Smith, M. (2017b). 3-D uncertainty-based topographic change detection with structure-from-motion photogrammetry: precision maps for ground control and directly georeferenced surveys. *Earth Surface Processes and Landforms, 42,* 1769–1788.

James, M. R., & Varley, N. (2012). Identification of structural controls in an active lava dome with high resolution DEMs:

Volcán de Colima, Mexico. *Geophysical Research Letters, 39,* L22303.

Jóźków, G., & Toth, C. (2014). Georeferencing experiments with UAS imagery. *ISPRS Annals of the Photogrammetry, Remote Sensing, and Spatial Information Sciences, II*(1), 25–29.

Linton, D. (1955). The problem of tors. *Geographical Journal, 121,* 470–487.

Lucieer, A., de Long, S. M., & Turner, D. (2014). Mapping landslide displacements using structure from Motion (SfM) and image correlation of multi-temporal UAV photography. *Progress in Physical Geography, 38,* 97–116.

Mancini, F., Dubbini, M., Gattelli, M., Stecchi, F., Fabbri, S., & Gabbianelli, G. (2013). Using unmanned aerial vehicles (UAV) for High-resolution reconstruction of topography: The structure from motion approach on coastal environments. *Remote Sensing, 5,* 6880–6898.

Marteau, B., Vericat, D., Gibbins, C., Batalla, R. J., & Green, D. R. (2017). Application of Structure-from-Motion photogrammetry to river restoration. *Earth Surface Processes and Landforms, 42,* 503–515.

Michniewicz, A., Jancewicz, K., Różycka, M. & Migoń, P. (2016). Rzeźba granitowego skalnego miasta Starościńskich Skał w Rudawach Janowickich (Sudety Zachodnie). *Landform Analysis, 31,* 17–33 (in Polish with Eng. abs.: Morphology of the granite rock city of Starościńskie Skały in the Rudawy Janowickie/Western Sudetes/).

Migoń, P. (1993). Kopułowe wzgórza granitowe w Kotlinie Jeleniogórskiej. *Czasopismo Geograficzne, 64(1),* 3–23 (in Polish with Eng. abs.: Granite domical hills (bornhardts) in the Jelenia Góra Basin).

Migoń, P. (1996). Granite landscapes of the Sudetes Mountains—some problems of interpretation: a review. *Proceedings of the Geologists' Association, 107,* 25–38.

Migoń, P. (2006). *Granite landscapes of the world.* Oxford: Oxford University Press.

Migoń, P. (2007). Granitoids in Poland. Archivum Mineralogiae Monograph 1. In A. Kozłowski & J. Wiszniewska (Eds.), *Geomorphology of granite terrains in Poland* (pp. 355–366). London: Komitet Nauk Mineralogicznych PAN & Wydział Geologii UW.

Migoń, P., Duszyński, F., & Goudie, A. (2017). Rock cities and ruiniform relief: forms—processes—terminology. *Earth-Science Reviews, 171,* 78–104.

Migoń, P. & Kasprzak, M. (2015). Analiza rzeźby stoliwa Szczelińca Wielkiego w Górach Stołowych na podstawie numerycznego modelu terenu z danych LiDAR, *Przegląd Geograficzny, 87(1),* 27–52 (in Polish with Eng. abs.: LiDAR DEM-based analysis of geomorphology of the Szczeliniec Wielki mesa in Poland's Stołowe Mountains).

Migoń, P., Kasprzak, M., & Traczyk, A. (2013). How high-resolution DEM based on airborne LiDAR helped to reinterpret landforms—examples from the Sudetes, SW Poland. *Landform Analysis, 22,* 89–101.

Miziński, B., & Niedzielski, T. (2017). Fully-automated estimation of snow depth in near real time with the use of unmanned aerial vehicles without utilizing ground control points. *Cold Regions Science and Technology, 138,* 63–72.

Phantom 3 Professional. User Manual V1.0 (2015). DJI. http://download.dji-innovations.com/downloads/phantom_3/en/Phantom_3_Professional_User_Manual_v1.0_en.pdf.

Ryan, J. C., Hubbard, A. L., Box, J. E., Todd, J., Christoffersen, P., Carr, J. R., et al. (2015). UAV photogrammetry and structure from motion to assess calving dynamics at Store Glacier, a large outlet draining the Greenland ice sheet. *Cryosphere, 9,* 1–11.

Sobczyk, A., Danišík, M., Aleksandrowski, P., & Anczkiewicz, A. (2015). Post-Variscan cooling history of the central Western Sudetes (NE Bohemian Massif, Poland) constrained by apatite fission-track and zircon (U-Th)/He thermochronology. *Tectonophysics, 649,* 47–57.

Szałamacha, J. (1969). Objaśnienia do Szczegółowej mapy geologicznej Sudetów 1:25 000. Arkusz Janowice Wielkie M 33—44 Bd. Wydawnictwo Geologiczne, Warszawa. (in Polish: Explanations to Detailed geological map of Sudetes 1:25 000, sheet Janowice Wielkie M33–44Bd).

Tonkin, T. N., & Midgley, N. G. (2016). Ground-control networks for image based surface reconstruction: An investigation of optimum survey designs using UAV derived imagery and structure-from-motion photogrammetry. *Remote Sensing, 8*(9), 786.

Twidale, C. R., & Vidal Romani, J. R. (2005). *Landforms and geology of the granite terrains.* Leiden: Balkema.

Vollgger, S. A., & Cruden, A. R. (2016). Mapping folds and fractures in basement and cover rocks using UAV photogrammetry, Cape Liptrap and Cape Paterson, Victoria, Australia. *Journal of Structural Geology, 85,* 168–187.

Westoby, M. J., Brasington, J., Glasser, N. F., Hambrey, M. J., & Reynolds, J. M. (2012). 'Structure-from-Motion' photogrammetry: A low-cost, effective tool for geoscience applications. *Geomorphology, 179,* 300–314.

Wężyk, P. (ed.) (2014). Podręcznik dla uczestników szkoleń z wykorzystania produktów LiDAR. Informatyczny System Osłony Kraju przed nadzwyczajnymi zagrożeniami. Główny Urząd Geodezji i Kartografii, Warszawa, 328 p. (in Polish: Manual for training participants on the use of LiDAR products).

(Received  April 17, 2017, revised  November 15, 2017, accepted  November 21, 2017, Published online  November 27, 2017)

Pure Appl. Geophys. 175 (2018), 3209–3221
© 2018 The Author(s)
https://doi.org/10.1007/s00024-017-1755-z

**❙ Pure and Applied Geophysics**

# Geovisualisation of Relief in a Virtual Reality System on the Basis of Low-Level Aerial Imagery

ŁUKASZ HALIK[1] [iD] and MACIEJ SMACZYŃSKI[1]

*Abstract*—The aim of the following paper was to present the geomatic process of transforming low-level aerial imagery obtained with unmanned aerial vehicles (UAV) into a digital terrain model (DTM) and implementing the model into a virtual reality system (VR). The object of the study was a natural aggretage heap of an irregular shape and denivelations up to 11 m. Based on the obtained photos, three point clouds (varying in the level of detail) were generated for the 20,000-$m^2$-area. For further analyses, the researchers selected the point cloud with the best ratio of accuracy to output file size. This choice was made based on seven control points of the heap surveyed in the field and the corresponding points in the generated 3D model. The obtained several-centimetre differences between the control points in the field and the ones from the model might testify to the usefulness of the described algorithm for creating large-scale DTMs for engineering purposes. Finally, the chosen model was implemented into the VR system, which enables the most lifelike exploration of 3D terrain plasticity in real time, thanks to the first person view mode (FPV). In this mode, the user observes an object with the aid of a Head- mounted display (HMD), experiencing the geovisualisation from the inside, and virtually analysing the terrain as a direct animator of the observations.

**Key words:** Unmanned aerial vehicles, virtual reality, geovisualisation, first person view, digital terrain model.

## 1. Introduction

Land relief is a key feature determining human activities in the field of space management. Its accurate representation, analysis, and visualisation in the digital environment makes it possible to identify the factors which may influence landscape and urban planning more precisely. Hence the importance of obtaining precise DTMs, which form a fundamental data-set in many applications, especially in geographic information systems (GIS). Unfortunately, the quality of these DTMs is rarely communicated to GIS users (Wood and Fisher 1993; Kraus et al. 2006). This may result in inaccurate geovisualisations of the analysed area, which do not correctly reflect spatial relations. A possible solution to this problem is to create DTMs on the basis of UAV-obtained data. This type of data might serve to create a DTM with centimetre-level accuracy, which may then be implemented into geovisualisations in the virtual reality (VR) system.

Unmanned aerial vehicles (UAVs) have played a significant role in the military for some time (Watts et al. 2012). They have been used, for example, in military activities such as intelligence, environmental surveillance, sea surveillance, and mine removal activities (Eisenbeiss 2004). It was this connection with the military sector that led to the significant development of unmanned aerial platforms and their wide use in the years 1960–1980 (Ahmad 2011). Nowadays, unmanned aerial vehicles are not only used for military purposes—their use in the civil sector is also expanding. In 2007, the European Commission published a report on the developing UAV technology in which the civil and commercial sectors were divided into six application segments: government; firefighting; energy; agriculture, forestry, and fisheries; Earth observation and remote sensing; communications and broadcasting (European Commission 2007). On the basis of the report, it might be concluded that unmanned aerial vehicles are a versatile tool for obtaining data on various aspects of everyday life. As explained by Mill et al. (2011, 2014), UAV-obtained imagery is used, among other things, in topographic surveying, land management, organization of road infrastructure, works

[1] Department of Cartography and Geomatics, Institute of Physical Geography and Environmental Planning, Faculty of Geographical and Geological Sciences, Adam Mickiewicz University in Poznań, Poznań, Poland. E-mail: lhalik@amu.edu.pl

connected with monitoring construction processes, technical inspection of engineering infrastructure, soil mass measurements, precise agriculture, and work connected with natural disasters. A detailed description of the development of UAV technology and its possible applications was given by Colomina and Molina (2014). In addition, Liu et al. (2014) focused on the usefulness of unmanned platforms in the field of land engineering, while Nex and Remondino (2014), on UAV-aided 3D mapping.

The possibility of using the UAV technology to monitor open pit mines, especially with the purpose of tracking the occurring changes and making cubature calculations on the basis of the point cloud, has been described, among others, by Esposito et al. (2017) based on the example of Sa Pigada Bianda in Sardinia. Due to the technological development of the miniaturisation process and the decreasing production costs of the components used in UAV construction, the vehicles are becoming more and more popular as surveying platforms in geophysical research. Their versatile use in geophysics, geodesy, and spatial planning has been described by Lin (2008), Remondino et al. (2012), Siebert and Teizer (2014), Li et al. (2016), Kršák et al. (2016) and Torres et al. (2016). The research material obtained with UAVs, in the form of images or point clouds (Axelsson 2000), can provide the basis for many analyses, such as analyses of visibility.

Horbiński and Medyńska-Gulij (2017) pointed out three main ways of visualising the state of natural aggregate pits and the changes affecting them: static two-dimensional (where the product is a traditional paper map), surface three-dimensional (a surface three-dimensional model of a single open pit), and interactive—created in an Internet browser. A new medium that may aid the process of visual terrain analysis in real time without the need to perform GIS analyses is the VR system. In the proposed reality–virtuality diagram, Milgram and Kishino (1994) defined several possible states such as: the real world, augmented reality, and virtual reality. The VR system might be seen as the opposite of the real world. This is connected with the fact that VR systems combine fast computer graphics systems with head-mounted display (HMD) and interface devices that provide the effect of immersion in an interactive virtual three-dimensional environment in which the objects have

spatial presence (Bryson 1995). VR systems allow real-time user feedback on landscape and urban planning. They enable users to manipulate the time dimension and discover spatial variation and patterns; they enable planners to experiment with objects irreversible in the real world; and they allow investigation of the interactions between multiple objectives (Orland et al. 2001). Sherman and Judkins (1992) have identified five characteristics essential to a VR system: Illustrative, Immersive, Interactive, Intensive, and Intuitive. Movement in VR in most cases is similar to movement in the real world (first-person view, FPV) and should result in a minimum of disorientation because humans on a daily basis navigate through three-dimensional space. That is why the VR system might aid the process of visual analysis, transporting the user directly into the analysed area and enabling its observation in the most natural way for a human, i.e., from the pedestrian perspective.

The main aim and contribution of the following paper was to present the geomatic process of transforming low-level aerial imagery obtained with unmanned aerial vehicles (UAV) into a digital terrain model (DTM) and implementing the model into a virtual reality system (VR). We verified the functionality of this process on the basis of one anthropogenically transformed landform (natural aggregate heap) generated from low-level aerial imagery. A subsequent aim was to verify the accuracy of the produced 3D model, i.e., to check whether the actual linear and angular dimensions between the control points were precisely represented in the 3D model. This was accomplished by comparing the coordinates of the control points in the 3D model with the results of those points' GNSS RTK measurements. The study was conducted using the geomatic method, combining a range of digital activities and calculations (carried out both in the office and in the field) which so far had been used separately in traditional geodesy, cartography, and photogrammetry.

## 2. Study Area

The study was conducted in the western part of Poland, in the Greater Poland Voivodeship: Poznań

County. The selected area of study features a great number of open gravel mining pits which significantly contribute to changes in land relief. The object chosen for the study was a 6500-m$^2$ heap of natural aggregate, located in the natural aggregate mine "Gołuń". The spatial extent of the aggregate mine is shown in Fig. 1, and the coordinates of its main border points (I, II, III, IV) are shown in Table 1. The selected heap had an irregular shape, with numerous hollows and protuberances. The difference in elevation between the highest and the lowest point of the heap was over 11 m.

Table 1

*Coordinates of the studied area*

| Point | X (EPSG: 2177) | Y (EPSG: 2177) |
|-------|----------------|----------------|
| I | 5814228.00 | 6455180.00 |
| II | 5813585.00 | 6455704.00 |
| III | 5813560.00 | 6454620.00 |
| IV | 5812934.00 | 6455241.00 |

## 3. Methods and Materials

### 3.1. Equipment and Software

The equipment used to obtain low-level aerial imagery included a multirotor unmanned aerial platform Tarot X6, equipped with a Panasonic DMC-GH4 camera with a 16.1 Mpx matrix and a DJI Zenmuse X5 lens (Fig. 2). The obtained images were processed using the Agisoft Lens and PhotoScan software, and the geovisualisation in the VR system was created with the aid of the Unity 3D software, which is a 3D game engine.

Figure 2
Unmanned aerial platform Tarot X6

### 3.2. Research Procedure

Creating a geovisualisation in the VR system entails a sequence of activities. The complexity of the conducted study required us to devise a specially designed geomatic workflow of the research process, illustrating the various research stages (Fig. 3). The individual components of the workflow will be described in more detail in later parts of the paper.

### 3.2.1 Planning the UAV Flight

At the office stage of planning the flight, an important role is played by cartometric map bases, which are helpful in the situational orientation of the studied area (Kędzierski et al. 2014). Moreover, in order to define the area of the flight and the ground control station, it is a good idea to use satellite photographs of the studied area (Ostrowski and Hanus 2016). During the office stage of flight planning, we

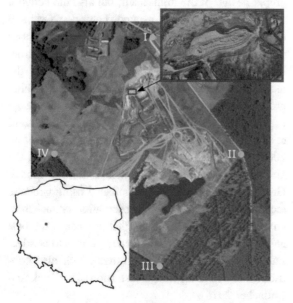

Figure 1
Location of the studied area (*background layer* orthophotomap from national geoportal www.geoportal.gov.pl)

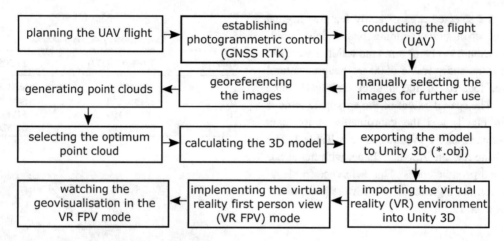

Figure 3
The workflow for the geomatic process of transforming low-level aerial imagery obtained with a UAV into a DTM uploaded into the VR system

designed the path of the UAV flight over the studied natural aggregate pit (Fig. 4). Based on the analysis of the area and the studied object, we decided to take oblique photographs of the pit from its every side (west, east, north, south) as well as vertical photographs from directly above it. The object was divided into five observation sides: northern, southern, eastern, western, and vertical. Eight oblique observation posts were defined for the UAV's flight around the object (two for each side) as well as two vertical observation posts. A flight conducted in such a way would yield a series of overlapping photographs of the studied object from every aspect.

In the conducted study, we used a cartometric map base as well as orthophotomaps downloaded from the national geoportal, which served to identify potential spots for establishing ground control. While planning the photogrammetric flight, it was decided that in order to obtain a model representing the relief with a sub-centimetre accuracy, photogrammetric control would be established in the area—a necessary step for conducting the georeferencing process, and eventually generating a numerical terrain model (Uysal et al. 2015). In addition, an intermediate aim of this paper was to conduct an independent check of the situational and height accuracy of the photogrammetric model obtained in the georeference process. That was why additional check points were established in the area of the study, constituting an independent control network (Gonçalves and Henriques 2015). It was assumed that the check of the

obtained photogrammetric model would consist in comparing the coordinates of the afore-mentioned control points measured with the Global Navigation Satellite System Real Time Kinematic (GNSS RTK) technique with the coordinates of the same points obtained from the numerical terrain model.

### 3.2.2 Establishing Ground Control

During the surveying activities performed with the help of UAVs, what is important is not only the proper design of the flight itself, but also the network of points constituting ground control (Nex and Remondino 2014). In a situation when the unmanned platform used in the flight is not equipped with an inertial measurement unit (IMU), or when the received satellite signal is interrupted, establishing ground control is a necessity, even at the office stage of planning the flight (Anai et al. 2012; Barazzetti et al. 2010; Eugster and Nebiker 2008; Wang et al. 2008). The process is especially necessary when the aim is to generate a cartometric terrain model. Ground control should be established using geodetic techniques such as satellite observation or tacheometry. It is believed, however, that nowadays the best and most effective measurement method is to conduct satellite observations in real time, which also contributes to reducing image distortions (de Kock and Gallacher 2016).

In the study in hand, ground control points (GCPs) were established using the GNSS RTK

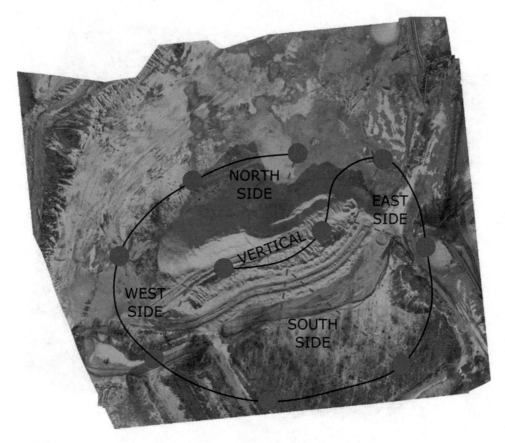

Figure 4
UAV flight path

technique, with the aid of the Trimble R4 Model 3 receiver. Taking into consideration the surveying technique used and the planar coordinate system EPSG:2177 (the one applicable in Poland, used in geodetic works) (Bosy 2014), it was decided that all the measurements made would be defined in this system for practical reasons. Due to the fact that UAV-obtained images are often oblique, it is recommended that a great number of control points be established in the surveyed area (Ruzgienė et al. 2015). Taking into account the size of the studied object and the planned altitude of the flight, four targeted GCPs were established (Fig. 5). They were located on every side of the observed aggregate heap. The ground control points were established in the field with the use of wooden poles, additionally marked with fluorescent orange paint so they could be easily identified in the photographs (Siebert and

Teizer 2014). Additionally, in order to perform a situational and height accuracy check of the obtained photogrammetric model, seven independent check points (ICPs) were established in the studied area and labelled with numbers from 1 to 7 (Fig. 5). The decision to establish seven independent check points (ICP) was motivated by several factors:

(a) Viewing the heap as a three-dimensional object, one might notice that (1) in terms of its vertical structure it consists of a foot and a summit. Hence, there should be points established in both those parts; (2) from the north, the object is shaded; from the south—lit by sunlight. Hence, establishing a single control point on each of those sides could be insufficient. Consequently, two points were established on each side (the shaded and the sunlit one). Two additional control points were established on the summit.

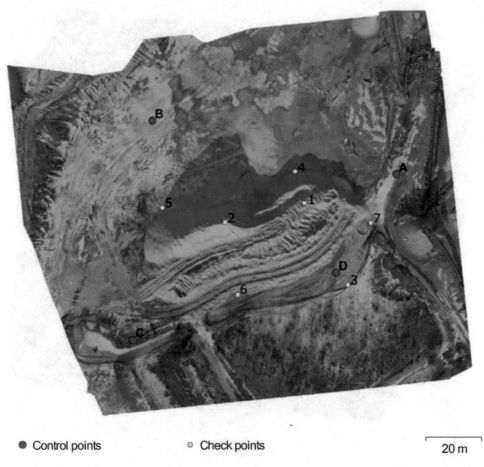

● Control points          ○ Check points          20 m

Figure 5
The location of ground control points (red dots) and independent check points (yellow dots) in the studied area

(b) One of the objectives of the study was to check whether for the applied research methodology, increasing the distance from which the photograph was taken would affect the accuracy of the obtained 3D model. To this end, one additional point (a third one) was established on the southern side, as photographs of this side were taken from a greater distance than those of the northern side (Fig. 6).

ICPs were not used in the process of georeferencing the images (Toutin and Chénier 2004).

### 3.2.3 Conducting the Flight

The platform used in the study was equipped with an inertial measurement unit (IMU) and a GPS receiver;

however, the coordinates were only used to stabilize the flight in the GPS mode and on the basis of available research (Gonçalves and Henriques 2015). It was assumed that the currently attainable accuracy levels are not sufficient for photogrammetric purposes. Hence, the flight was carried out in the manual mode, thus acquiring photographs on the basis of image obtained in real time. Taking into account the size of the studied object and the legal restrictions in Poland related to UAV flights, we decided that a sufficient elevation of the flight would be 100 m AGL (above ground level) and that the photographs for further analysis would be taken from all sides of the object. In accordance with the currently binding regulations in Poland (Regulation of the Minister of Infrastructure and Construction, Dz. U. 2016, Item 1993), carrying out UAV flights with a platform not

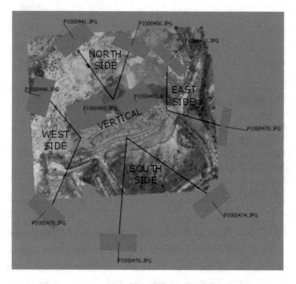

Figure 6
Location of 10 selected images for further analysis

exceeding 25 kg of take-off weight is allowed up to an altitude of 100 m above ground level (AGL) and at a distance of more than 6 km away from the nearest airport.

The weather conditions during the flight were favourable: wind speed approximately 3 m/s and a clear sky. During the 30-min flight, a total of 103 images of the object were obtained. The given time of the flight included 10 min taken to prepare the platform for flight and run a checklist ensuring flight safety.

### 3.2.4 Manual Selection of the Images for Further Analysis

Before proceeding to further stages of the study, the researchers performed a manual harmonisation of the obtained set of images, reducing their number while making sure to retain the ones which presented the studied object in the best possible way. As a result of this stage, ten photographs were chosen for further analysis—two images from every direction and two vertical images presenting the natural aggregate heap (Fig. 6).

### 3.2.5 Georeferencing the Selected Images

The next stage consisted in georeferencing the chosen images into a metric coordinate system. The first step

towards georeferencing them into a coordinate system, identical to the one in which the ground control points had been defined, was to recreate their relative internal orientation. This process was performed on the basis of the EXIF file containing metadata about the photographs which made it possible to estimate the location of the images in space. Owing to the necessity of generating a cartometric DTM, the next stage was to recreate the relative orientation of the points, but already on the basis of the established GCPs. The task was completed using Agisoft PhotoScan Professional software. This programme is especially useful for processing UAV-obtained images, as it enables its user to generate a DTM or an orthophotomap in any chosen coordinate system.

It must be added, however, that before the calculation process was performed in the above mentioned software, a twin programme, Agisoft Lens, had been used to calibrate the camera used during the UAV flight. Brown's distortion model implemented in this software made it possible to determine and factor into further calculations the radial and tangential distortion of the images.

After performing the georeferencing process on the basis of the established ground control, the RMSE (root-mean-square error) was calculated—a value specifying the deviation of the original tie-up points from the corresponding points calculated on the basis of the generated model. For calculating the RMSE, the following formula was used (Smaczynski and Medynska-Gulij 2017):

$$\text{RMSE} = \sqrt{\sum_{i=1}^{n} \frac{\left(X_{i,\text{est}} - X_{i,\text{in}}\right)^2 + \left(Y_{i,\text{est}} - Y_{i,\text{in}}\right)^2 + \left(Z_{i,\text{est}} - Z_{i,\text{in}}\right)^2}{n}},$$

where $X/Y/Z_{i,\text{in}}$—the actual value of a particular coordinate, $X/Y/Z_{i,\text{est}}$—the estimated value of this coordinate.

On the basis of the obtained results, it was concluded that the ground control point whose total error of georeferencing had the lowest value was point D, and the point burdened with the greatest error—point B (Table 2). The further part of the study involved a check on the basis of independent check points.

Table 2

*The calculated georeferencing errors of the GCPs*

| Label | X error (cm) | Y error (cm) | Z error (cm) | Total (cm) |
|-------|--------------|--------------|--------------|------------|
| B | 0.21 | 1.46 | − 0.29 | 1.50 |
| D | − 0.04 | − 0.10 | − 0.74 | 0.74 |
| A | − 1.00 | − 0.51 | 0.79 | 1.38 |
| C | 0.75 | − 0.96 | 0.43 | 1.29 |
| Total | 0.64 | 0.91 | 0.60 | 1.26 |

Table 3

*Parameters of the generated 3D models*

| Variable | Model I | Model II | Model III |
|----------|---------|----------|-----------|
| Density of the point cloud (pts.) | 23,509,622 | 1,537,422 | 381,379 |
| Point density (point/m$^2$) | 1420 | 90.1 | 22.7 |
| Total RMSE (cm) | 2.05 | 1.26 | 1.58 |
| Number of model faces | 4,690,164 | 101,128 | 19,438 |
| Number of model vertices | 2,346,456 | 50,806 | 9835 |
| Generation time (h:m:s) | 03:53:26 | 00:03:41 | 00:00:57 |
| File size (MB) | 363 | 6.83 | 1.21 |

### 3.2.6 Generating Point Clouds

The next step towards the creation of the terrain model was to use the obtained images as the basis for generating a set of three-dimensional point clouds of the studied object. At this stage, three point clouds representing the landform were created, differing in their construction parameters. The first point cloud was created on the basis of photographs in their original size, i.e., 4608 × 2592 pixels (Model 1). The second cloud (Model 2) was generated after reducing the size of the images 4 times (which meant that the length of each side of the photo was reduced twice). The last, smallest point cloud (Model 3), was created from images reduced 16 times in size, i.e., 4 times on every side.

### 3.2.7 Choosing the Optimum Point Cloud and Calculating the 3D Model

Table 3 presents the characteristics of the three point clouds and the photogrammetric 3D models generated from them. As expected, the time needed to calculate Model I with the greatest precision in representing the terrain was the longest, amounting to nearly four hours. A disadvantage of this model, directly connected with the size of the created point cloud (over 23.5 mln points), was the very large size of the output file (363 MB), which could render it impossible for implementation into the VR system. Bearing in mind also the fact that Model I yielded the greatest value of the RMSE error (2.05 cm), the researchers decided to exclude it from further stages of the study. In Model II, the point cloud was much smaller (over 1.5 mln points), which reduced the file size to only 6.83 MB. The calculation time was also greatly reduced, to below 4 min, yet the model provided a very accurate representation of the terrain

(RMSE 1.26 cm). Model III was constructed from the lowest number of points (approximately 0.4 mln), which, as in the previous cases, influenced file size (1.21 MB). However, it yielded worse results in terms of the RMSE error (1.58 cm). Taking into consideration the much worse parameters of Model III and the insignificant difference in the size of the output files, the model chosen by the authors as more appropriate for further geovisualisation in the VR system was Model II.

### 3.2.8 Exporting the 3D Model into the Unity 3D Environment

Based on the results obtained in the previous stage, the model chosen to undergo further analyses was Model II, calculated from a cloud of 1,537,422 points (90 points/m$^2$). On the basis of the cloud, a 3D model of the studied area was generated, consisting of 50,806 vertices, translating into 101,128 triangles making up the area of the model. Next, a texture was applied to the generated model, to reflect the real appearance of the natural aggregate heap (Fig. 7). The last step was to export the resultant model into the *.OBJ format, which enabled its subsequent implementation into Unity 3D.

### 3.2.9 Importing the Model into the VR Environment in Unity 3D

Activities at this stage were carried out in the programming environment of Unity 3D, which is an engine for the creation of computer games. The Unity project was first upgraded with the Google VR package, which boosts the functionality of the Unity

Figure 7
The generated terrain model with the applied texture

engine, enabling it to display content in the VR system. The Google VR package provides additional features such as: user head tracking, side-by-side stereo rendering, detection of user interaction with the system, distortion correction for a VR viewer's lenses. After the upgrade, the 3D model of the aggregate heap was imported and set to a proper scale. Because by default the model was displayed without any texture, texture was added in order to give the user an impression of greater realism while visually analysing the terrain. Finally, the virtual world was made complete by adding the sky as the background. The authors used Google Cardboard, which is the cheapest and easiest VR platform to obtain. This smartphone-based headset uses the user's smartphone as a device to display the VR content.

### 3.2.10 Implementing the Virtual Reality First-Person View Mode (VR FPV)

A key element of the geovisualisation in the VR system is the FPV mode which enables the observation of objects from the pedestrian perspective. This works by displaying on the HMD two slightly different angles of the scene for each eye, which generates an illusion of depth (Fig. 8). This, along with other methods of simulating depth, such as parallax (objects farther from the observer seen to move more slowly) and shading, creates an almost lifelike experience. In this mode, the user becomes the main animator of the geovisualisation; all his/her movements are registered by sensors in the HMD and faithfully rendered in the geovisualisation. The VR system generates two stereoscopic images, one for each eye. This creates an impression that the user is immersed in the geovisualisation; in other words, the geovisualisation reaches its recipient directly. An additional advantage of the VR FPV mode is the possibility of adjusting the height of the observed picture to the actual height of the observer.

### 4. Results

The RMSE values obtained in the georeferencing process constitute only a mean squared error that occurred while georeferencing the photogrammetric model calculated on the basis of the adopted tie-up points (GCPs). Results obtained in a study of this kind should be treated in accordance with the basic principles of geodesy, such as the necessity of performing independent checks. This activity should be carried out on the basis of previously established and

Figure 8
Location of the VR FPV viewer (left), virtual reality first-person view mode (right)

surveyed independent check points (ICPs) (Fig. 1), which were not directly used in the georeferencing process. In this study, the values that the authors treated as a reliable reference were the coordinates of check points 1–7, obtained via a GNSS RTK survey. Points from the field satellite survey were marked with the affix "gnss"; the ones from the photogrammetric Model II—with the affix "uav". A correlation of both those types of coordinates resulted in obtaining the coordinate deviations $\Delta X$, $\Delta Y$, $\Delta H$, as well as the Euclidean distance $\Delta L$ (Table 4).

In accordance with the arrangements concerning the distribution of ICPs (Fig. 5), two of them (points 4 and 5) were located at the shaded foot of the aggregate heap, while the three points 3, 6, and 7, were located at its sunlit foot. Comparing the obtained accuracy values $\Delta X$, $\Delta Y$, $\Delta H$, $\Delta L$ for the seven ICPs surveyed with a GNSS receiver and the values obtained from the 3D Model II generated from UAV images, one must conclude that in the study in hand, shading did not influence the situational coordinates ($X$, $Y$) or the Euclidean distance ($L$) of the analysed points. There was, however, a visible height ($H$) deviation in the case of point 4, which was the most shaded of all points.

What might have caused such a big deviation was the location of point 4 on the northern, shaded side of the natural aggregate heap. The pixels in its proximity

Table 4

*Differences in the coordinates of the independent check points—ICPs*

| Points | $\Delta X$ (m) | $\Delta Y$ (m) | $\Delta H$ (m) | $\Delta L$ (m) |
|---|---|---|---|---|
| 1gnss–1uav | 0.000 | 0.031 | 0.045 | 0.031 |
| 2gnss–2uav | − 0.019 | − 0.004 | − 0.016 | 0.019 |
| 3gnss–3uav | − 0.026 | − 0.004 | − 0.020 | 0.026 |
| 4gnss–4uav | − 0.019 | − 0.002 | − 0.209 | 0.019 |
| 5gnss–5uav | 0.013 | − 0.006 | − 0.041 | 0.014 |
| 6gnss–6uav | − 0.008 | − 0.017 | − 0.018 | 0.019 |
| 7gnss–7uav | − 0.015 | 0.011 | 0.029 | 0.019 |

were mostly of very similar colour. This might have caused image noise, resulting in incorrect results. In addition, it should be noted that despite establishing ground control on ground level, check points 1 and 2 located in two mutually independent spots at the top of the aggregate heap were afflicted with very small georeferencing errors. The second experimental aim was to analyse whether a twofold increase in the photographing distance from the studied aggregate heap would affect the obtained accuracy values $\Delta X$, $\Delta Y$, $\Delta H$, $\Delta L$. After studying the results, it was found that increasing the distance had not affected the obtained accuracy of the points' location on the 3D model used to create the VR geovisualisation.

In spite of the one outlying elevation value, the results obtained in this study testify to the fact that the generated relief model has a high situational and height accuracy. Therefore, it can be used for visibility analysis in the VR system.

## 5. Discussion and Conclusions

The conducted study was inspired by the paper of Esposito et al. (2017), who outlined the possibilities of using UAV technology for monitoring open pit mines, especially recording the changes taking place and performing cubature calculations on the basis of the point cloud. In addition, this paper aimed to further the research done by Slater et al. (2010) concerning the possibility of using VR for a more immersive exploration of 3D geovisualisation in the VR FPV mode. Authors believe that the integration of UAV-based DTMs into the VR system may be very important in anti-collision systems, in manual navigation of the UAV and in UAV simulators. The example work showing anti-collision system is the paper by Fong and Thorpe (2001). The other application of UAV-based terrain reconstruction within the VR system are simulators. VR FPV mode interfaces can improve the operator situational awareness and provide valuable tools to help understand and analyse the vehicle surroundings and plan command sequences Nguyen et al. (2000), Postal et al. (2016) and Smolyanskiy and Gonzalez-Franco (2017).

The obtained deviations in the location of check points surveyed with GNSS RTK and the same points calculated on the basis of UAV-obtained images enable the following conclusions to be drawn: (1) selecting two overlapping UAV images for each of the observed sides of the object (west, east, north, south, top) is sufficient to create a highly accurate 3D model; (2) using UAV technology is much more effective, both in terms of accuracy and the time required to carry out the measurements, than the use of traditional geodetic techniques: tacheometry, levelling, GNSS. Besides, the use of UAV for data acquisition does not cause any disturbance to the unstable material structures such as aggregate heaps or dunes, and so it is a very good option for studying such objects; (3) with the help of the workflow proposed in the paper, it is possible to generate a relief model with a centimetre-level accuracy. Such a 3D model might be used in precise, large-scale

engineering projects. Thanks to such accurate models, it is possible to create an interactive geovisualisation in the VR system, in the FPV mode, which lets the user move around in the digital 3D environment and analyse the visibility in real time. An advantage of using the VR system for geovisualisation purposes is the possibility of analysing terrain changes in time, in a way most closely resembling natural time, without the need to be physically present in the studied area.

Low-level aerial imagery obtained by monitoring open pit mining areas at specified time intervals might become the basis for designing an animated visualisation presenting changes affecting those areas with the use of visual variables (Halik and Medyńska-Gulij 2017; Medyńska-Gulij and Cybulski 2016). The workflow for geomatic data transformation and geovisualisation as proposed in the paper might have a wide application in many areas of life, raising the level of spatial awareness and serving to clarify spatial relations between the geospheres. Nevertheless, we are mindful of the fact that in the future, with the increasing accuracy of IMU and the more widespread use of RTK GPS in UAV platforms, it will be possible to modify the proposed geomatic process of transforming low-level aerial imagery in such a way as to render the establishment of GCPs in the field unnecessary. Such a solution would enable georeferencing without the need to design, establish, and survey a ground control network, thus accelerating the process of creating the geovisualisation.

## References

Ahmad, A. (2011). Digital mapping using low altitude UAV. *Pertanika Journal of Science and Technology, 19*(S), 51–58.

Anai, T., Sasaki, T., Osaragi, K., Yamada, M., Otomo, F., & Otani, H. (2012). Automatic exterior orientation procedure for low-cost UAV photogrammetry using video image tracking technique and

GPS information. *ISPRS International Archives of the Photogrammetry, Remote Sensing and Spatial Information Sciences, XXXIX-B7,* 469–474.

Axelsson, P. (2000). DEM generation from laser scanner data using adaptive TIN models. *International Archives of Photogrammetry and Remote Sensing., XXXIII*(4B), 203–210.

Barazzetti, L., Remondino, F., Scaioni, M., Brumana, R. (2010). Fully automatic UAV image-based sensor orientation. *International Archives of Photogrammetry Remote Sensing and Spatial Information Sciences Vol XXXVIII Part 5 Commission V Symposium,* p. 6. http://www.isprs.org/proceedings/XXXVIII/part1/12/12_02_Paper_75.pdf.

Bosy, J. (2014). Global, regional and national geodetic reference frames for geodesy and geodynamics. *Pure and Applied Geophysics, 171*(6), 783–808.

Bryson, S. (1995). Approaches to the successful design and implementation of VR applications. In E. Earnshaw, J. Vince, & H. Jones (Eds.), *Virtual reality applications* (pp. 3–15). London: Academic Press.

Colomina, I., & Molina, P. (2014). Unmanned aerial systems for photogrammetry and remote sensing: A review. *ISPRS Journal of Photogrammetry and Remote Sensing, 92,* 79–97.

de Kock, M.E., Gallacher, D. (2016). *From drone data to decisions: Turning images into ecological answers.* Conference: Innovation Arabia 9, (February)

Eisenbeiss, H. (2004). A mini unmanned aerial vehicle (UAV): System overview and image acquisition. *Proceedings of the International Workshop on Processing and Visualization using High-Resolution Imagery.* https://doi.org/10.1017/S0003598X00047980.

Esposito, G., Mastrorocco, G., Salvini, R., Oliveti, M., & Starita, P. (2017). Application of UAV photogrammetry for the multitemporal estimation of surface extent and volumetric excavation in the Sa Pigada Bianca open-pit mine, Sardinia, Italy. *Environmental Earth Sciences, 76*(3), 103. https://doi.org/10.1007/s12665-017-6409-z.

Eugster, H., & Nebiker, S. (2008). UAV-based augmented monitoring—real-time georeferencing and integration of video imagery with virtual globes. *Archives, 37,* 1229–1236.

European Commission, (2007). Study analysing the current activities in the field of UAV, ENTR/2007/065

Fong, T., & Thorpe, Ch. (2001). Vehicle teleoperation interfaces. *Autonomous Robots, 11*(1), 9–18.

Gonçalves, J. A., & Henriques, R. (2015). UAV photogrammetry for topographic monitoring of coastal areas. *ISPRS Journal of Photogrammetry and Remote Sensing, 104,* 101–111.

Halik, Ł., & Medyńska-Gulij, B. (2017). The differentiation of point symbols using selected visual variables in the mobile augmented reality system. *The Cartographic Journal, 54*(2), 147–156. https://doi.org/10.1080/00087041.2016.1253144.

Horbiński, T., Medyńska-Gulij, B. (2017). Geovisualisation as a process of creating complementary visualisations: Enhanced/static two-dimensional, surface three-dimensional, and interactive; *Geodesy & Cartography, 66*(1), 45–58.

Kędzierski, M., Fryśkowska, A., & Wierzbicki, D. (2014). *Opracowania fotogrametryczne z niskiego pułapu.* Warszawa: Wojskowa Akademia Techniczna.

Kraus, K., Karel, W., Briese, C., & Mandlburger, G. (2006). Local accuracy measures for digital terrain models. *The Photogrammetric Record, 21,* 342–354. https://doi.org/10.1111/j.1477-9730.2006.00400.x.

Kršák, B., Blištan, P., Pauliková, A., Puškárová, P., Kovanic, L., Palková, J., et al. (2016). Use of low-cost UAV photogrammetry to analyse the accuracy of a digital elevation model in a case study. *Measurement, 91,* 276–287.

Li, M., Nan, L., Smith, N. & Wonka, P. (2016). Reconstructing building mass models from UAV images. *Computers and Graphics, 54,* 84–93. Available at: http://www.sciencedirect.com/science/article/pii/S0097849315001077

Lin, Z. (2008). UAV for mapping—low altitude photogrammetric survey. *The International Archives of the Photogrammetry, Remote Sensing and Spatial Information Sciences, XXXVII*(Part B1), 1183–1186. http://citeseerx.ist.psu.edu/viewdoc/download?doi=10.1.1.150.9698&rep=rep1&type=pdf.

Liu, P., Chen, A. Y., Yin-Nan, H., Jen-yu, H., Jihn-Sung, L., Shil-Chung, K., et al. (2014). A review of rotorcraft unmanned aerial vehicle (UAV) developments and applications in civil engineering. *Smart Structures and Systems, 13*(6), 1065–1094.

Medyńska-Gulij, B., Cybulski, P., (2016). Spatio-temporal dependencies between hospital beds, physicians and health expenditure using visual variables and data classification in statistical table. *Geodesy and Cartography, 65*(1), 67–80. http://www.degruyter.com/view/j/geocart.2016.65.issue-1/geocart-2016-0002/geocart-2016-0002.xml.

Milgram, P., & Kishino, F. (1994). Taxonomy of mixed reality visual displays. *IEICE Transactions on Information and Systems, 12,* 1321–1329.

Mill, T., Ellmann, A., Aavik, A., Horemuz, M., & Sillamäe, S. (2014). Determining ranges and spatial distribution of road frost heave by terrestrial laser scanning. *The Baltic Journal of Road and Bridge Engineering, 9*(3), 225–234. https://doi.org/10.3846/bjrbe.2014.28.

Mill, T., Ellmann, A., Uuekula, K., & Joala, V. (2011). Road surface surveying using terrestrial laser scanner and total station technologies. In D. Čygas & K. D. Froehner (Eds.), *Proceedings of 8th International Conference "Environmental engineering: Selected papers"* (Vol. 3, pp. 1142–1147). Technika: Vilnius.

Nex, F., & Remondino, F. (2014). UAV for 3D mapping applications: A review. *Applied Geomatics, 6*(1), 1–15.

Nguyen, L. A., Bualat, M., Edwards L. J., Flueckiger L., Neveu, Ch, Schwerh, K., Wagner, M. D. & Zbinden E. (2000). Virtual reality interfaces for visualization and control of remote vehicles. in *IEEE International Conference on Robotics and Automation 2000, San Francisco, California.*

Orland, B., Bedthimedhee, K., & Uusitalo, J. (2001). Considering virtual worlds as representations of landscape realities and as tools for landscape planning. *Landscape and Urban Planning, 54,* 139–148.

Ostrowski, W., & Hanus, K. (2016). Budget UAV systems for the prospection of small- and medium- scale archaeological sites. *ISPRS Archives of Photogrammetry, Remote Sensing and Spatial Information Sciences, XLI*(July), 971–977.

Postal, G. R., Pavan W. & Rieder, R., (2016). A virtual environment for drone pilot training using VR devices. *2016 XVIII Symposium on Virtual and Augmented Reality (SVR), Gramado,* pp 183–187. https://doi.org/10.1109/svr.2016.39

Remondino, F., Barazzetti, L., Nex, F., Scaioni, M., Sarazzi, D. (2012). UAV Photogrammetry for mapping and 3D modeling—current status and future perspectives. *ISPRS—International Archives of the Photogrammetry, Remote Sensing and Spatial Information Sciences, XXXVIII-1*(September), 25–31. http://

www.int-arch-photogramm-remote-sens-spatial-inf-sci.net/XXX VIII-1-C22/25/2011/.

Ruzgienė, B., Berteška, T., Gecyte, S., Jakubauskiene, E. & Aksamitauskas, V. C. (2015). The surface modelling based on UAV photogrammetry and qualitative estimation. *Measurement, 73,* 619–627. http://www.sciencedirect.com/science/article/pii/ S0263224115002316

Sherman, B., & Judkins, P. (1992). *Glimpses of heaven, vision of hell: Virtual reality and its implications.* UK: Hodder and Stoughton.

Siebert, S., & Teizer, J. (2014). Mobile 3D mapping for surveying earthwork projects using an unmanned aerial vehicle (UAV) system. *Automation in Construction, 41,* 1–14. https://doi.org/10. 1016/j.autcon.2014.01.004.

Slater, M., Spanlang, B., Sanchez-Vives, M. V., & Blanke, O. (2010). First person experience of body transfer in virtual reality. *PLoS ONE, 5*(5), e10564. https://doi.org/10.1371/journal.pone. 0010564.

Smaczyński, M. & Medyńska-Gulij, B. (2017). Low aerial imagery—an assessment of georeferencing errors and the potential for use in environmental inventory. *Geodesy and Cartography, 66*(1), 89–104.

Smolyanskiy, N., & Gonzalez-Franco, M. (2017). Stereoscopic first person view system for drone navigation. *Frontiers in Robotic and AI, 4,* 11. https://doi.org/10.3389/frobt.2017.00011.

Torres, M., Pelta, D. A., Verdegay, J. L., & Torres, J. C. (2016). Coverage path planning with unmanned aerial vehicles for 3D terrain reconstruction. *Expert Systems with Applications, 55,* 441–451.

Toutin, T. & Chénier, R. (2004). GCP requirement for high-resolution satellite mapping. XXth ISPRS Congress, pp. 12–23. http://www.cartesia.org/geodoc/isprs2004/comm3/papers/385. pdf.

Uysal, M., Toprak, A. S., & Polat, N. (2015). DEM generation with UAV photogrammetry and accuracy analysis in Sahitler hill. *Journal of the International Measurement Confederation, 73,* 539–543. https://doi.org/10.1016/j.measurement.2015.06.010.

Wang, J., Garratt, M., Lambert, A., Han, S. & Sinclair, D. (2008). Integration of Gps/Ins/vision sensors to navigate unmanned aerial vehicles. *The International Archives of the Photogrammetry, Remote Sensing and Spatial Information Sciences, 37,* 963–970.

Watts, A. C., Ambrosia, V. G., & Hinkley, E. A. (2012). Unmanned aircraft systems in remote sensing and scientific research: Classification and considerations of use. *Remote Sensing, 4,* 1671–1692. https://doi.org/10.3390/rs4061671.

Wood, J., & Fisher, P. (1993). Assessing interpolation accuracy in elevation models. *IEEE Computer Graphics and Applications, 13*(2), 48–56.

(Received  April 11, 2017, revised  December 4, 2017, accepted  December 9, 2017, Published online  December 18, 2017)

Pure Appl. Geophys. 175 (2018), 3223–3245
© 2018 Springer International Publishing AG, part of Springer Nature
https://doi.org/10.1007/s00024-018-1874-1

**Pure and Applied Geophysics**

# Detection and Mapping of the Geomorphic Effects of Flooding Using UAV Photogrammetry

JAKUB LANGHAMMER[1] and TEREZA VACKOVÁ[1]

*Abstract*—In this paper, we present a novel technique for the objective detection of the geomorphological effects of flooding in riverbeds and floodplains using imagery acquired by unmanned aerial vehicles (UAVs, also known as drones) equipped with an panchromatic camera. The proposed method is based on the fusion of the two key data products of UAV photogrammetry, the digital elevation model (DEM), and the orthoimage, as well as derived qualitative information, which together serve as the basis for object-based segmentation and the supervised classification of fluvial forms. The orthoimage is used to calculate textural features, enabling detection of the structural properties of the image area and supporting the differentiation of features with similar spectral responses but different surface structures. The DEM is used to derive a flood depth model and the terrain ruggedness index, supporting the detection of bank erosion. All the newly derived information layers are merged with the orthoimage to form a multiband data set, which is used for object-based segmentation and the supervised classification of key fluvial forms resulting from flooding, i.e., fresh and old gravel accumulations, sand accumulations, and bank erosion. The method was tested on the effects of a snowmelt flood that occurred in December 2015 in a montane stream in the Sumava Mountains, Czech Republic, Central Europe. A multi-rotor UAV was used to collect images of a 1-km-long and 200-m-wide stretch of meandering stream with fresh traces of fluvial activity. The performed segmentation and classification proved that the fusion of 2D and 3D data with the derived qualitative layers significantly enhanced the reliability of the fluvial form detection process. The assessment accuracy for all of the detected classes exceeded 90%. The proposed technique proved its potential for application in rapid mapping and detection of the geomorphological effects of flooding.

**Key words:** UAV, structure from motion, flood, accumulations, bank erosion, OBIA, classification, textural features.

[1] Department of Physical Geography and Geoecology, Faculty of Science, Charles University in Prague, Albertov 6, 128 43 Prague 2, Czech Republic. E-mail: jakub.langhammer@natur.cuni.cz

## 1. Introduction

Unmanned aerial vehicles (UAVs, also known as UASs or drones) represent a rapidly evolving field of technology allowing geoscientists to obtain spatial data products that are suitable for rapid mapping and analysis of dynamic landscape processes.

The recent surge in the use of UAVs in different geoscientific disciplines has been facilitated by the rapid progress in the methods of photogrammetric processing (Anderson and Gaston 2013; Hirschmüller 2011; Westoby et al. 2012). The growing number of applications of UAV-based photogrammetry is apparent mostly in geomorphological research. As topography is a key element, the accuracy and reproducibility of the data based on UAV imagery (Clapuyt et al. 2016; Tonkin et al. 2014) are issues of primary interest. The areas of application are rapidly growing and include the fields of geomorphic mapping (Hackney and Clayton 2015), landslide research (Lucieer et al. 2013), erosion studies (Eltner et al. 2015; Smith and Vericat 2015), and sedimentation (Wheaton et al. 2010).

The UAV applications in hydrological research are most developed in the areas of stream and riverscape mapping, where the ability to rapidly acquire accurate and detailed spatial data, represented by orthomaps and digital elevation models (DEMs), represents a substantial shift in field surveying methods (Dietrich 2016; Flener et al. 2013). Detailed mapping of stream properties by means of UAVs is of special importance for hydromorphological research, because it provides a means to obtain objective spatial information for the classification of physical river habitats (Woodget et al. 2017), dynamics of water stages (Witek et al. 2014), and limited (but efficient) detection of hydromorphological features (Casado et al. 2015).

The importance of UAV-based photogrammetry is increasing in fluvial geomorphology, because it allows researchers to perform quantitative analysis of changes in stream and riparian zones at multitemporal scales, thereby enabling volumetric analyses of bank erosion and fluvial deposition (Miřijovský and Langhammer 2015; Tamminga et al. 2015a). However, hydrological applications of UAV-based mapping, typically based on RGB sensors, are substantially restricted by the limited ability to detect and reconstruct the properties of the submerged stream channel (Lejot et al. 2007; Thumser et al. 2017). This limitation applies mostly to streams that feature turbulent flow and riffle-pool sequences with significant differences in channel depth or water turbidity (Tamminga et al. 2015b). UAV applications in the shallow riverbeds have, however, exhibited good performance in mapping of stream hydromorphology enabling to partially cover the submerged zones of the channel (Woodget et al. 2015).

Detection of changes in the fluvial morphology of stream and riparian zones and the operability of data acquisition are key factors in the analysis of flood effects. Recent studies have thus frequently focused on stream planform changes, and most have been limited to the detection and analysis of fluvial morphologic changes in streams after flooding (Langhammer et al. 2017; Tamminga et al. 2015) or delimiting flood-prone areas based on DEM analysis (Şerban et al. 2016). To better separate flood-related features from anthropogenic features in an urban area, a method for identifying the extent and effects of flooding was tested by Feng et al. (2015).

The aim of this study was to develop an objective protocol enabling rapid mapping, identification, and classification of the geomorphological effects of floods in natural streams. Our goal was to develop a framework based on RGB imagery (as RGB sensors are the standard sensor in common commercial UAVs) that can detect the key fluvial forms resulting from floods in stream channels and riparian zones. The particular research goals were (i) to define a framework that uses UAV-based imagery for detection and classification of flood effects (ii) to identify indices based on 2D and 3D photogrammetric processing products that enhance the performance of classification of fluvial forms, and (iii) to test the framework on a case study to identify the limits of its applicability.

## 2. Materials and Methods

### 2.1. Identification and Classification of Flood Effects

Floods exert complex effects on the landscape geomorphology that act at different temporal and spatial scales. Floods shape valley and river systems and drive the dynamics of geomorphic processes (Baker et al. 1988; Magilligan 1992). On the other hand, geomorphologic controls limit and affect the course and consequences of floods in a given environment (Křížek 2008; Poole et al. 2002). Floodplains, as the natural environment for the fluvial processes, also represent cultural landscapes and are frequently remodeled by a variety of societal impacts, ranging from settlement, industry, agriculture, transport, and flood protection measures, thereby heavily altering the hydrological properties and fluvial dynamics (Langhammer and Vilímek 2008; Magilligan 1992; Wohl 2000). Knowledge of the geomorphic response to observed flood pulses in a given environment is thus essential for understanding the behavior of the fluvial system and for designing efficient risk-reduction measures.

Objective, accurate, and readily available spatial information on the distribution of the fluvial forms resulting from floods is thus an essential input for further geomorphologic and hydrologic analysis and modeling. Methods for geomorphic mapping of fluvial processes are evolving with the development of new field survey and remote sensing technologies. Field mapping and surveying methods, although based on the conventional approaches, are still frequently used for mapping of flood effects in different environments (Křížek 2008; Wolman 1971). The increasing spatial resolution of DEMs allows them to be used for flood inundation mapping (Papaioannou et al. 2016), delineation of the extent of the floodplain (Hartvich and Jedlicka 2008), and hydrodynamic modeling (Caviedes-Voullième et al. 2014; Sanders 2007/2008). For such applications, detailed DEMs based on aerial LiDAR data have become a standard resource applicable at different spatial scales (Cook and Merwade 2009; Geerling et al. 2009; Hooshyar et al. 2015). Progress in remote data acquisition techniques and data availability has enabled the use of remote sensing spatial data products in the assessment of complex stretches of stream floodplains. In addition to detailed DEMs, multispectral satellite imagery (Hamilton et al. 2007; Sanyal and Lu 2004; Tralli et al. 2005) and aerial photogrammetry (Bryant and Gilvear 1999; Mertes 2002; Poole et al. 2002) have also contributed to the newly emerging UAV photogrammetry methods (Flener et al. 2013; Hervouet et al. 2011; Miřijovský and Langhammer 2015).

The acquired imagery has an ultra-high spatial resolution, which is suitable for the identification of flooding-induced fluvial forms and changes in streams and floodplains (Miřijovský and Langhammer 2015; Tamminga et al. 2015a). Although high-resolution spatial data products provide information of sufficient geometric resolution for the identification of the fluvial forms resulting from flooding, the automated detection of fluvial features is difficult (Dietrich 2016) as the spectral resolution of most common types of UAV cameras is limited to the visible spectral bands (RGB).

The limited spectral resolution results in a similar spectral response for objects with different origins, making it difficult to separate the appropriate functional classes. This problem is of special significance in flooded areas, where the objects are coated in fine-grained material delivered by the flood, because this coating diminishes differences in the natural color of the objects. The inability to separate classes with a similar spectral appearance in the visible bands can be bypassed using additional information, either by applying a multi- or hyperspectral sensor or by deriving specific indices to support the identification of the fluvial features based on textural characteristics (Feng et al. 2015). In addition to the information derived from 2D orthoimagery, UAVs can also be used to obtain other information derived from a 3D point cloud or digital surface model (DSM). Such information can be vital for the separation of features according to their relative vertical position in the riverscape and their variability.

The proposed solution (Fig. 1) extends the number of bands available from UAV imagery based on

supplementary information suitable for the identification and separation of objects in a floodplain. The RGB bands are supplemented with information on the vertical position of the cells relative to the stream and on the terrain ruggedness index (TRI), which are derived from the 3D DEM, and with information on object texture, which is derived from the 2D imagery.

The flood depth model derived from the DEM can help to distinguish pixels with similar spectral responses but different vertical positions in the fluvial system. This process can be used to separate objects located in different functional zones of the fluvial system, such as the river channel, floodplain, or terrace.

The object texture information can help to distinguish objects of similar spectral responses but different structures, suggesting different origins. This information is typically related to the differences between pixels corresponding to sand or clasts and pixels corresponding to turbid water. The information on object structure can significantly improve distinguishing among different functional classes. This approach is enabled by the unique ability of UAV photogrammetry to simultaneously provide both a 2D orthoimage and a 3D DEM for a study area. These two data sources are used to generate new layers that are added to the RGB imagery as extra bands to assist in supervised classification (Fig. 1). The basic categories of fluvial forms, resulting from flooding in unregulated streams, are identified as follows: recent and old gravel accumulations, fresh sand accumulations, and bank erosion.

## 2.2. Study Area

The case study focuses on Javoří brook (Šumava Mountains, Czech Republic), which is a small montane stream with a catchment area of 11 km². The basin is located in a mid-mountain environment and preserves the natural dynamics of the fluvial processes, which have recently accelerated in response to climate change and extensive forest disturbance (Borrelli et al. 2016; Langhammer et al. 2015b). Despite the position of the headwaters at an elevation of 1172 ±139 m above sea level (a.s.l.), the basin slope is moderate, and the stream in the study area meanders extensively (Fig. 2). The study site is a

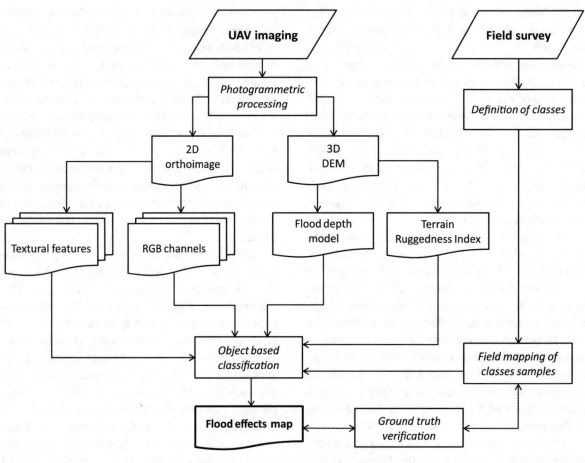

Figure 1
Workflow of the objective identification of the geomorphological effects of floods

1-km-long stretch of the river channel and riparian zone, and includes several meanders with active bank erosion and accumulation point bars, consisting of gravel material of various grain sizes (Langhammer et al. 2017).

The hydrological regime of the upper Vydra basin is dominated by significant spring snowmelt occurring in April–May (Vlasák 2003). In the study basin, low-magnitude flood events occur frequently. As a headwater catchment, this basin is the source area of large floods in the Vltava and Elbe river basins, including the extreme flood in August 2002 (Langhammer et al. 2015a). However, due to the catchment's position at the basin divide and relatively flat topography, the flood magnitudes in the catchment are usually lower than those in the lowland segments of the streams.

The latest recent significant flood, resulting from a rain-on-snow event in December 2015, was used as the reference event for the analysis of the geomorphological effects of flooding in the meandering montane stream. The flood event was driven by snowmelt that was accelerated by intense precipitation at the end of November 2015 and that completely washed out the first compact snow cover of the winter season 2015–2016 accumulated prior the event (Fig. 3). The snow cover had reached a depth of 20–30 cm and melted over the course of a single day. The peak flow of the flood reached a peak discharge value of 51.2 m$^3$/s. The magnitude of the flood at Vydra-Modrava station in the basin outlet corresponded to a 5–10-year flood, and the peak flow at the lowland downstream station Otava-Písek corresponded to a 50-year flood (CHMI 2008).

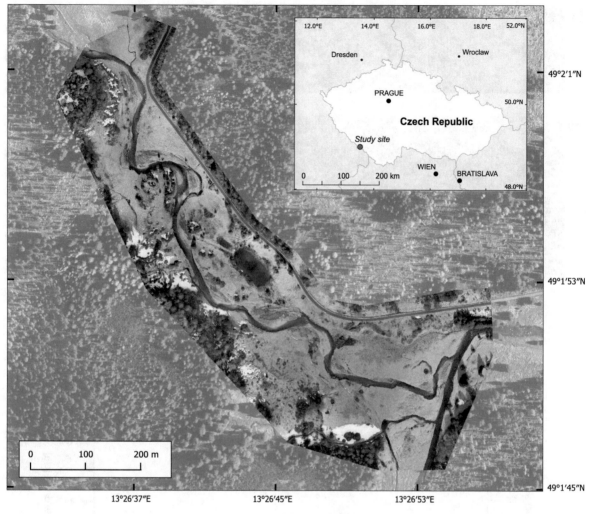

Figure 2
Study area

## 2.3. UAV Imagery Acquisition

The UAV imagery was acquired with a DJI Inspire 1 Pro imaging platform with a Zenmuse X5 camera equipped with the Olympus M.ZUIKO ED 12-mm prime lens (Fig. 4). The Zenmuse X5 camera has a resolution of 4608 × 3456 pixels (16 MP) and is equipped with a 17.3 × 13.0-mm CMOS sensor to capture panchromatic imagery that is separable into the RGB spectral bands.

The flight was performed at an altitude of 70 m and acquired 352 images with 60–70% front and side overlap. The internal GPS of the drone was used to obtain the location of each image. The study site featured 12 fixed points distributed unevenly over the area. The positions of these points were determined using GNSS Topcon HiperSR device, achieving centimeter accuracy. UAV imagery was acquired on December 4th, 2015, 3 days after the peak of the flood.

## 2.4. Photogrammetric Processing

Photogrammetric processing of the UAV imagery is designed to derive 2D orthomosaic and 3D DEM data from a set of overlapping images (Flener et al. 2013). The applied image processing is based on the structure from motion (SfM) algorithm (Fonstad et al.

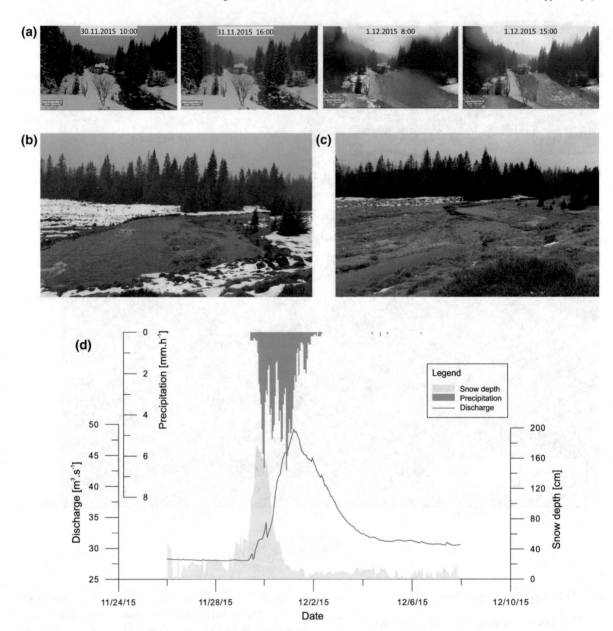

Figure 3

Progress of the flood in December 2015. **a** Snowmelt at the outlet of Roklanský brook, as captured by the webcam at Modrava (http://www.sumavanet.cz), **b** flood spill in the study site 3 h after peak flow, **c** study site during the UAV imaging campaign (photo by J. Langhammer, 2015), **d** hydrograph of the flood event at Javoří brook station, approximately 1 km upstream of the monitoring site, and **e** hydrograph at the Roklanský brook station, 2 km downstream of the monitoring site. Data Charles University, 2015

2013; Westoby et al. 2012), which allows alignment and image matching of sets of unstructured images acquired in nadir or oblique directions. The whole image processing chain consists of a set of sequential steps that comprise the calculation and application of camera corrections, alignment of the imagery based on the SfM algorithm (Fig. 5a), generation of a sparse point cloud from the matching points in the imagery, and densification of the point cloud (Fig. 5b). The dense point cloud is then classified to separate the points representing the ground from those representing vegetation and structures (Fig. 5c). Then, based

**(a)**

**(b)**

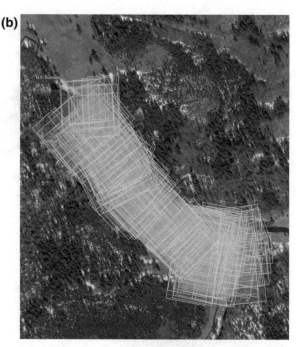

Figure 4
Imagery acquisition in the study area. **a** Imaging campaign, and **b** coverage of the study area by the UAV imagery. Photo by J. Langhammer, 2015

on the resulting point model, two key data products of the photogrammetric reconstruction—the DEM (Fig. 5d) and the orthomosaic (Fig. 5e)—are derived (Miřijovský and Langhammer 2015).

From the UAV imaging, a total of 352 images were collected and used for processing. After image alignment, automatic detection, and calibration of the camera parameters, this set of images was used to generate the sparse point cloud. The processing was done with a high alignment accuracy and automatic detection of the overlapping points. The dense point cloud, created in the next step, consisted of 26,460,000 points. The total area covered by the reconstruction of the terrain was 0.13 km$^2$, while the ground sampling distance of the resulting orthoimage was 1.7 cm and the average error was 0.942 pixels. For further analysis, the orthoimage and the DEM were sampled to the same resolution of 2 cm per pixel (Table 1).

The photogrammetric reconstruction of the riverbed and floodplain was done using Agisoft Photoscan Pro 1.31 software. The data processing was performed on a PC-based workstation featuring a Core i7 6700 4.0 GHz processor, 64 GB DDR4 RAM, and two NVIDIA GTX 960 GPU acceleration units.

### 2.5. Image Textural Features

Texture is an important characteristic that enables the identification of objects in an image based on gray-toned spatial dependencies. The textural features are statistical descriptors that measure the gray-scale distribution in an image and consider the spatial interactions between pixels (Porebski et al. 2008). The features relate to specific textural characteristics such as homogeneity, contrast, and the presence of organized structure within the image and characterize the complexity and nature of tone transitions in the image (Haralick et al. 1973). Texture analysis is a general image analysis technique (Vijayalakshmi and Subbiah Bharathi 2011) with applications in various disciplines, e.g., medical image analysis (Feng 2011), bioimaging (Miyamoto and Merryman 2005) and industrial control (Mäenpää et al. 2003).

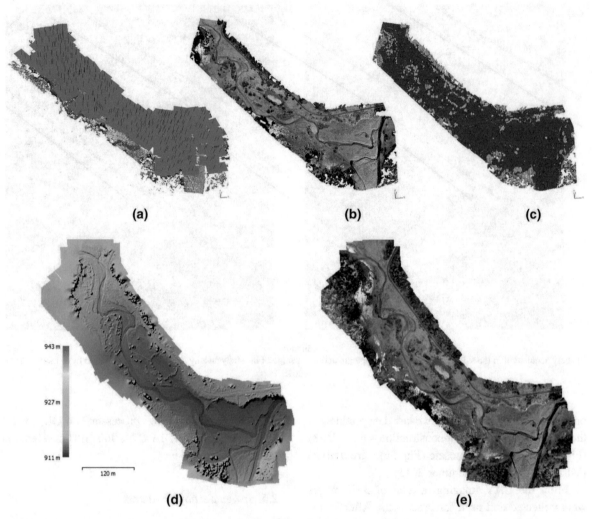

Figure 5

Photogrammetric processing chain for deriving the orthomap and DEM from UAV imagery in Agisoft Photoscan Pro. **a** Alignment of the imagery and generation of sparse point cloud, **b** generation of the dense point cloud, **c** classification of the ground points from the dense point cloud, **d** derived DEM, and **e** derived orthoimage

Table 1

*Parameters of the flight, imagery, and terrain reconstruction*

| Parameter | Value |
|---|---|
| Number of images | 352 |
| Flight altitude | 70 m |
| Ground sampling distance | 1.7 cm/pix |
| Total area covered | 0.13 km$^2$ |
| Average error | 0.898 pix |
| Point density | 2691.6 points per sq. m |

For the texture analysis in this study, the textural features introduced by Haralick et al. (1973) were applied. To calculate the textural features from the UAV-based imagery, we used the SAGA GIS 4.0 software (Conrad et al. 2015). The RGB composite images were used to calculate two sets of textural features using two approaches: the selected local statistical measures (Zhang 2001), i.e., Contrast, Energy, Entropy, and Variance, and the original set of textural features based on (Haralick et al. 1973), i.e., Entropy, Contrast, Correlation, Variance, Inverse Diff Moment, Sum Average, Sum Entropy, Sum Variance, Difference Variance, Difference Entropy, Angular Second Moment, Measure of Correlation-1, and Measure of Correlation-2. All these textural features express the different aspects of the distribution and homogeneity of the pixels in a gray-tone image. The measure of Contrast reflects the sharpness of images and the depth of textures. Energy, calculated as the sum of each matrix element, reflects the gray-scale distribution homogeneity of images and texture crudeness. Entropy is a measure of a non-uniformity and complexity of image texture as it represents a statistical measure of randomness of the input image. The Angular second moment feature reflects the intra-regional changes in image textures. The Correlation feature is a measure of linear dependencies of gray levels in a matrix to the neighboring pixels and expresses the consistency of image texture (Kumar and Sreekumar 2014). The potential for use of the given measures largely depends on the given environment, image source, and quality.

The resulting patterns of local textures, calculated on the applied imagery, are highly heterogeneous (Fig. 6). The visual assessment of the calculated textures at selected representative spots in the study area indicated that only two of the textural features deliver a spatial distribution of values relevant to the studied features, i.e., Energy (Fig. 6d) and Entropy (Fig. 6e). Some of the textural features reflect only the basic landscape or fluvial features (i.e., Variance and Angular Second Moment—Fig. 6g, h), while some even introduce artificial noise that could hinder the interpretation rather than increasing its accuracy (i.e., Contrast, Inverse Diff Moment, Difference Variance, and Sum Average—Fig. 6f, i, j, k).

The selection of the textural textures that can act as supporting layers with the potential to further enhance the classification was thus based on their ability to reflect the key physical features of the riverscape and to support separation of the stream, fluvial forms, and vegetation. Because it generated the most complex results, the Entropy feature (Fig. 6e) was used in further classification.

### 2.6. DEM Analysis

The use of a DEM as a supporting layer in the identification of fluvial features has two goals. First, the DEM was used to derive a reconstruction of flood depths as a measure of the vertical position and the exposure of the given riverscape elements to the stream flow energy. Second, the DEM was used to detect areas with abrupt changes in elevation resulting from bank erosion.

The flood depth reconstruction was based on a simplified model calculating the vertical distance of the DEM grid cells to the observed peak water level during the flood event. The DEM was first aligned according to the water levels, expressing the hydraulic gradient in the assessed stream stretch and then flood filled up to the levels, corresponding the peak flow marks, observed by the field survey. This flood depth reconstruction was done in SAGA GIS (Conrad et al. 2015) using the Grid Tools module for DEM alignment and the Lake Flood module for the reconstruction of the extent and depth of the flood spill in the floodplain based on the flood level marks observed by the field survey.

The TRI was then used to support the identification of bank erosion. This index, based on the gradient of values in a set of raster cells, was originally developed as a tool for spatial ecology studies, in which the ability to identify linear features and edges is vital for the prediction of habitats with heterogeneous environmental conditions (Böhner and Selige 2006; Riley 1999). The techniques used to detect linear features in the DEM are also of great importance in terrain analysis, because they can be used to detect structures with steep elevation gradients, i.e., faults, breaks, or linear structures (Quackenbush 2004). In our study, we have used the TRI to identify linear features with high elevation

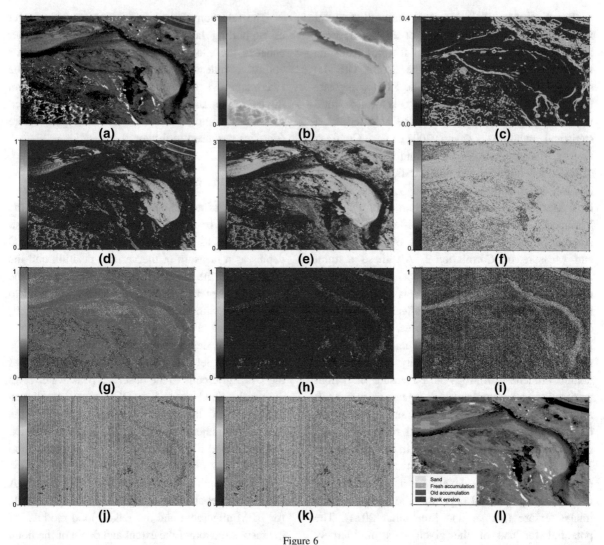

Figure 6

Examples of textural features derived from the UAV imagery. **a** RGB composite image, **b** aligned DEM, **c** terrain ruggedness index, **d** energy, **e** entropy, **f** contrast, **g** variance, **h** angular second moment, **i** inverse diff moment, **j** difference variance, **k** sum average, and **l** gray-tone composite image with sample of training areas

gradients in the ultra-high-resolution DEM of a riverscape to potentially identify flooding-induced bank erosion. We have used SAGA GIS with a TRI module (Conrad et al. 2015) with the cell radius parameter for TRI calculation set to ten grid cells.

### 2.7. Multi-band Image Segmentation and Classification

The multi-band raster data set was then prepared as a base for the image segmentation and

classification. The following raster layers were included as the input bands: the RGB channels from the orthoimage, the Entropy textural features, the aligned DEM, and the TRI data.

Based on this multi-band data set, object-based image segmentation was performed. The core idea of the object-based image analysis (OBIA), representing, recently, a rapidly evolving image analysis technique, is to extract the real-world objects from the raster imagery, independently on the scale (Blaschke 2010). The object extraction stems on the assumption that the resulting internal heterogeneity of the created segment shall be less than the heterogeneity when compared with its neighbors (Burnett and Blaschke 2003). The object extraction typically applies one of the two main approaches: region-based and edge-based segmentation (Haralick and Shapiro 1985). The region-based algorithms are seeking for homogeneous regions by either fast one-pass thresholding or more complex iterative seed growing methods. The edge-based detection approach is then based on a search for discontinuities in the image as the boundaries of regions, typically by the first-order differential operators, e.g., the Sobel or Roberts operator (Yuheng and Hao 2017).

In our study, the object-based image analysis (OBIA) module in SAGA GIS was used (Conrad et al. 2015), and the segmentation, using the seed region growing algorithm, resulted in 2.6 million polygons covering the study area. The set of OBIA polygons has been applied as a base for training the supervised classification of the flood-related fluvial forms and other key elements of the riverscape. The supervised classification aimed to identify the spatial distribution of recent and old traces of fluvial activity in the stream and floodplain. The following categories were distinguished: (i) fresh sand accumulations, (ii) fresh gravel accumulations, (iii) old gravel accumulations, and (iv) bank erosion.

For each category, we defined multiple training areas to include representative samples of each class. The training areas were selected to capture the variability within a given class while maintaining the major differences among the classes. The locations of the training areas were determined based on the field survey after the flood event, which was performed simultaneously with the UAV imaging campaign. For each class, there were selected multiple samples from the different parts of the study area to cover the potentially heterogeneous conditions in the assessed stream segment (Fig. 6l). From the total of 2,327,000 polygons, created by the OBIA segmentation, 16,070 of them was used as training samples of the classified categories. The distribution of categories among the training samples was reflecting the frequency of the categories (Table 2). Hence, the most frequent categories belong to the new (58.4%) and old coarse accumulations (36.9%),

while the number of samples of the sand accumulations (3.2%) and bank erosion (1.6%) is limited by their naturally scarce occurrence in the assessed stream segment. The training areas were further tested for homogeneity, as reflected by the unimodal distribution of the digital number (DN) values in the given bands. Training areas with overlapping distributions that contain pixels belonging to different categories were excluded or corrected prior to further classification.

Based on these training areas, the supervised classification of the OBIA polygons was performed. The maximum-likelihood classifier was used with equal weights assigned to the classes and a threshold of 0.1% of the unclassified image area. The classification was performed using the SAGA GIS supervised classification module (Conrad et al. 2015).

## 2.8. Accuracy Assessment

From the training and classification data sets, a confusion matrix was derived to analyze the distribution of training samples among the resulting classification categories. Based on the results, the basic metrics of classification performance commonly used in remote sensing were calculated. The set of basic measures of classification precision was calculated, comprising the basic indicators of true and false positives and negatives. The true positives indicate the correctly identified samples of a given class, while the false positives, called also as false alarms, are the incorrectly identified instances, i.e., the samples, assigned to a given class, although, in fact, they should not belong to it. The true negatives represent the correctly rejected cases, i.e., samples, that are not belonging to a given class, and that are also not assigned to it. Hence, the false negatives represent the incorrectly rejected cases, i.e., the samples, not assigned to a given class, while they, in fact, belong to it (Fawcett 2006).

Based on these basic indices, the measures of classification precision, sensitivity, and specificity (Foody 2002) were derived to analyze the different aspects of classification performance. The precision, one of the most frequently used metrics, represents the proportions of the true and false positives (Powers 2011), expressed as a fraction of relevant instances to

Table 2

*Distribution of DN values in training samples, used for the supervised classification*

| Band | Measure | Fresh coarse accumulation | Old coarse accumulation | Sand accumulation | Bank erosion |
|------|---------|---------------------------|-------------------------|-------------------|--------------|
| Red | Min | 136.00 | 90.93 | 34.77 | 24.91 |
| | Median | 167.67 | 139.78 | 85.71 | 57.63 |
| | Max | 175.69 | 191.60 | 159.88 | 99.92 |
| Green | Min | 129.50 | 66.14 | 19.65 | 2.26 |
| | Median | 172.31 | 139.47 | 76.47 | 49.40 |
| | Max | 179.31 | 195.70 | 163.06 | 95.29 |
| Blue | Min | 131.50 | 67.56 | 0.00 | 0.20 |
| | Median | 170.76 | 146.58 | 62.68 | 56.35 |
| | Max | 180.20 | 202.80 | 175.66 | 95.35 |
| DEM | Min | 0.29 | 0.12 | 0.33 | 0.13 |
| | Median | 0.54 | 0.47 | 0.63 | 0.96 |
| | Max | 0.70 | 0.70 | 1.13 | 2.09 |
| TRI | Min | 0.00 | 0.00 | 0.00 | 0.28 |
| | Median | 0.02 | 0.01 | 0.01 | 0.48 |
| | Max | 0.06 | 0.05 | 0.05 | 0.69 |
| Entropy | Min | 1.35 | 2.06 | 1.61 | 2.15 |
| | Median | 1.95 | 2.62 | 2.61 | 2.64 |
| | Max | 2.46 | 2.75 | 2.76 | 2.74 |
| Training polygons | | 5085 | 2948 | 177 | 101 |
| Training area (sq. m) | | 207.83 | 131.33 | 11.29 | 5.51 |

the total number of instances. The sensitivity (or recall or true positive rate) is the proportion of correctly identified positives and is calculated as the ratio of correctly classified positives to the sum of the true positives and false negatives (Fawcett 2006). The specificity (or the true negative rate) represents the correctly identified negatives and is calculated as the ratio of true negatives to the sum of the false positives and true negatives (Fawcett 2006).

The accuracy assessment and confusion matrix was calculated in the KNIME 3.2.0 using the Scorer node (Morent et al. 2011).

## 3. Results

### 3.1. Reconstruction of the Flooded Area

The 3D reconstruction of a riverscape based on optical imagery has significant limitations in terms of reconstructing the submerged zone. As mentioned in the methodological part, the optical data cannot provide a complete reconstruction of the channel bottom, especially in segments with considerable water depths, turbulent flow, or high turbidity. UAV imagery can, however, be used to reconstruct the bathymetry of shallow streams with an unobstructed view, low turbidity, and non-turbulent stream flow. With respect to these limitations, we have performed the 3D reconstruction of the floodplain, which was further aligned according to the water level to calculate the flood spill model (Fig. 7).

The reconstruction of the flooded area, derived from the DEM, proved a solid fit with the extent of the flood spill, observed during the flood event (Fig. 7). The good fit of the model, although calculated by the simplified approach, was supported by a flat topography of the area, while the superelevation of the stream bottom between the start point and end point of the assessed stream segment reaches 4 m.

From the extent of the flooded zone and the distribution of water depths in the floodplain, there is apparent that the flood spill exceeded significantly the stream channel and reached the full area of the floodplain. Highest observed flood depths are detected in the area of the stream channel. For the dynamics of the fluvial processes, there is, however, critical the distribution of the water depth over the active zones of the meanders. In most significant point bars of the meandering system featuring large

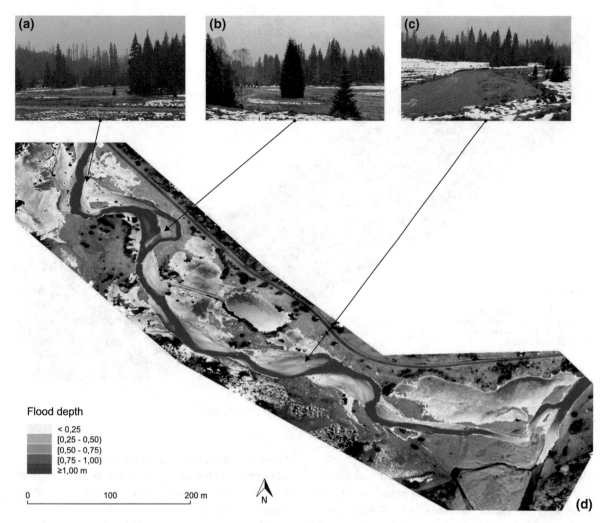

Figure 7

Flooded area and depths in the assessed stream stretch based on the 3D riverscape reconstruction from the UAV imagery. The images display the flood extent in selected sites of the study area during the flood event: **a** Wide-spread flood spill in the riparian zone at the entrance to the meandering belt, **b** Water filling the cut-off meander, **c** Extent of the flood spill in the most active meander of the system. Photo J. Langhammer, 2015

gravel and sand depositions, the flood spill depth exceeded 0.5 m (Fig. 7), which represents and appropriate driver for fluvial processes. The spill of the flood reached most of the floodplain area, including the abandoned channels as well as the local depressions at the fringes of the floodplain which enabled redistribution of the fine alluvial depositions relatively far from the active point bars and stream channel.

### 3.2. Spatial Distribution of the Flood Effects

The geomorphological effects of the flood are concentrated in the active belt that crosses the meandering zone of the Javoří creek. The fresh gravel and sand accumulations are located dominantly at the active point bars. However, a significant part of the fresh fluvial accumulations can be found also in the cut-off meanders and local depressions in the floodplain, where the course of the flood has relocated the fluvial depositions during the flood event (Fig. 8).

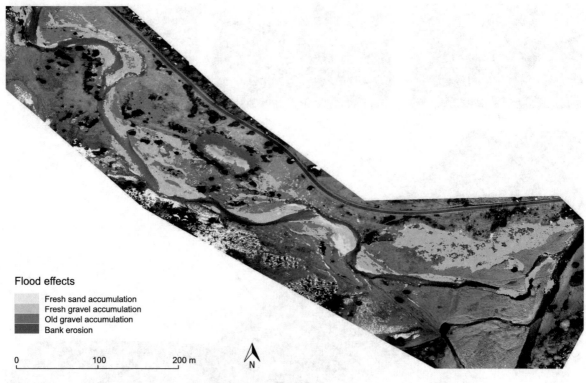

Figure 8
Distribution of the flood effects in the study area

The flood effects are dominated by the accumulation forms, as only 0.95% of the detected forms belong to bank erosion. The most extensive fluvial forms are fresh gravel accumulations (70.62% of the area) and old gravel accumulations (24% of the area) because of the nature of a headwater montane stream. Sand accumulations, covering only 4.43% of the assessed area, are detected only in isolated spots located on the back sides of meanders (Table 3).

The spatial extent of fluvial accumulations and bank erosion fits with the extent of the area flooded by the studied flood event. Almost 98% of the identified flood effects are located within the extent of the flooded area (Fig. 8).

Two categories of fluvial forms were detected outside the active area of the flood: old gravel accumulations covered 95.67% of this area, and bank erosion accounted for 4.33% of this area (Table 3). The old gravel accumulations located outside the area of recent flooding are remnants of the previous floods with magnitudes greater than that of the studied

event. The detected spots of bank erosion outside the active zone of the flood are related to the edge of the stream bank that collapsed following the flood-induced scouring.

The spatial distribution of the flood effects is the direct effect of the extent and depth of the flood spill in the floodplain. Based on the aligned DEM and observed flood marks, it was possible to reconstruct the extent and depth of flooding in the study area to analyze the relations between the occurrence of the individual categories of flood effects and the depth of flooding.

The distribution of the flood effects reflects variations in the stream power, the nature of the fluvial processes, the flood depth, and the resulting fluvial forms (Fig. 9). The most intense fluvial activity occurred in the peripheral part of the active zone of the flood. For example, 87% of the flood effects were detected within the first two water depth categories up to 0.6 m, and 98.47% of the flood effects were found at flood depths of less than 1.2 m;

Table 3

*Structure of flood effects*

|  | Area (sq. m) | Share (%) |
|---|---|---|
| Distribution of flood effects in respect to flooded zone |  |  |
| Effects inside flooded area | 12494.0 | 97.96 |
| Effects outside flooded area | 259.5 | 2.04 |
| Total | 12753.5 | 100 |
| Structure of flood effects |  |  |
| Fresh gravel accumulations | 9006.2 | 70.62 |
| Old gravel accumulations | 3060.6 | 24.00 |
| Fresh sand accumulations | 565.1 | 4.43 |
| Bank erosion | 121.7 | 0.95 |
| Total | 12753.5 | 100 |
| Structure of flood effects outside the flooded zone |  |  |
| Old gravel accumulations | 248.3 | 95.67 |
| Bank erosion | 11.3 | 4.33 |
| Total | 259.5 | 100 |

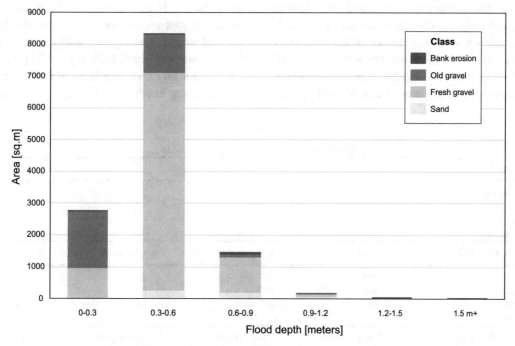

Figure 9

Distribution of flood effects in relation to the depth of flooding

hence, very little activity occurred in areas with depths greater than 1.2 m.

The shallowest zone of flooding, with water depths up to 30 cm, is dominated by old gravel accumulations that are partially covered by fresh gravel accumulations (Fig. 9). Together with the remnants in the areas outside the flooded area and their partial occurrence in the second water depth category, the old gravel accumulations are clearly concentrated in the outer part of the active zone of the flood. The fresh gravel accumulations are predominant in the water depth category of 0.3–0.6 m and represent 65.45% of all flood effects. In this flood depth category, the fresh gravel deposits represent 81% of the flood effects, the old gravel accumulations represent 15%, and sand deposits represent 4%. Sand accumulations are generally sparse in this montane headwater stream and are most common in water depths of 0.6–1.2 m. In contrast to the accumulation forms, bank erosion forms are concentrated in the highest water depth categories, corresponding to the location of the bank scouring of the convex banks of the meanders. Bank erosion thus fully dominates the water depth category greater than 1.5 m, where it is the only identified fluvial form (Fig. 9).

### 3.3. Classification Accuracy

The statistical reliability of the classification, tested by the confusion matrix and the derived overall and per-class indices of classification accuracy, revealed a solid performance of the supervised classification (Table 4). From the total number of 5526 training samples in the four classes (bank erosion, new gravel accumulations, old gravel

accumulations, and new sand accumulations), 4928 were correctly classified, reaching an overall classification accuracy of 89.18%.

There are, however, apparent differences in the classification performance for the erosion and accumulation forms. The bank erosion was the only fluvial form for which all of the training samples were correctly identified by the classification in the appropriate class as the true positives (Table 4).

Among the accumulation forms, the new gravel depositions, as the largest category, featured the most reliable results. From the 3478 tested elements, only 161 were assigned to an incorrect category, particularly fresh sand accumulations. The precision of this classification, as a fraction of true and false positives, is 0.954, while the sensitivity is 0.888. The reliability appears to be generally better for the detection of fresh fluvial forms, benefiting from the distinct spectral responses of the fresh accumulations and the background. This high reliability is also apparent for sand accumulations, featuring a solid precision of 0.822 and a very high specificity of 0.990, indicating a good ability to correctly identify negative samples

Table 4

*Confusion matrix and classification accuracy*

| (a) Total accuracy | | | | Value |
|---|---|---|---|---|
| Correct classified | | | | 4928 |
| Wrong classified | | | | 598 |
| Accuracy | | | | 89.18 |
| Cohen's kappa | | | | 0.7822 |

| (b) Training/classified | Sand | New gravel | Old gravel | Bank erosion |
|---|---|---|---|---|
| Sand | 235 | 51 | 0 | 0 |
| New gravel | 161 | 3317 | 0 | 0 |
| Old gravel | 19 | 367 | 1319 | 0 |
| Bank erosion | 0 | 0 | 0 | 57 |

| (c) Classification accuracy | True positives | False positives | True negatives | False negatives |
|---|---|---|---|---|
| Sand | 235 | 51 | 5060 | 180 |
| New gravel | 3317 | 161 | 1630 | 418 |
| Old gravel | 1319 | 386 | 3821 | 0 |
| Bank erosion | 57 | 0 | 5469 | 0 |

| (d) Sensitivity and specificity | Precision | Sensitivity | Specificity | f-Measure |
|---|---|---|---|---|
| Sand | 0.822 | 0.566 | 0.990 | 0.670 |
| New gravel | 0.954 | 0.888 | 0.910 | 0.920 |
| Old gravel | 0.774 | 1.000 | 0.908 | 0.872 |
| Bank erosion | 1.000 | 1.000 | 1.000 | 1.000 |

that are not sand accumulations. The old gravel accumulations have the weakest performance, with partial overlaps with new gravel and sand accumulations (Table 3) and a precision of 0.774. A correct identification of this category from the imagery is not always straightforward. The distinction of the new and old accumulations is based on comparing samples with almost identical textural properties, elevation, and ruggedness; and is thus dependent on the color response. The outer boundary of the fresh fluvial deposits can gradually transition into old gravel, resulting in a mixed color response.

## 3.4. Ground Truth

The accuracy of the principal fluvial forms was further verified by a field survey of the selected spots featuring typical examples of the studied features, i.e., fresh gravel accumulations, fresh sand accumulations, and bank erosion, as well as the ability to distinguish between recent and old gravel accumulation in a point bar. In a meander with extensive traces of fluvial activity, four segments featuring the identified fluvial forms were selected (Fig. 10).

To test the detection of fresh gravel accumulation, a point bar on the inner side of a meander near start of the most active zone in the meandering belt was selected (Fig. 10a). Gravel accumulations were detected appropriately both in the core zone of the point bar and on its margins. At this spot, detection of bank erosion resulting from scouring was also tested. The classification here correctly identified the presence of bank erosion on the perpendicular bank slopes. The upper edges of these slopes are covered with vegetation; thus, bank erosion would have been difficult to detect based solely on the orthoimagery. The detection of old gravel accumulations (Fig. 10b) was tested on the largest point bar with apparent traces of gravel accumulations resulting from the recent and past floods. The identification of the old and new (Fig. 10c) gravel accumulations in the classification was distinct. Although the deposits have the same granulometric structure, the old accumulations have a significantly different color response and can thus be separated by the supervised classification. The most critical aspect is the separation of the gravel accumulations along the bank from the same material in the submerged zone. Here, the different textural features of the water surface help to distinguish between those two categories. The identification of sand accumulations is aided by distinct color response differences and different surface textures (Fig. 10c). Extensive bank erosion (Fig. 10d) was detected correctly along banks both fully exposed and partially covered by the remnants of vegetation. The field verification of the classification results thus confirmed the calculated high performance of the protocol (Table 3).

## 4. Discussion

The combined application of the spatial data products based on UAV photogrammetric reconstruction of the riverscape—the 2D orthoimage, the 3D DEM, and the derived information, i.e., textural features, TRI values, and flood depth model—were used to bypass the limitations of conventional data sources in the identification of fluvial forms resulting from floods. Orthoimages based on RGB data are the most common spatial data source used for the analysis of riverscapes, hydromorphological mapping (Casado et al. 2015), and the assessment of fluvial morphologic features (Bryant and Gilvear 1999) based on aerial mapping or UAV-based imagery (Dietrich 2016). However, RGB imagery has a limited ability to distinguish among categories with similar spectral responses in the visible bands but with different physical properties and origins but also due to the occurrence of shadows, shifting the spectral response (Casado et al. 2015). Use of RGB imagery as the only data source for automated detection may thus result in misclassifications.

Adding information on the texture, vertical position, and the terrain ruggedness enables us to distinguish qualitative features that allow us to distinctly separate similar-looking features with different physical properties. The effect of additional data sources on the accuracy of classification was tested by detecting the selected fluvial forms using different combinations of input layers, i.e., the RGB orthoimage, the RGB orthoimage with textural features, and the RGB orthoimage with textural features plus the layers derived from the DEM.

Figure 10
Test spots for ground truth testing of the detection of the flood effects. **a** Fresh sand accumulation, **b** fresh gravel accumulation, **c** old gravel accumulation, and **d** bank erosion

The increase in accuracy with the addition of textural features and DEM-derived layers is substantial (Table 5). In particular, the effect of the additional layers is apparent in the determination of gravel accumulations and bank erosion. The most substantial increase in the classification accuracy is apparent in the case of bank erosion, where the accuracy increased from 60.1% to 98.3%, due largely to the addition of information derived from the DEM. Adding the elevation data enables us to capture

Table 5

*Classification accuracy*

| Class | RGB | RGB + textural features | RGB + textural features + TRI | RGB + textural features + TRI + DEM |
|---|---|---|---|---|
| Fresh sand accumulation | 93.9 | 95.1 | 95.4 | 95.9 |
| Fresh gravel accumulation | 80 | 83.7 | 86.7 | 95.7 |
| Old gravel accumulation | 75.5 | 76 | 76 | 93.2 |
| Bank erosion | 61.7 | 71.1 | 98.1 | 98.3 |

abrupt changes in elevation values at the location of bank erosion resulting from bank scouring. The accuracy of determination of the old gravel accumulations also significantly increased with the addition of supplementary data from an initial value of 75.5% to 93.7%. Here, the correct location of the spots with old gravel accumulations is improved by determining the vertical position of the site relative to the stream. The detection of fresh gravel accumulation was also improved, resulting in an accuracy increase from 80% to 95.7%. The detection of fresh sand accumulations, which feature a color response that is distinctly different from those of the other classes, remained reliable under all configurations, and the accuracy in all cases exceeded 93% (Table 5).

The ultra-high-resolution spatial data resulting from UAV digital photogrammetry can be used to seamlessly merge 2D and 3D information layers and have been shown to have a significant potential for identification of flood effects. Fusion of the orthoimagery with qualitative data, combining textural features and elevation indices, is helpful for distinguishing among riverscape categories with different structures or locations in the floodplain but with similar color responses in the visible bands.

The general limitation of the approaches, based on optical sensing, is the need for visibility of the assessed phenomena to enable the imaging and further correct photogrammetric reconstruction of the scene (Casado et al. 2015). As the UAV imaging is based on passive sensing in the visual or near-infrared spectrum, the coverage of the stream by a dense riparian vegetation, being a natural and frequent element of the riverscape, can hinder the mapping (Woodget et al. 2017). In part, such limitation can be minimized by proper scheduling the imaging

campaigns off the vegetation season, which is also in line with the recommendations for hydromorphological mapping of streams and floodplains.

The proposed protocol was tested in a natural environment of a montane stream, featuring the full range and typical distributions of fluvial geomorphological features resulting from flooding. As the methodology is based on general principle, the protocol could be applied to modified streams in the cultural landscapes as well. Here, the additional information bands, derived from DEM, could be useful to distinguish the artificial structures from fluvial features and improve the classification performance in urban areas (Zhang 2001).

To accurately assess the effects of natural hazards, there is a growing need for easily operable and reliable techniques for rapid mapping and quantitative assessment of phenomena, ranging from the mapping of the geomorphic impacts of flooding in a headwater basins (Magilligan et al. 1998; Phillips 2002) to the assessment of stream channel changes (Baker et al. 1988). Quantitative analyses are substantially limited by the fact that some fluvial features are difficult to directly detect based on the conventional data sources and acquisition techniques, i.e., aerial imaging or satellite remote sensing, or by the spatial scale and related level of detail. However, field surveys, which offer the most detailed information, are time-consuming and dependent on skilled surveyors. Hence, areas affected by flooding and covered by reliable mapping of the flood effects remain sparse even in cultural landscapes.

Based on the experience gained by testing of the protocol, there is emerging efficiency of the UAV-based approach for rapid mapping and detection of flood effects. To give an example, the detailed mapping of 1-km-long stretch of the riverscape by a multi-rotor imaging platform has typically a time span of 1 h of flight with the adequate extra time necessary for the pre-flight planning and post-flight field survey, in particular, the GCP positioning. After photogrammetric processing, either by desktop or cloud-based tools, the campaign results in a seamless detailed map of the riverscape, which could, besides the discussed protocol, be used as supporting tool for hydromorphological assessment and as a basis for multitemporal analysis. The application of UAVs and digital photogrammetric analysis techniques, which feature high operability, high spatial resolutions, and high cost efficiency, can thus provide a reliable protocol for rapid mapping and classification of the geomorphic changes resulting from natural hazard processes.

## 5. Conclusions

This study has presented a novel approach that enables the objective detection of the geomorphological effects of flooding occurring in riverbeds and floodplains using imagery acquired by UAVs with panchromatic cameras. We have proposed and tested a workflow that applies the methods of photogrammetric analysis, texture analysis, and supervised classification based on the two key data products of the UAV photogrammetry—the 3D DEM and the 2D orthoimage. From these data inputs, a multi-band data set was derived and served as a base for the subsequent classification of flood effects.

The RGB channels of the derived orthoimage were supplemented with the image textural features and the information on the 3D properties of the stream derived from the DEM, in particular, the flood depth model and the terrain ruggedness index. Based on this data set, a supervised classification was performed to identify certain fluvial features, i.e., fresh and old gravel accumulations, sand accumulations, and bank erosion. The results of the classification clearly indicated that the classification accuracy of the fluvial features improved when the additional information layers were added to the RGB bands. Among the layers derived from the DEM, the TRI enabled the distinct detection of bank erosion, which is usually difficult to detect in orthoimagery. The

flood depth model was used to separate features with similar color and textural responses but different vertical positions relative to the stream, which assisted mainly with the separation of old and fresh gravel accumulations. Furthermore, identification of the observed flood traces enabled us to simulate the extent of the flooding and the distribution of flood depth, which were used to analyze the distribution of the given types of flood effects according to the depth of flooding. The textural features obtained from the orthoimage were then used to better distinguish between features with similar color responses but with different surface structures, i.e., the gravel deposits in a point bar and in the shallow submerged zone.

Adding the supporting analytical layers to the classification process increased its accuracy. The most significant increase was apparent for bank erosion, for which the accuracy increased from 61.7% using only RGB channels to 98.3% using all available layers. The accuracy of the assessment of gravel accumulations increased from 75.2% for old accumulations and 80.1% for new accumulations to 93.2 and 95.7%, respectively. The most stable results were observed for fresh sand accumulations, which feature a distinct spectral response and structural properties; the accuracy shifted from 93.9% using only RGB layers to 95.9% using all data bands.

This study proved that the proposed method of fusing RGB imagery with 3D information derived from UAV imagery and the derived analytical layers significantly extends the potential for using common UAV imaging platforms for supervised classification and detection of fluvial forms. The proposed protocol represents a reliable, operable, and cost-efficient method for the rapid mapping and analysis of flood effects in streams and floodplains.

## Acknowledgements

The research was supported by the EU COST Action 1306 project LD15130 "Impact of landscape disturbance on the stream and basin connectivity" and Czech Science Foundation project 13-32133S "Retention potential of headwater areas".

## REFERENCES

Anderson, K., & Gaston, K. J. (2013). Lightweight unmanned aerial vehicles will revolutionize spatial ecology. *Frontiers in Ecology and the Environment, 11*(3), 138–146.

Baker, V. R., Kochel, R. C., & Patton, P. C. (1988). *Flood Geomorphology.* New York: Wiley.

Blaschke, T. (2010). Object based image analysis for remote sensing. *The ISPRS Journal of Photogrammetry and Remote Sensing, 65,* 2–16.

Böhner, J., & Selige, T. (2006). Spatial prediction of soil attributes using terrain analysis and climate regionalisation. *Gottinger Geographische Abhandlungen, 115,* 13–28.

Borrelli, P., Panagos, P., Langhammer, J., Apostol, B., & Schütt, B. (2016). Assessment of the cover changes and the soil loss potential in European forestland: First approach to derive indicators to capture the ecological impacts on soil-related forest ecosystems. *Ecological Indicators, 60*(January 2016), 1208–1220.

Bryant, R. G., & Gilvear, D. J. (1999). Quantifying geomorphic and riparian land cover changes either side of a large flood event using airborne remote sensing: River Tay, Scotland. *Geomorphology, 29*(3), 307–321.

Burnett, C., & Blaschke, T. (2003). A multi-scale segmentation/object relationship modelling methodology for landscape analysis. *Ecological Modelling, 168,* 233–249.

Casado, M. R., Gonzalez, R. B., Kriechbaumer, T., & Veal, A. (2015). Automated identification of river hydromorphological features using UAV high resolution aerial imagery. *Sensors, 15*(11), 27969–27989.

Caviedes-Voullième, D., Morales-Hernández, M., López-Marijuan, I., & García-Navarro, P. (2014). Reconstruction of 2D river beds by appropriate interpolation of 1D cross-sectional information for flood simulation. *Environmental Modelling & Software, 61,* 206–228.

CHMI. (2008). *Precipitation and runoff database.* Prague: CHMI.

Clapuyt, F., Vanacker, V., & Van Oost, K. (2016). Reproducibility of UAV-based earth topography reconstructions based on structure-from-motion algorithms. *Geomorphology, 260,* 4–15. https://doi.org/10.1016/j.geomorph.2015.05.011.

Conrad, O., Bechtel, B., Bock, M., Dietrich, H., Fischer, E., Gerlitz, L., et al. (2015). System for automated geoscientific analyses (SAGA) v. 2.1.4. *Geoscientific Model Development, 8*(7), 1991–2007.

Cook, A., & Merwade, V. (2009). Effect of topographic data, geometric configuration and modeling approach on flood inundation mapping. *Journal of Hydrology, 377*(1), 131–142.

Dietrich, J. T. (2016). Riverscape mapping with helicopter-based structure-from-motion photogrammetry. *Geomorphology, 252,* 144–157.

Eltner, A., Baumgart, P., Maas, H.-G., & Faust, D. (2015). Multitemporal UAV data for automatic measurement of rill and interrill erosion on loess soil. *Earth Surface Processes and Landforms, 40*(6), 741–755.

Fawcett, T. (2006). An introduction to ROC analysis. *Pattern Recognition Letters, 27*(8), 861–874.

Feng, D. D. (2011). *Biomedical information technology.* Amsterdam: Elsevier.

Feng, Q., Liu, J., & Gong, J. (2015). Urban flood mapping based on unmanned aerial vehicle remote sensing and random forest classifier—a case of Yuyao, China. *Water, 7*(4), 1437–1455.

Flener, C., Vaaja, M., Jaakkola, A., Krooks, A., Kaartinen, H., Kukko, A., … Alho, P. (2013). Seamless mapping of river channels at high resolution using mobile liDAR and UAV-photography. Remote Sensing, *5*(12), 6382–6407.

Fonstad, M. A., Dietrich, J. T., Courville, B. C., Jensen, J. L., & Carbonneau, P. E. (2013). Topographic structure from motion: a new development in photogrammetric measurement. *Earth Surface Processes and Landforms, 38*(4), 421–430.

Foody, G. M. (2002). Status of land cover classification accuracy assessment. *Remote Sensing of Environment, 80*(1), 185–201.

Geerling, G. W., Vreeken-Buijs, M. J., Jesse, P., Ragas, A., & Smits, A. (2009). Mapping river floodplain ecotopes by segmentation of spectral (CASI) and structural (LiDAR) remote sensing data. *River Research and Applications, 25*(7), 795–813.

Hackney, C., & Clayton, A. (2015). 2.1. 7. Unmanned Aerial Vehicles (UAVs) and their application in geomorphic mapping. Retrieved from https://eprints.soton.ac.uk/376639/1/2.1.7_UAV.pdf.

Hamilton, S. K., Kellndorfer, J., Lehner, B., & Tobler, M. (2007). Remote sensing of floodplain geomorphology as a surrogate for biodiversity in a tropical river system (Madre de Dios, Peru). *Geomorphology, 89*(1), 23–38.

Haralick, R. M., Shanmugam, K., & Dinstein, I. (1973). Textural features for image classification. *IEEE Transactions on Systems, Man, and Cybernetics, SMC, 3*(6), 610–621.

Haralick, R. M., & Shapiro, L. G. (1985). Image segmentation techniques. *Computer Vision, Graphics, and Image Processing, 29,* 100–132.

Hartvich, F., & Jedlicka, J. (2008). Progressive increase of inputs in floodplain delineation based on the DEM: application and evaluation of the model in the catchment of the Opava River. *AUC Geographica, 53*(1–2), 87–104.

Hervouet, A., Dunford, R., Piégay, H., Belletti, B., & Trémélo, M.-L. (2011). Analysis of post-flood recruitment patterns in braided-channel rivers at multiple scales based on an image series collected by unmanned aerial vehicles, ultra-light aerial vehicles, and satellites. *GIScience and Remote Sensing, 48*(1), 50–73.

Hirschmüller, H. (2011). Semi-global matching-motivation, developments and applications. *Photogrammetric Week, 11,* 173–184.

Hooshyar, M., Kim, S., Wang, D., & Medeiros, S. C. (2015). Wet channel network extraction by integrating LiDAR intensity and elevation data. *Water Resources Research, 51*(12), 10029–10046.

Křížek, M. (2008). Erosion and accumulation flood landforms in Sázava River in spring 2006. *AUC Geographica, 53*(1–2), 163–181.

Kumar, R. M., & Sreekumar, K. (2014). A survey on image feature descriptors. *Computers & Electrical Engineering, 5,* 7847–7850.

Langhammer, J., Hartvich, F., Kliment, Z., Jeníček, M., Bernsteinová, J., Vlček, L., … Miřijovský, J. (2015). The impact of disturbance on the dynamics of fluvial processes in mountain landscapes. Silva Gabreta, *21*(1), 105–116.

Langhammer, J., Lendzioch, T., Miřijovský, J., & Hartvich, F. (2017). UAV-based optical granulometry as tool for detecting changes in structure of flood depositions. *Remote Sensing, 9*(3), 240.

Langhammer, J., Su, Y., & Bernsteinová, J. (2015b). Runoff response to climate warming and forest disturbance in a mid-mountain basin. *Water, 7,* 3320–3342.

Langhammer, J., & Vilímek, V. (2008). Landscape changes as a factor affecting the course and consequences of extreme floods in the Otava river basin. *Czech Republic. Environmental Monitoring and Assessment, 144*(1–3), 53–66.

Lejot, J., Delacourt, C., Piégay, H., Fournier, T., Trémélo, M.-L., & Allemand, P. (2007). Very high spatial resolution imagery for channel bathymetry and topography from an unmanned mapping controlled platform. *Earth Surface Processes and Landforms, 32*(11), 1705–1725.

Lucieer, A., de Jong, S. M., & Turner, D. (2013). Mapping landslide displacements using structure from motion (SfM) and image correlation of multi-temporal UAV photography. *Progress in Physical Geography, 38*(1), 97–116.

Mäenpää, T., Turtinen, M., & Pietikäinen, M. (2003). Real-time surface inspection by texture. *Real-Time Imaging, 9*(5), 289–296.

Magilligan, F. J. (1992). Thresholds and the spatial variability of flood power during extreme floods. *Geomorphology, 5*(3), 373–390.

Magilligan, F. J., Phillips, J. D., James, L. A., & Gomez, B. (1998). Geomorphic and sedimentological controls on the effectiveness of an extreme flood. *The Journal of Geology, 106*(1), 87–96.

Mertes, L. A. K. (2002). Remote sensing of riverine landscapes. *Freshwater Biology, 47*(4), 799–816.

Miřijovský, J., & Langhammer, J. (2015). Multitemporal monitoring of the morphodynamics of a mid-mountain stream using UAS photogrammetry. *Remote Sensing, 7*(7), 8586–8609.

Miyamoto, E., & Merryman, T. (2005). Fast calculation of Haralick texture features. Human Computer Interaction Institute. Retrieved from https://www.inf.ethz.ch/personal/markusp/teaching/18-799B-CMU-spring05/material/eizan-tad.pdf.

Morent, D., Stathatos, K., Lin, W.-C., & Berthold, M. R. (2011). Comprehensive PMML preprocessing in KNIME. In Proceedings of the 2011 workshop on predictive markup language modeling (pp. 28–31). San Diego, CA: ACM.

Papaioannou, G., Loukas, A., Vasiliades, L., & Aronica, G. T. (2016). Flood inundation mapping sensitivity to riverine spatial resolution and modelling approach. *Natural Hazards, 83*(1), 117–132.

Phillips, J. D. (2002). Geomorphic impacts of flash flooding in a forested headwater basin. *Journal of Hydrology, 269*(3), 236–250.

Poole, G. C., Stanford, J. A., Frissell, C. A., & Running, S. W. (2002). Three-dimensional mapping of geomorphic controls on flood-plain hydrology and connectivity from aerial photos. *Geomorphology, 48*(4), 329–347.

Porebski, A., Vandenbroucke, N., & Macaire, L. (2008). Haralick feature extraction from LBP images for color texture classification. In 2008 First Workshops on Image Processing Theory, Tools and Applications (pp. 1–8).

Powers, D. M. (2011). Evaluation: from precision, recall and F-measure to ROC, informedness, markedness and correlation. Retrieved from http://dspace2.flinders.edu.au/xmlui/handle/2328/27165.

Quackenbush, L. J. (2004). A review of techniques for extracting linear features from imagery. *Photogrammetric Engineering & Remote Sensing, 70*(12), 1383–1392.

Riley, S. J. (1999). Index that quantifies topographic heterogeneity. *Intermountain Journal of Sciences: IJS, 5*(1–4), 23–27.

Sanders, B. F. (2007/2008). Evaluation of on-line DEMs for flood inundation modeling. Advances in Water Resources, 30(8), 1831–1843.

Sanyal, J., & Lu, X. X. (2004). Application of remote sensing in flood management with special reference to Monsoon Asia: A review. *Natural Hazards, 33*(2), 283–301.

Şerban, G., Rus, I., Vele, D., Breţcan, P., Alexe, M., & Petrea, D. (2016). Flood-prone area delimitation using UAV technology, in the areas hard-to-reach for classic aircrafts: case study in the north-east of Apuseni Mountains, Transylvania. *Natural Hazards, 82*(3), 1817–1832.

Smith, M. W., & Vericat, D. (2015). From experimental plots to experimental landscapes: topography, erosion and deposition in sub-humid badlands from structure-from-motion photogrammetry. *Earth Surface Processes and Landforms, 40*(12), 1656–1671.

Tamminga, A., Eaton, B., & Hugenholtz, C. H. (2015a). UAS-based remote sensing of fluvial change following an extreme flood event. *Earth Surface Processes and Landforms, 40*(11), 1464–1476.

Tamminga, A., Hugenholtz, C., Eaton, B., & Lapointe, M. (2015b). Hyperspatial remote sensing of channel reach morphology and hydraulic fish habitat using an unmanned aerial vehicle (UAV): A first assessment in the context of river research and management. *River Research and Applications, 31*(3), 379–391.

Thumser, P., Kuzovlev, V. V., Zhenikov, K. Y., Zhenikov, Y. N., Boschi, M., Boschi, P., et al. (2017). Using structure from motion (SfM) technique for the characterization of riverine systems—case study in the headwaters of the Volga River. *Geography, Environment, Sustainability, 10*(3), 31–43.

Tonkin, T. N., Midgley, N. G., Graham, D. J., & Labadz, J. C. (2014). The potential of small unmanned aircraft systems and structure-from-motion for topographic surveys: A test of emerging integrated approaches at Cwm Idwal. *North Wales. Geomorphology, 226*(Supplement C), 35–43.

Tralli, D. M., Blom, R. G., Zlotnicki, V., Donnellan, A., & Evans, D. L. (2005). Satellite remote sensing of earthquake, volcano, flood, landslide and coastal inundation hazards. *ISPRS Journal of Photogrammetry and Remote Sensing: Official Publication of the International Society for Photogrammetry and Remote Sensing, 59*(4), 185–198.

Vijayalakshmi, B., & Subbiah Bharathi, V. (2011). A novel approach to texture classification using statistical feature. arXiv [cs.CV]. Retrieved from http://arxiv.org/abs/1111.2391.

Vlasák, T. (2003). Overview and classification of historical floods in the Otava river basin. *Acta Universitatis Carolinae—Geographica, 38*(2), 49–64.

Westoby, M. J., Brasington, J., Glasser, N. F., Hambrey, M. J., & Reynolds, J. M. (2012). "Structure-from-Motion" photogrammetry: A low-cost, effective tool for geoscience applications. *Geomorphology, 179,* 300–314.

Wheaton, J. M., Brasington, J., Darby, S. E., & Sear, D. A. (2010). Accounting for uncertainty in DEMs from repeat topographic surveys: Improved sediment budgets. *Earth Surface Processes and Landforms, 35*(2), 136–156.

Witek, M., Jeziorska, J., & Niedzielski, T. (2014). An experimental approach to verifying prognoses of floods using an unmanned aerial vehicle. *Meteorology Hydrology and Water Management. Research and Operational Applications, 2*(1), 3–11.

Wohl, E. E. (2000). *Geomorphic effects of floods. Inland flood hazards: Human, riparian, and aquatic communities* (pp. 167–193). Cambridge, UK: Cambrige University Press.

Wolman, M. G. (1971). Evaluating alternative techniques flood-plain mapping. *Water Resources Research, 7*(6), 1383–1392.

Woodget, A. S., Austrums, R., Maddock, I. P., & Habit, E. (2017). Drones and digital photogrammetry: from classifications to continuums for monitoring river habitat and hydromorphology. *Wiley Interdisciplinary Reviews: Water.* https://doi.org/10.1002/wat2.1222.

Woodget, A. S., Carbonneau, P. E., Visser, F., & Maddock, I. P. (2015). Quantifying submerged fluvial topography using hyperspatial resolution UAS imagery and structure from motion photogrammetry. *Earth Surface Processes and Landforms, 40*(1), 47–64.

Yuheng, S., & Hao, Y. (2017). Image Segmentation Algorithms Overview. arXiv preprint arXiv:1707.02051.

Zhang, Y. (2001). Texture-integrated classification of urban treed areas in high-resolution color-infrared imagery. *Photogrammetric Engineering and Remote Sensing, 67*(12), 1359–1366.

(Received  December 23, 2017, revised  March 21, 2018, accepted  April 19, 2018, Published online  April 23, 2018)

Pure Appl. Geophys. 175 (2018), 3247–3261
© 2018 Springer International Publishing AG, part of Springer Nature
https://doi.org/10.1007/s00024-018-1929-3

**❚ Pure and Applied Geophysics**

# Assessment of Riverine Morphology and Habitat Regime Using Unmanned Aerial Vehicles in a Mediterranean Environment

ELIAS DIMITRIOU[1] ⓘ and ELENI STAVROULAKI[2]

*Abstract*—The use of Unmanned Aerial Vehicles (UAVs) for monitoring environmental parameters presents an increasing trend in the last few years since it comprises a low cost, rapid and high spatial resolution alternative to classic remote sensing monitoring techniques. One of the most common applications of UAVs today is the development of Digital Terrain Models through photogrammetric processes which can then be used for geomorphological and habitat mapping. Monitoring the sedimentation regime in a riverine ecosystem, focusing on the fluvial habitat availability and changes is a necessary approach to assess water and soil management practices. The aim of the particular study is to assess the spatiotemporal erosion and deposition regime in a Mediterranean river using UAV photogrammetric riverbed measurements and apply habitat mapping, to quantify the suitability of the ecohydrologic conditions for a typical fish species. The output of this effort illustrated the highly dynamic sediment transfer conditions of the examined river which, however, maintained suitable fish habitats during the study period (06/2016 and 08/2017). Moreover, the results indicated that using UAVs to acquire spatial ecohydrologic and sediment relevant information in a very detailed resolution can be an efficient and accurate approach for managing vulnerable aquatic ecosystems.

**Key words:** Unmanned Aerial Vehicle, sediment, erosion, habitat, geomorphology, ecohydrology.

## 1. Introduction

Rivers are complex ecosystems where several ecological, hydrological and geomorphological parameters interact (Malanson 1995). Monitoring the spatial and temporal variations in river functions is important for understanding and improving habitat quality and distribution, (Kingsford 2011; Newson and Large 2006; Jimenez Cisneros et al. 2014). For this purpose, data on the physical conditions of rivers, including geomorphological and hydrological, are often collected (Maddock 1999; Orr et al. 2008; Vaughan et al. 2009). The associated river habitat parameters (e.g. depth and velocity) vary in space and time, depending on the underlying geological, climatic and anthropogenic conditions and modifications. These alterations affect the basic, physical hydromorphologic regime of rivers and, therefore, influence the quality and availability of fluvial habitats (Kingsford 2011; Vaughan et al. 2009; Nilsson and Berggren 2000).

Large-scale, coarse mapping and discrete point or transect sampling are not enough to represent the detailed spatial structure of fluvial habitats (Vaughan et al. 2009). Fausch et al. (2002) suggest that spatially continuous measurements (rather than sampling points or lines) should be applied in rivers, to allow precise quantification of habitat properties and improve our understanding regarding the size, distribution, and connectivity of different habitat types.

Remote sensing methods are extensively used for surveying and monitoring physical river habitats and hydromorphological parameters (Fausch et al. 2002; Carbonneau et al. 2012). A growing number of research efforts worldwide demonstrate the use of digital photogrammetry (Winterbottom and Gilvear 1997; Westaway et al. 2001) and spectral depth correlations (Legleiter et al. 2004, 2009; Legleiter 2012) for quantifying fluvial topography, substrate properties (Carbonneau et al. 2004; Heritage and Milan 2009; Hodge et al. 2009; Brasington et al. 2012; Rychov et al. 2012), and the mapping of hydrogeomorphic units (Wright et al. 2000).

¹ Hellenic Centre for Marine Research, Institute of Marine Biological Resources and Inland Waters, 46.7 km of Athens-Sounio Ave, 19013 Anavissos, Greece. E-mail: elias@hcmr.gr
² School of Environmental Engineering, Technical University of Crete, Chania-Crete, Greece. E-mail: estavroulaki@isc.tuc.gr

Using stereoscopic images of UAV to build 3D topographic models for monitoring displacement (Pierzchala et al. 2014; Clapuyt et al. 2016) and soil erosion processes (Neugirg et al. 2016) is a quite common practice today. Several studies support that this approach offers rapid and flexible data acquisition, which is relatively inexpensive and provides very high spatial resolutions (Cheek et al. 2015; Woodget 2015) that often reach centimeter levels (Anderson and Gaston 2013; Zarco-Tejada et al. 2014; Neugirg et al. 2016).

Application of conventional remote sensing platforms often has cost, resolution, and flexibility limitations (Whitehead et al. 2014). The emerging industry of the Unmanned Aerial Systems (UAS) offers the possibility to study environmental processes at spatiotemporal scales that would be extremely difficult using traditional remote sensing techniques (Whitehead et al. 2014).

The purpose of this study is to use UAV acquired photogrammetric data to investigate the riverbed morphology changes, within a year, in a Mediterranean river due to erosional and depositional processes and assess the fluvial habitat type availability and suitability for a typical fish species. This study is important for monitoring and managing ecological changes in riverine ecosystems since it provides valuable information for assessing and conserving aquatic habitats in an efficient manner at a high spatiotemporal scale.

## 2. Methodology

### 2.1. Case Study Area

Spercheios river basin is in the Region of Sterea Ellada, at Greece (Fig. 1) and covers an area of 1660 km$^2$ with an average and maximum altitudes of 640 and 2315 m, respectively. The main land cover types are forest and other natural vegetation land (65%), followed by agricultural land (32%) and urban areas/bare land (3%, Mentzafou et al. 2017). Spercheios river's riparian zone and delta is an EU protected area (Natura 2000 site), while there are no dams in its main course. The section of the river that was monitored for the particular research effort is located approximately in the middle of the basin, at the lowland part and was selected because it is an open area without dense vegetation and, therefore, appropriate for aerial monitoring. The length of the river section monitored (near Loutra Ypatis village) was approximately 500 m, while the active riverbed width was about 70 m. The surrounding land cover type is mainly agricultural land and limited anthropogenic construction, including roads (Fig. 1).

### 2.2. Topographic Survey with UAV

The field surveys in the study area (Sperchios river main branch) were conducted in June 2016 and August 2017 using a DJI Phantom 3 Professional UAV and a Spectra Precision SP60 RTK GPS. The software used for photo capturing on the field was 'Pix4D capture', while the photogrammetric process in the office was realized with the Pix4D mapper'.

The area covered in the field surveys was approximately 500 × 200 m along the river course, the flight altitude was 50 m and the percentage overlap of the photos was set to 80%. The resolution of each photo was 4000 × 3000 pixels, while Ground Control Points (GCPs) were captured in six different points widely spread in the study area and were imported to the Pix4D mapper software to improve the geospatial accuracy of the produced Digital Terrain Model.

The Pix4D photogrammetric process includes the extraction of homologous image points (tie points) between the overlapping images captured, the estimation of camera parameters for image calibration and bundle adjustment as well as the creation of a Digital Surface Model which is georeferenced by using Ground Control Points (GCPs, Stumpf et al. 2015). The photogrammetry algorithms are analytically described in Unger et al. (2009), while the mean error of a reconstructed surface is approximately 1–3 times the Ground Sampling Distance (Küng et al. 2011).

The cell size of the produced DTM was approximately 2.5 cm and it was resampled by selecting 20 cm cell size and the 'Nearest neighborhood' algorithm in ArcMap 10.4 GIS software. The specific resampling algorithm was used due to the low error margins (ESRI ArcMap manual, http://desktop.

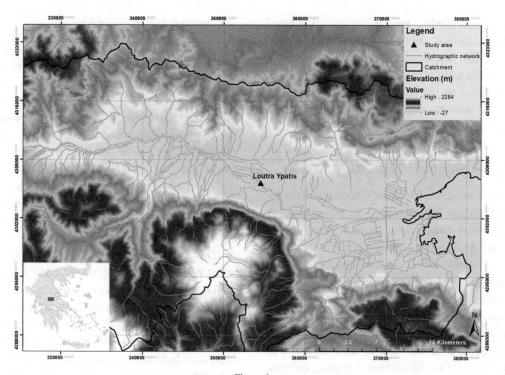

Figure 1
Digital elevation model of the Spercheios river catchment and the exact study area

arcgis.com/en/arcmap/) that are less than half of the cell size (1.2 cm). Such low error margins are necessary due to the high spatial resolution of the data required for this case study. Moreover, the selection of the 20 cm cell size was made by considering the computational demands of the ArcMap software to elaborate the respective raster files as well as the potential magnitude of photogrammetric error (cell size is approx. two times the maximum error). The official National coordinate system (Greek Grid, GGRS 87) was used in the entire elaboration process.

The data collection period was selected to be in the summer when the river flow is very low (few cm only in a limited part of the riverbed area) and, therefore, the photogrammetric process would be able to record most of the riverbed topographic features. The Pix4D software was also able to estimate the riverbed altitude at these shallow, clear waters since the riverbed features (cobbles, gravel and boulders) were visible in the images and, therefore, detectable by the keypoint matching, photogrammetric algorithm. Several other studies have also indicated that there is the potential to include the submerged portions of the study area in the DTM through optical bathymetric modeling, for which the UAV-based aerial photos are appropriate (e.g. Lejot et al. 2007; Williams et al. 2014; Javernick et al. 2014).

### 2.3. Mapping Erosion and Deposition Patterns

The DTMs of the years 2016 and 2017 were developed in the Pix4D software and imported to ArcMap for further elaboration. The DTMs potential errors were assessed by selecting 130 random points from a stable, common feature in both DTMs (a road), for which the altitude differences were analyzed. A mask layer delineating the active riverbed was created through manual digitization using the orthophoto maps produced in the Pix4D. The significant vegetation features observed in the riverbed were excluded from the mask which was used to extract the Region of Interest (RoI) from the DTMs. The Raster Calculator component of ArcMap was then used to subtract the 2016 DTM from the 2017 one and the resulting differences were classified into nine classes, following the 'Jenks natural breaks' classification method, depicting areas of high erosion to areas of high deposition.

The results were explained in physical terms and validated based on the morphology of the river and the recorded flow properties in the field surveys and in the respective orthophotos.

## 2.4. Hydromorphic Units (Mesohabitat) Mapping

For the Hydromorphic Units (Mesohabitat) mapping, the DTMs of 2016 and 2017 were used together with a reference surface raster for each year that was created in ArcMap by introducing the Absolute Elevation (meters above sea level, m.a.s.l.) of the water surface in the upstream and downstream ends of the examined river section. The referenced surface raster was subtracted from the respective DTM and the negative values were considered as the bathymetry of the river in the date of field survey. The Hydromorphic Units (Mesohabitat) types: pool-run-riffle-very shallow waters have been mapped by reclassifying the bathymetry maps according to the relevant classification types (Table 1) presented in the River Habitat Survey protocol (EA 2003). The respective habitat types depend on both water depth and velocity but for this study due to the lack of velocity measurements, a relevant approximation has been used (Hydromorphic Units Mesohabitat, Hauer et al. 2009).

A comparative assessment followed with the resulted Hydromorphic Units (Mesohabitat) maps in which compliance with the visual characteristics of the respective orthophotos was evaluated while the identified changes in the habitat types and extents over the study period were quantified.

## 2.5. Habitat Suitability Index

The Habitat Suitability Index (HSI) is extensively used to describe the appropriateness of a Mesohabitat type for the vital functions of a specific species (Papadaki et al. 2016). The HSI is expressed separately for each habitat parameter such as water depth, velocity and substrate and takes values from zero (unsuitable habitat) to one (very suitable, McMahon 1982). In this study, the HSI for water depth regarding Chub Fish species that can be found in the particular part of the river (Barbieri et al. 2015) has been estimated based on a relevant model developed by McMahon (1982, Fig. 2).

The bathymetric maps of the river for the two study periods have been transformed to suitability indices in ArcMap software using the raster calculator module and the aforementioned model (Fig. 2), while the results were classified into five categories (with equal intervals). The area of the different HSI categories has been quantified for the 2016 and 2017 maps and a comparative assessment followed.

## 3. Results and Discussion

### 3.1. Photogrammetric Process and Uncertainties

The Quality Report of the Pix4D software indicated that the estimated error of the computed GCPs/3D points by the reconstructed model in relation to their original location as provided by the RTK GPS was very low in all directions (Table 2). The mean error was always less than 3 mm, while the Root-Mean-Square (RMS) error was less than 10 cm. The RTK GPS used in the study provided in situ, online

Table 1

*Hydromorphic Units (Mesohabitat) types and depths*

| Habitat type | Depth (m) |
| --- | --- |
| Pool | > 0.6 |
| Run | 0.4–0.6 |
| Riffle | 0.2–0.4 |
| Very shallow waters | < 0.2 |

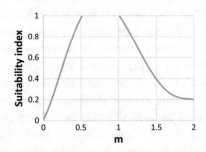

Figure 2
Habitat Suitability Index vs water depth (summer months) for Chub fish species, developed by McMahon (1982)

Table 2

*Localization accuracy and GCPs mean errors in the three coordinate directions*

| Type of error | Error X (m) | | Error Y (m) | | Error Z (m) | |
|---|---|---|---|---|---|---|
| | 2016 | 2017 | 2016 | 2017 | 2016 | 2017 |
| Mean (m) | − 0.001 | − 0.003 | − 0.002 | − 0.002 | − 0.013 | 0.014 |
| Sigma (m) | 0.104 | 0.028 | 0.081 | 0.027 | 0.066 | 0.019 |
| RMS error (m) | 0.104 | 0.028 | 0.081 | 0.027 | 0.067 | 0.024 |

corrections through a 3G connection with a nearby base station, offering fixed positions with less than 1 cm accuracy.

The average Ground Sampling Distance was 2.44 cm in 2016 and 2.35 cm in 2017, while the mean reprojection error was 0.24 and 0.21 pixels, respectively (Table 3). The number of calibrated images reached 259 in 2016 and 214 in 2017 while the number of 3D points that used for Bundle adjustment overcame 1.78 million points in 2016 and 967,000 points in 2017. Therefore, the accuracy of the reconstructed surfaces can be considered satisfactory for this study since it is estimated as lower than three times the GSD (less than 8 cm, Küng et al. 2011).

The DTMs inherent potential errors were assessed by comparing the altitude differences in the two study years at four stable control points (building roofs and cement constructions), which were available in the area. Moreover, to assess the potential maximum errors due to the lack of adequate number of stable control points, 380 randomly distributed points were selected along two dirt roads located on the river banks. These roads are prone to erosion and deposition processes and, therefore, cannot be used as reference points for the accurate estimation of the

potential DTMs errors but since these roads are much less dynamic environments in relation to the riverbed, they were used as an indication of the results reliability.

This approach indicated that in the four stable control points the differences in the DTMs' altitudes fluctuate between 0.08 and 0.18 m (Fig. 3). Moreover, the analysis of the elevation difference in the two dirt roads indicates an average difference between the two DTMs of 0.12 m, a median of 0.16 and a standard deviation of 0.24 m. Moreover, 25% of the road points have differences of less than 0.04 m, while 75% have differences less than 0.28 m.

Regarding potential horizontal offsets between the 2 DTMs, an alignment procedure in ArcGIS has been followed to co-register both DTMs based on the aforementioned four stable control points. This process ended up in horizontal differences at the control points of less than 5 cm. Based on the resampling process that followed (the pixel size converted from 2.4 to 20 cm) and the above described error probability analysis, the DTM differences of less than 0.2 m are considered within the potential error limits of the process.

Table 3

*Quality and technical details of the photogrammetric process*

| | Acquisition year | |
|---|---|---|
| | 2016 | 2017 |
| Average ground sampling distance (GSD—cm) | 2.44 | 2.35 |
| Mean reprojection error (pixels) | 0.242 | 0.214 |
| Median matches per calibrated image | 17,700 | 11,094 |
| Number of 3D points for bundle block adjustment | 178,0231 | 96,7443 |
| Number of calibrated images | 259 | 214 |

Figure 3
Difference in altitude at four control points between the 2016 and 2017 DTMs (the blue points are randomly selected points along two dirt roads on the river banks)

## 3.2. Erosion and Deposition Patterns

The Digital Terrain Model (DTM) of the year 2016 illustrates slightly higher fluctuations on the elevation values in relation to the year 2017 (Fig. 4). This pattern has led to the maintenance of the riverbed at a significantly lower altitude in 2016 and the elevation values' fluctuations at a higher level. This is because the topographically higher parts of the study area coincide with agricultural areas that have stable elevation values, at an interannual basis, while the riverbed is a highly dynamic environment that changes every year depending on the hydrodynamic conditions that are affected by the annual precipitation regime. Moreover, the topographic anomaly recorded in the eastern part of the study area with the significant deepening in the year 2017 represents a soil extraction site for anthropogenic purposes that operated for the first time in 2017.

The differences in the elevation values between the 2017 and 2016 DTMs fluctuate from − 2 to 2 m (Fig. 5). The largest positive differences are illustrated in the upstream part of the river section and some of the highest elevation changes can be assigned to submerged vegetation riverbed changes over the study period (2016–2017). Moreover, the lowest negative values, observed mainly, in the downstream part of the river can be attributed to anthropogenic sediment extraction which occurred in the 2017 summer.

Regarding the volumetric aspect of the changes in the riverbed sediment between 2016 and 2017, a net sediment deficit has been estimated if both natural and anthropogenic processes are considered (507 m$^3$, Table 4), while a significant sediment surplus is estimated if only the natural processes are considered (2874 m$^3$). The net volume of sediment that corresponds to the elevation differences from − 0.2 to 0.2 m which are within the potential error levels is 200 m$^3$ only (392 m$^3$ eroded and 592 m$^3$ deposited). The deposition rate in the study area reaches 0.27 m$^3$/m$^2$ while the erosion rate is approximately 0.19 m$^3$/m$^2$ (without the human abstractions). Therefore, the behavior of the river system in the particular section

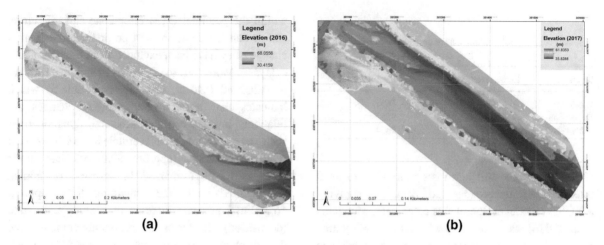

Figure 4
Digital Terrain Model (elevation, m.a.s.l): **a** 06/2016 and **b** 08/2017

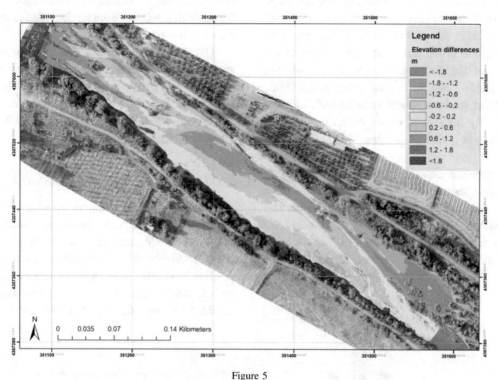

Figure 5
Elevation difference (m a.s.l.) between the 2016 and 2017 DTMs

of the hydrographic network is dominated by the deposition of sediment (Table 4). This is in accordance with the physiographic characteristics of this part of the river, which is in the lowland part of the catchment, approximately at the middle of it and with relatively low hydraulic slopes (approx. 0.2%). Thus, the erosion rate in this segment of the river ranges between 1855 and 3150 tons/year/ha, depending on the type of soil materials (1 m$^3$ of soil is approx. 1–1.7 tons), which is in agreement with Sigalos et al.

113

Table 4

*Sediment volumes (m³) that have been eroded and deposited in the study area*

| Sediment | Entire riverbed (m³) | Natural riverbed (m³) |
|---|---|---|
| Erosion | − 10,131 | − 6677 |
| Deposition | 9624 | 9551 |
| Total | − 507 | 2874 |

(2010) that found soil loss rates from 1300 to 4750 tons/year/ha in the broader study area, after applying a RUSLE model at Spercheios catchment. Mertzanis et al. (2016) state that the lowland sedimentary area of river Spercheios compose a gentle to flat landscape with significant volumes of eroded materials in the area of low discharge, due to the presence of erosion prone flysch in the specific catchment. The discharge levels during the summer period are frequently well below 1 m³/s, in this part of the river (Kormas et al. 2003), while recent studies (Markogianni et al. 2017) indicate that the discharge during dry periods are often negligible due to human water abstractions. Nevertheless, the intra-annual variations of discharge are relatively high with values exceeding 10 m³/s during the wet period of the year (Mentzafou et al. 2017).

The elevation differences frequency distribution diagram (Fig. 6) accredit also that deposition is more intense and important in the study area in comparison to erosion, while approximately 70% of the area have elevation differences from − 0.2 to 0.6 m. A large part of the studied area (approx. 25%) indicates elevation differences from − 0.2 to 0.2 m which are

within the potential DTM error zone and, therefore, can be considered as areas of no or limited elevation change (they have been excluded from the volumetric analysis). Moreover, the 25th and 75th percentiles are − 0.26 and 0.46 m, respectively (Table 5) which indicates that most of the riverbed changes are relatively low for a highly seasonal, dynamic river such as Spercheios. Nevertheless, 38% of the riverbed coverage (approx. 9980 m²) has lower altitudes in 2017 than 2016 (erosion or human abstractions) while the remaining 62% (approx. 16,390 m²) presents increase of altitude with time (deposition). In terms of sediment volumes, the deposition processes dominate, as already mentioned, if only natural mechanisms are considered. Approximately, 9630 m² of the area (27% of the total) have elevation changes within the error limits (− 0.2 to 0.2 m) and, therefore, were not included in the above-mentioned estimations.

The above stated information is essential for the sustainable management of the river and sediment resources to achieve a balance between the anthropogenic needs and the hydroecological processes that are crucial for maintaining the riverine ecosystem services (Gurnell et al. 2016).

### 3.3. Hydromorphic Units (Mesohabitat) Mapping

The Mesohabitat mapping considered only the riverbed areas that were below the water surface during the field surveys and the resulting maps (Fig. 7) indicate significant differences in the composition and extent of habitats during the study period. August is a dry month in this area and in combination with the significant water abstractions

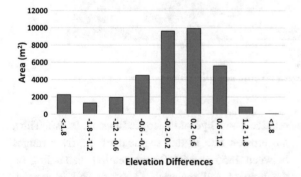

Figure 6
Frequency distribution of the estimated elevation differences in the riverbed during the study period

Table 5

*Statistical indices for the elevation differences in the riverbed during the study period*

| Statistical indices | Elevation differences (m) |
|---|---|
| Average | − 0.009 |
| Median | 0.148 |
| Standard deviation | 0.799 |
| Skewness | − 1.152 |
| Kurtosis | 1.350 |
| 25th percentile | − 0.263 |
| 75th percentile | 0.455 |

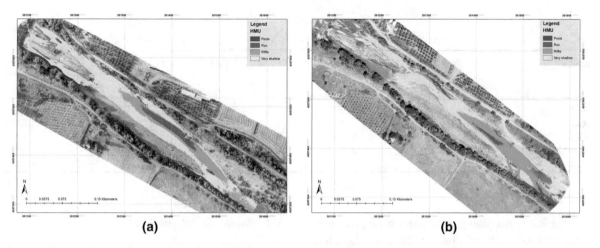

**(a)**     **(b)**

Figure 7
Habitat maps in the study area: **a** 06/2016 and **b** 08/2017

for irrigation, the water volumes remaining in the river are very low (Mentzafou et al. 2017). On the contrary, the water discharge during June is still relatively high in this lowland part of the catchment due to the high travel time of the snowmelt (particularly through groundwater) and the lower irrigation demands (Mentzafou et al. 2017). This is accredited by the results of this effort which indicate a higher wetted area in June 2016 (approx. 20,600 m$^2$) than in August 2017 (approx. 15,500 m$^2$, Table 6). According to an operational hydrologic modeling platform (http://hydro-data.hcmr.gr/) developed by the Hellenic Centre for Marine Research after a recent research project in the area (Mentzafou et al. 2017), the average discharge in June 2016 from Spercheios river to Maliakos Gulf is approximately 1.8 m$^3$/s while in August 2017 the respective value is 1.3 m$^3$/s. This difference in the discharge together with the changes in the geomorphology of the riverbed causes significant alterations in the composition of the habitats in the examined river reach.

The percentage of pools in relation to the total extent of the wetted area is lower in 06/2016 than in 08/2017 which is probably due to the relatively high erosion conditions that dominated during the winter of 2017 (12% more rainfall than in the previous year). The extent of the riffle habitats was slightly lower in 06/2016 than in 08/2017, while the very shallow areas followed the opposite pattern (Table 6). This is

again due to the aforementioned geomorphological changes during the study area which transformed very shallow to riffle habitats as a result of erosional processes. Comparing the habitat changes from June 2016 to August 2017, a reduction of the Run and very shallow habitats has been observed, mainly as a result of the lower discharge in August in relation to June. The extent of pools has been increased during August 2017 which is against the hydrologic conditions, but this change is mainly assigned to the significant erosional processes between the two field surveys.

The accuracy of the habitat mapping was assessed visually by comparing the resulted habitat maps with the respective orthophoto maps. The results from this assessment indicated that the boundaries of the habitat maps followed the water surface limits in the orthophoto maps except for some very shallow parts in the habitat maps in 2016 that appear dry in the orthophoto maps. These few inconsistencies exist due to the inherent error range of the DTMs that can cause misclassification close to the wetted area boundaries. Thus, these zones were not considered as suitable habitats, since depths below 20 cm in rivers with fish fauna are forbidden by the Greek Ecological Flow legislation and illustrate suitability index below 0.5 for Chub species according to McMahon (1982). The remaining available area for fish fauna in this part of the river (500 m length) was approximately 6500 m$^3$ in June 2016 and 7060 m$^3$ in

Table 6

*Coverage of different habitat types in the study area during 06/2016 and 08/2017*

| HMUs | June 2016 | | August 2018 | | Difference | |
|---|---|---|---|---|---|---|
| | m$^2$ | % | m$^2$ | % | % | % |
| Pool | 53 | 0.26 | 178 | 1.15 | | 235.85 |
| Run | 1563 | 7.58 | 1183 | 7.63 | | − 24.31 |
| Riffle | 4831 | 23.44 | 5699 | 36.75 | | 17.97 |
| Very shallow | 14167 | 68.73 | 8448 | 54.47 | | − 40.37 |
| Total | 20614 | | 15,508 | | | − 24.77 |

August 2017. According to a recent ecological flow estimation study, in this part of the river, for Chub species (Stamou et al. 2018), the Weighted Useable Area (WUA) when the river discharge varies from 1 to 2 m$^3$/s ranges between 10,000 m$^2$/1000 m (length of river) to 13,000 m$^2$/1000 m. These values are very similar to the above-mentioned relevant estimations of the present study (13,000–14,120 m$^2$/1000 m).

The depths of the habitats mapped in this effort range from below 0.2 m to slightly above 0.6 m with a large percentage of the area having depths from 0.2 to 0.6 m. Excluding the very shallow HMUs which are within the potential DTMs error magnitude, the Riffle zones (0.2–0.4 m depth) dominate in both study periods followed by the Run zones (0.4–0.6 m depth). This result is in accordance with the Stamou et al. (2018) study in which a hydraulic model, for this part of the river, indicated dominant depth ranges from 0.1 to 0.6 m for a discharge of 1.8 m$^3$/s (similar to June 2016 discharge).

The above-mentioned information is of high importance since it facilitates the physical habitat monitoring in riverine systems which is a key aspect of the European Union's Water Framework Directive (EC/2000/60) and a necessary basis for the development and implementation of water resources management and restoration plans. The development of new methods to acquire topographic data (through UAVs) provides the opportunity to operationally investigate the spatiotemporal changes of the geomorphological features and processes (Nield et al. 2011) which allow mitigation measures to be undertaken at a timely basis.

### 3.4. Habitat Suitability Index

The HSI maps (Fig. 8) illustrate some similarities to the habitat maps (Fig. 7) as expected since both follow a depth categorization approach but the HSI maps present significantly higher spatial variability than the habitat maps. Most of the very shallow waters are classified as of unsuitable for Chub species while the optimum depth fluctuates from 0.6 to 1.3 m. The Pools illustrate a moderate habitat suitability, while the Riffles and mainly the Runs present the highest suitability in the study area (Fig. 8).

The categories of low and very low habitat suitability covered significantly smaller area in June 2016 (7% of the study area) than in August 2017 (15% of the study area, Table 7).

In June 2016, the high HSI areas (0.6–0.8) had a larger extent (36% of the total) than in August 2017 (21%), while the very high HSI levels (0.8–1) illustrated only minor fluctuations between the two study periods (Table 7). The moderate suitability area (0.4–0.6) is higher in August 2017 than in June 2016 while more than 50% of the study area is characterized as of high and very high habitat suitability for Chub species in both study periods. The differences between June 2016 and August 2017 can be partially explained by the higher discharge values and water depth observed in June, while the morphological changes through erosion/deposition processes play an important role mostly for the very low HSI values.

As a general conclusion of the HSI maps, it can be stated that during both study periods, the particular part of the river incorporates adequate areas with suitable habitats for Chub species, while in August 2017 the most important problem seems to be the

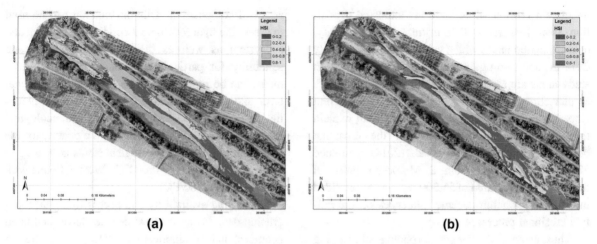

(a)                                                            (b)

Figure 8
HSI maps in the study area: **a** 06/2016 and **b** 08/2017

Table 7

*Coverage of different HSI classes in the study area during 06/2016 and 08/2017*

| HSI | June 2016 | | August 2017 | |
|---|---|---|---|---|
| | m³ | % | m³ | % |
| 0–0.2 | 1433 | 6.97 | 2315 | 14.94 |
| 0.2–0.4 | 16 | 0.08 | 7 | 0.04 |
| 0.4–0.6 | 4791 | 23.30 | 4690 | 30.27 |
| 0.6–0.8 | 7463 | 36.29 | 3239 | 20.91 |
| 0.8–1 | 6862 | 33.37 | 5240 | 33.83 |

connectivity of the habitats due to the relatively extensive dry zones of the riverbed. Based on these observations, important conclusions can be drawn for the implementation of Ecological flow principles that target on maintaining adequate areas of suitable habitat types in the river. The seasonal quantification of the available habitats through UAVs and photogrammetry efforts can enhance the assessment of the applied water management scheme in the area and offer improvements for the preservation of the economically and ecologically important fish stocks of the area (Papadaki et al. 2016).

## 4. Discussion and Conclusions

During the last few decades, satellite images are commonly used for mapping and monitoring environmental dynamics and processes, in riparian, forest and coastal ecosystems. Nevertheless, high spatial resolution satellite images are very expensive and not easily accessible for the scientific community (Ventura et al. 2016). For this purpose, the use of UAVs is increasingly expanding since they offer a low cost and high spatial resolution alternative to the satellite data.

A series of recent environmental studies that utilize UAVs focus on water quality monitoring (Su 2017; Zeng et al. 2017), underwater habitat mapping (Ventura et al. 2016), estimation of water levels and ecological flows in rivers (Bandini et al. 2017; Zhao et al. 2017) as well as erosion sediments quantification (Pineux et al. 2017; Neugirg et al. 2016; Peter et al. 2014).

Zhao et al. (2017) attempted to measure river cross-section areas using UAV-based photogrammetric approaches in a similar manner to the one used in the present research effort and commented on the satisfactory accuracy of the results (RMSE of cross section less than 0.25 m). This study indicated that the combination of UAV acquired data (topographic) and in situ (bathymetric) measurements can offer an optimal and accurate approach in river environments.

Neugirg et al. (2016) estimated a 5.3 cm average lowering of the study catchment due to sediment loss (6300 m³) estimated based on UAV multitemporal, high-resolution surveys. A total area of 17,900 m² underwent erosion while 4400 m$^2$ indicated deposition (soil loss ratio: 0.28 m$^3$/m$^2$). In this study, deposition dominates with a volume of approximately 2874 m$^3$ and a soil gain ratio of a 0.11 m$^3$/m$^2$. Both study areas are Mediterranean river catchments with similar climatic conditions but significantly different slope and altitude characteristics which can explain the aforementioned differences on the erosional behavior. Moreover, Peter et al. (2014) estimated 720 m$^3$ soil erosion from a Moroccan gully of 869 m$^2$, in a study area of 35,300 m$^2$, which provided indications for anthropogenic influences on the gully's erosional processes.

Thus, using UAVs for geomorphological mapping incorporates significant advantages with respect to cost, effort applied, output accuracy and availability of data at the required spatial and temporal scale. The development of small lidar sensors that can be fitted to UAVs overcomes the limitation of data capturing in densely vegetated areas (Rusnák et al. 2018).

A potential obstacle for the expanded use of UAVs is the legislative framework that illustrates significant differences amongst the various countries and progressively imposes restrictions and permit granting processes that can affect, especially the use of larger, professional UAVs (Rusnák et al. 2018). For this reason, a homogenization of the relevant laws at the European level could be an optimal solution for achieving the necessary safety standards and allow the wide use of the UAVs from all relevant stakeholders (public authorities, researchers, NGOs, rescue teams, mail companies, etc.) and individuals.

In this research effort, photogrammetric data acquired by a UAV were used to estimate the DTM of a riverbed during different time periods, to assess the spatiotemporal changes in its geomorphology and fish habitat availability. Pix4D software was used for this purpose and this process created elevation rasters for an area of approximately 10 ha with a 2.4 cm resolution and up to 8 cm spatial accuracy. This amount of topographic data in such resolution and accuracy would be impossible to collect within a reasonable time span and cost for a typical, field survey.

The aforementioned processes allowed the estimation of sediment deposits changes in the riverbed and the spatiotemporal alterations of Mesohabitat area and composition. This information can be used to assess the hydroecological conditions in an aquatic ecosystem as well as the habitat availability and suitability for particular organisms. Such measurements can be repeated on a monthly and/or seasonal basis and provide valuable information about potential hydromorphological alterations and ecological flow disturbances that are important elements for the assessment of the river ecological status according to the requirements of the EU Water Framework Directive (2000/60/EC).

Increasing availability and access to digital topographic data during the past decades have facilitated geomorphometry applications (Hengl and Reuter 2009), including erosion studies (Stumpf et al. 2015). However, the use of UAV capabilities for habitat suitability mapping is not yet commonly applied.

Relatively common environmental UAV applications include vegetation mapping (Hugenholtz 2010) as well as creation of flow velocity maps in riverine environments (Hauet et al. 2008). Small UAVs have also been used to monitor the woodlands of Poland (Zmarz 2014) and the Himalayan glacier dynamics (Immerzeel et al. 2014), and are considered a revolutionary tool for studying spatial ecology (Anderson and Gaston 2013).

The output of this study offered satisfactory results that can be used for various purposes including Environmental Impact Assessment studies downstream of hydraulic works as well as ecological flow requirement estimations. Further improvements can be achieved with incorporating RTK GPS capabilities on the UAV itself as well as using LiDAR equipment on board which is steadily decreasing in cost. Thus, hydroecological studies including Mesohabitat mapping will be significantly facilitated by these technological advances which will improve the amount and accuracy of information that is necessary for optimal management and preservation of the water resources and ecosystem services.

## REFERENCES

Anderson, K., & Gaston, K. J. (2013). Lightweight unmanned aerial vehicles will revolutionize spatial ecology. *Frontiers in Ecology and the Environment, 11*(3), 138–146. https://doi.org/10.1890/120150.

Bandini, F., Jakobsen, J., Olesen, D., Reyna-gutierrez, J. A., & Bauer-gottwein, P. (2017). Measuring water level in rivers and lakes from lightweight Unmanned Aerial Vehicles. *Journal of Hydrology, 548,* 237–250. https://doi.org/10.1016/j.jhydrol.2017.02.038.

Barbieri, R., Zogaris, S., Kalogianni, E., Stoumboudi, M. Th., Chatzinikolaou, Y., Giakoumi, S., Kapakos, Y., Kommatas, D., Koutsikos, N., Tachos, V., Vardakas, L., & Economou, A. N. 2015. *Freshwater fishes and lampreys of Greece: An annotated checklist. Monographs on Marine Sciences No. 8.* Hellenic Centre for Marine Research: Athens, Greece, p. 130. http://epublishing.ekt.gr/el/11080. Accessed 1 Feb 2018.

Brasington, J., Vericat, D., & Rychov, I. (2012). Modeling river bed morphology, roughness, and surface sedimentology using high resolution terrestrial laser scanning. *Water Resources Research, 48,* W11519. https://doi.org/10.1029/2012WR012223.

Carbonneau, P. E., Fonstad, M. A., Marcus, W. A., & Dugdale, S. J. (2012). Making riverscapes real. *Geomorphology, 123,* 74–86. https://doi.org/10.1016/j.geomorph.2010.09.030.

Carbonneau, P. E., Lane, S. N., & Bergeron, N. (2004). Catchmentscale mapping of surface grain size in gravel bed rivers using airborne digital imagery. *Water Resources Research, 40,* W07202. https://doi.org/10.1029/2003WR002759.

Cheek, B. D., Grabowski, T. B., Bean, P. T., Groeschel, J. R., & Magnelia, S. (2015). Evaluating habitat associations of a fish assemblage at multiple spatial scales in a minimally disturbed stream using low-cost remote sensing. *Aquatic Conservation: Marine and Freshwater Ecosystems, 26,* 20–34. https://doi.org/10.1002/aqc.2569.

Clapuyt, F., Vanacker, V., & Van, O. K. (2016). Reproducibility of UAV-based earth topography reconstructions based on Structure-from-Motion algorithms. *Geomorphology, 260,* 4–15.

Environment Agency. (2003). *River habitat survey in Britain and Ireland.* Field Survey Guidance Manual: 2003. Bristol.

Fausch, K. D., Torgersen, C. E., Baxter, C. V., & Hiram, L. W. (2002). Landscapes to riverscapes: Bridging the gap between research and conservation of stream fishes. *BioScience, 52,* 483–498. https://doi.org/10.1641/0006-3568(2002)052[0483:LTRBTG]2.0.CC.

Gurnell, A. M., Rinaldi, M., Belletti, B., Bizzi, S., Blamauer, B., Braca, G., et al. (2016). A multi-scale hierarchical framework for developing understanding of river behaviour to support river management. *Aquatic Sciences, 78*(1), 1–16. https://doi.org/10.1007/s00027-015-0424-5.

Hauer, C., Mandlburger, G., & Habersack, H. (2009). Hydraulically related hydro-morphological units: Description based on a new conceptual mesohabitat evaluation model (MEM) using LiDAR data as geometric input. *River Research and Applications, 25,* 29–47. https://doi.org/10.1002/rra.1083.

Hauet, A., Kruger, A., Krajewski, W., Bradley, A., Muste, M., Creutin, J., et al. (2008). Experimental system for real-time discharge estimation using an image based method. *Journal of Hydrologic Engineering, 13*(2), 105–110.

Hengl, T., & Reuter, H. I. (2009). *Geomorphometry: Concepts, software, applications.* Amsterdam: Elsevier.

Heritage, G. L., & Milan, D. J. (2009). Terrestrial laser scanning of grain roughness in a gravel-bed river. *Geomorphology, 113,* 4–11. https://doi.org/10.1016/j.geomorph.2009.03.021.

Hodge, R., Brasington, J., & Richards, K. (2009). In situ characterization of grain-scale fluvial morphology using terrestrial laser scanning. *Earth Surface Processes and Landforms, 34,* 954–968. https://doi.org/10.1002/esp.1780.

Hugenholtz, C. H. (2010). Topographic changes of a supply-limited inland parabolic sand dune during the incipient phase of stabilization. *Earth Surface Processes and Landforms, 35,* 1674–1681.

Immerzeel, W. W., Kraaijenbrink, P. D. A., Shea, J. M., Shrestha, A. B., Pellicciotti, F., Bierkens, M. F. P., et al. (2014). High resolution monitoring of Himalayan glacier dynamics using unmanned aerial vehicles. *Remote Sensing of Environment, 150,* 93–103. https://doi.org/10.1016/j.rse.2014.04.025.

Javernick, L., Brasington, J., & Caruso, B. (2014). Modeling the topography of shallow braided rivers using structure-from-Motion photogrammetry. *Geomorphology, 213,* 166–182.

Jimenez Cisneros, B. E., Oki, T., Arnell, N. W., Benito, G., Cogley, J. G., Doll, P., et al. (2014). Freshwater resources. In C. B. Field, V. R. Barros, D. J. Dokken, K. J. Mach, M. D. Mastrandrea, T. E. Bilir, M. Chatterjee, K. L. Ebi, Y. O. Estrada, R. C. Genova, B. Girma, E. S. Kissel, A. N. Levy, S. MacCracken, P. R. Mastrandrea, & L. L. White (Eds.), *Climate change 2014: Impacts, adaptations and vulnerability. Part A: Global and sectoral aspects. Contribution of Working Group II to the Fifth Assessment Report of the Intergovernmental Panel on Climate Change.* Cambridge: Cambridge University Press.

Kingsford, R. T. (2011). Conservation management of rivers and wetlands under climate change—a synthesis. *Marine and Freshwater Research, 62,* 217–222. https://doi.org/10.1071/MF11029.

Kormas, K., Nicolaidou, A., & Thessalou-legaki, M. (2003). Variability of environmental factors of an Eastern Mediterranean river influenced coastal system. *Mediterranean Marine Science, 4/1,* 67–77. https://doi.org/10.12681/mms.242.

Küng, O., Strecha, C., Beyeler, A., Zufferey, J.-C., Floreano, D., Fua, P., & Gervaix, F., 2011. The accuracy of automatic photogrammetric techniques on ultra-light UAV Imagery. UAV-G 2011—Unmanned Aerial Vehicle in Geomatics. Zürich, CH. https://infoscience.epfl.ch/record/168806/files/uavg.pdf. Accessed 1 Feb 2018.

Legleiter, C. J. (2012). Remote measurement of river morphology via fusion of LiDAR topography and spectrally based bathymetry. *Earth Surface Processes and Landforms, 37,* 499–518. https://doi.org/10.1002/esp.2262.

Legleiter, C. J., Roberts, D. A., & Lawrence, R. L. (2009). Spectrally based remote sensing of river bathymetry. *Earth Surface Processes and Landforms, 34,* 1039–1059. https://doi.org/10.1002/esp.1787.

Legleiter, C. J., Roberts, D. A., Marcus, W. A., & Fonstad, M. A. (2004). Passive optical remote sensing of river channel morphology and instream habitat: Physical basis and feasibility. *Remote Sensing of Environment, 93,* 493–510. https://doi.org/10.1016/j.rse.2004.07.019.

Lejot, J., Delacourt, C., Piégay, H., Fournier, T., Trémélo, M. L., & Allemand, P. (2007). Very high spatial resolution imagery for channel bathymetry and topography from an unmanned mapping controlled platform. *Earth Surface Processes and Landforms, 32,* 1705–1725.

Maddock, I. (1999). The importance of physical habitat assessment for evaluating river health. *Freshwater Biology, 41,* 373–391. https://doi.org/10.1046/j.1365-2427.1999.00437.x.

Malanson, G. P. (1995). *Riparian landscapes.* Cambridge: Cambridge University Press.

Markogianni, V., Varkitzi, I., Pagou, K., Pavlidou, A., & Dimitriou, E. (2017). Nutrient flows and related impacts between a

Mediterranean river and the associated coastal area. *Continental Shelf Research, 134,* 1–14. https://doi.org/10.1016/j.csr.2016.12.014.

McMahon, T. E. (1982). Habitat suitability index models: Creek chub. USDL Fish and Wildlife Service. FWS/OBS-82/10.4. p. 23. https://www.nwrc.usgs.gov/wdb/pub/hsi/hsi-004.pdf. Accessed 1 Feb 2018.

Mentzafou, A., Vamvakaki, Ch., Zacharias, I., Gianni, A., & Dimitriou, E. (2017). Climate change impacts on a Mediterranean river and the associated interactions with the adjacent coastal area. *Environmental Earth Sciences, 76,* 259. https://doi.org/10.1007/s12665-017-6572-2.

Mertzanis A., Marabini F., Mertzanis K., Angeli M. G., Pontoni F., & Gasparetto P. 2016. Environmental management of aquatic resources and manmade ecoenvironmental impacts: Lakes, ponds and wetlands management in central Greece. *Ecology and Safety, 10* (**ISSN 1314-7234**).

Neugirg, F., Stark, M., Kaiser, A., et al. (2016). Erosion processes in calanchi in the Upper Orcia Valley, Southern Tuscany, Italy based on multitemporal high-resolution terrestrial LiDAR and UAV surveys. *Geomorphology, 269,* 8–22.

Newson, M. D., & Large, A. R. G. (2006). 'Natural' rivers, 'hydromorphological quality' and river restoration: A challenging new agenda for applied fluvial geomorphology. *Earth Surface Processes and Landforms, 31,* 1606–1624. https://doi.org/10.1002/esp.1430.

Nield, J. M., Wiggs, G. F. S., & Squirrell, R. S. (2011). Aeolian sand strip mobility and protodune development on a drying beach: Examining surface moisture and surface roughness patterns measured by terrestrial laser scanning. *Earth Surface Processes and Landforms, 36,* 513–522.

Nilsson, C., & Berggren, K. (2000). Alterations of riparian ecosystems caused by river regulation. *BioScience, 50,* 783–792. https://doi.org/10.1641/0006-3568(2000)050[0783:AORECB]2.0.CO;2.

Orr, H. G., Large, A. R. G., Newson, M. D., & Walsh, C. L. (2008). A predictive typology for characterising hydromorphology. *Geomorphology, 100,* 32–40. https://doi.org/10.1016/j.geomorph.2007.10.022.

Papadaki, C., Soulis, K., Munoz-Mas, R., Martinez-Capel, F., Zogaris, S., Ntoanidis, L., et al. (2016). Potential impacts of climate change on flow regime and fish habitat in mountain rivers of the south-western Balkans. *Science of the Total Environment, 540,* 418–428.

Peter, K. D., d'Oleire-Oltmanns, S., Ries, J. B., Marzolff, I., & Hssaine, A. A. (2014). Soil erosion in gully catchments affected by land-levelling measures in the Souss Basin, Morocco, analysed by rainfall simulation and UAV remote sensing data. *CATENA, 113,* 24–40.

Pierzchala, M., Talbot, B., & Astrup, R. (2014). Estimating soil displacement from timber extraction trails in steep terrain: Application of an unmanned aircraft for 3D modelling. *Forests, 5*(6), 1212–1223.

Pineux, N., Lisein, J., Swerts, G., Bielders, C. L., Lejeune, P., Colinet, G., et al. (2017). Geomorphology Can DEM time series produced by UAV be used to quantify diffuse erosion in an agricultural watershed? *Geomorphology, 280,* 122–136.

Rusnák, M., Sládek, J., Kidová, A., & Lehotský, M. (2018). Template for high-resolution river landscape mapping using UAV technology. *Measurement, 115*(October 2017), 139–151. https://doi.org/10.1016/j.measurement.2017.10.023.

Rychov, I., Brasington, J., & Vericat, D. (2012). Computational and methodological aspects of terrestrial surface analysis based on point clouds. *Computers and Geosciences, 42,* 64–70. https://doi.org/10.1016/j.cageo.2012.02.011.

Sigalos, G., Loukaidi, V., Dasaklis, S., & Alexouli-Livaditi, A. 2010. Assessment of the quantity of the material transported downstream of Sperchios River, Central Greece, Bulletin of the Geological Society of Greece. *Proceedings of the 12th International Congress, Patras, May, 2010.* https://ejournals.epublishing.ekt.gr/index.php/geosociety/article/view/11239/11285. Accessed 1 Feb 2018.

Stamou, A., Polydera, A., Papadonikolaki, G., Martínez-Capel, F., Muñoz-Mas, R., Papadaki, Ch., et al. (2018). Determination of environmental flows in rivers using an integrated hydrological-hydrodynamic-habitat modelling approach. *Journal of Environmental Management.* https://doi.org/10.1016/j.jenvman.2017.12.038. (**ISSN 0301-4797**).

Stumpf, A., Malet, J.-P., Allemand, P., Pierrot-Deseilligny, M., & Skupinski, G. (2015). Ground-based multi-view photogrammetry for the monitoring of landslide deformation and erosion. *Geomorphology, 231,* 130–145. https://doi.org/10.1016/J.GEOMORPH.2014.10.039.

Su, T. (2017). A study of a matching pixel by pixel (MPP) algorithm to establish an empirical model of water quality mapping, as based on unmanned aerial vehicle (UAV) images. *International Journal of Applied Earth Observations and Geoinformation, 58,* 213–224. https://doi.org/10.1016/j.jag.2017.02.011.

Unger, M., Pock, T., Grabner, M., Klaus, A., & Bischof, H. (2009). A variational approach to semiautomatic generation of digital terrain models. In G. Bebis, et al. (Eds.), *Advances in visual computing. ISVC 2009. Lecture notes in computer science* (Vol. 5876). Berlin: Springer.

Vaughan, I. P., Diamond, M., Gurnell, A. M., Hall, K. A., Jenkins, A., Milner, N. J., et al. (2009). Integrating ecology with hydromorphology: A priority for river science and management. *Aquatic Conservation: Marine and Freshwater Ecosystems, 19,* 113–125. https://doi.org/10.1002/aqc.895.

Ventura, D., Bruno, M., Jona, G., & Belluscio, A. (2016). Estuarine, coastal and shelf science a low-cost drone based application for identifying and mapping of coastal fish nursery grounds. *Estuarine, Coastal and Shelf Science, 171,* 85–98. https://doi.org/10.1016/j.ecss.2016.01.030.

Westaway, R. M., Lane, S. N., & Hicks, D. M. (2001). Remote sensing of clear-water, shallow, gravel-bed rivers using digital photogrammetry. *Photogrammetric Engineering and Remote Sensing, 67,* 1271–1281.

Whitehead, K., Hugenholtz, C. H., Myshak, S., Brown, O., LeClair, A., Tamminga, A., et al. (2014). Remote sensing of the environment with small unmanned aircraft systems (UASs), part 2: Scientific and commercial applications 1. *Journal of Unmanned Vehicle Systems, 02*(03), 86–102. https://doi.org/10.1139/juvs-2014-0007.

Williams, R. D., Brasington, J., Vericat, D., & Hicks, D. M. (2014). Hyperscale terrain modelling of braided rivers. Fusing mobile terrestrial laser scanning and optical bathymetric mapping. *Earth Surface Processes and Landforms, 39,* 167–183.

Winterbottom, S. J., & Gilvear, D. J. (1997). Quantification of channel bed morphology in gravel-bed rivers using airborne multispectral imagery and aerial photography. *Regulated Rivers*

*Research and Management, 13,* 489–499. https://doi.org/10.1002/(SICI)1099-1646(199711/12)13:6<489:AID-RRR471>3.0.CO;2-X.

Woodget, A.S. (2015). Quantifying physical river habitat parameters using hyperspatial resolution UAS imagery and SfM-photogrammetry. Unpublished PhD Thesis, University of Worcester, UK.

Wright, A., Marcus, W. A., & Aspinall, R. (2000). Evaluation of multispectral, fine scale digital imagery as a tool for mapping stream morphology. *Geomorphology, 33,* 107–120. https://doi.org/10.1016/S0169-555X(99)00117-8.

Zarco-Tejada, P. J., Diaz-Varela, R., Angileri, V., & Loudjani, P. (2014). Tree height quantification using very high-resolution imagery acquired from an unmanned aerial vehicle (UAV) and automatic 3D photo-reconstruction methods. *European Journal of Agronomy, 55,* 89–99.

Zeng, C., Richardson, M., & King, D. J. (2017). The impacts of environmental variables on water reflectance measured using a lightweight unmanned aerial vehicle (UAV)-based spectrometer system. *ISPRS Journal of Photogrammetry and Remote Sensing, 130,* 217–230. https://doi.org/10.1016/j.isprsjprs.2017.06.004.

Zhao, C. S., Zhang, C. B., Yang, S. T., Liu, C. M., Xiang, H., Sun, Y., et al. (2017). Calculating e-flow using UAV and ground monitoring. *Journal of Hydrology, 552,* 351–365. https://doi.org/10.1016/j.jhydrol.2017.06.047.

Zmarz, A. (2014). UAV – a useful tool for monitoring woodlands. *Miscellanea Geographica-Regional Studies on Development, 18*(2), 46–52.

(Received  December 10, 2017, revised  June 11, 2018, accepted  June 15, 2018, Published online  June 20, 2018)

Pure Appl. Geophys. 175 (2018), 3263–3283
© 2018 The Author(s), corrected publication July 2018
This article is an open access publication
https://doi.org/10.1007/s00024-017-1707-7

**Pure and Applied Geophysics**

# Application of Low-Cost Fixed-Wing UAV for Inland Lakes Shoreline Investigation

Tomasz Templin,[1] Dariusz Popielarczyk,[1] and Rafał Kosecki[1]

*Abstract*—One of the most important factors that influences the performance of geomorphologic parameters on urban lakes is the water level. It fluctuates periodically, causing shoreline changes. It is especially significant for typical environmental studies like bathymetric surveys, morphometric parameters calculation, sediment depth changes, thermal structure, water quality monitoring, etc. In most reservoirs, it can be obtained from digitized historical maps or plans or directly measured using the instruments such as: geodetic total station, GNSS receivers, UAV with different sensors, satellite and aerial photos, terrestrial and airborne light detection and ranging, or others. Today one of the most popular measuring platforms, increasingly applied in many applications is UAV. Unmanned aerial system can be a cheap, easy to use, on-demand technology for gathering remote sensing data. Our study presents a reliable methodology for shallow lake shoreline investigation with the use of a low-cost fixed-wing UAV system. The research was implemented on a small, eutrophic urban inland reservoir located in the northern part of Poland—Lake Suskie. The geodetic TS, and RTK/GNSS measurements, hydroacoustic soundings and experimental aerial mapping were conducted by the authors in 2012–2015. The article specifically describes the UAV system used for experimental measurements, the obtained results and the accuracy analysis. Final conclusions demonstrate that even a low-cost fixed-wing UAV can provide an excellent tool for accurately surveying a shallow lake shoreline and generate valuable geoinformation data collected definitely faster than when traditional geodetic methods are employed.

**Key words:** UAV, DEM, bathymetry, morphometry, GNSS, Lake Suskie.

## 1. Introduction

Water environment is a complex phenomenon, dynamic in both space and time. To understand its

nature a methodology that allows to accurately describe the objects and analyze the spatio-temporal changes is required. Over the years, different approaches have been employed to accurately model the topography of water reservoirs, lakes and rivers. They introduced new procedures, the processing chain, sensors, and algorithms (Popielarczyk et al. 2015).

The aquatic environment changes are the result of natural and anthropogenic forces, as well as the protection and restoration process of water reservoirs. The study of the state of this environment usually begins with a thorough analysis of the bathymetry and morphometry (Popielarczyk and Templin 2014; Lopata et al. 2014). The quality of morphometric parameters largely depends on the accuracy and quality of the collected data. One of the biggest challenges in the processing chain is the effective methodology to quickly and efficiently collect accurate and up-to-date data. To realize this idea the authors started a research program with the goal of developing a dedicated, open unmanned aerial system for semi-automated shallow water morphometric and bathymetric data acquisition.

Recent advances in surveying technology allow to construct a novel, automated measuring platform using modern sensors and processing algorithms. (Toth and Jóźków 2016). Traditional geodetic survey techniques (total station and RTK/GNSS surveys) are nowadays more frequently replaced with high-resolution laser sensors and semi-automatic platforms: aerial light detection and ranging systems (ALS), terrestrial light detection and ranging systems (TLS) or mobile light detection and ranging systems (MLS), personal laser scanning (PLS) (Marshall et al. 2016).

Water environment research usually requires continuous mapping of underwater and land topography. In most reservoirs, bathymetry can be measured with instruments such as sonar, single-

The original version of this article was revised: the first name of the third author the small letter has been used instead to be capitalized. Furthermore the version of 'RAFAŁ' is incorrect. It should be 'RAFAŁ'.

[1] Institute of Geodesy, University of Warmia and Mazury in Olsztyn, Oczapowskiego 1, 10-719 Olsztyn, Poland. E-mail: tomasz.templin@uwm.edu.pl

beam echo sounder (SBES) or multi-beam systems (MBES) and bathymetric airborne LiDAR (International Hydrographic Organization 2005). Additional analysis of the adjacent topography requires remote sensing products like aerial photography, satellite images or radar imagery (Heine et al. 2015; Szostak et al. 2014; Szulwic et al. 2015).

One of the most important factors that influences the result of geomorphologic parameters on urban lakes is the water level. It fluctuates periodically, causing shoreline changes. The shape and length of the shoreline determine the bathymetry and morphometric parameters of the lake. It is especially significant for typical environmental studies like bathymetric surveys, sediment depth changes, thermal structure, water quality monitoring, etc. (Shintani and Fonstad 2017). In most reservoirs, the water level can be obtained from digitized historical maps and plans or direct measurements using: geodetic total station, GNSS receivers, satellite and aerial photos, and even UAV with different sensors.

Unmanned aerial vehicles (UAVs), known as drones, offer significant advantages in geodata collection and are a low-cost alternative to the classical manned aerial photogrammetry or complementary solution to terrestrial acquisition (Nex and Remondino 2014; Niedzielski et al. 2016). The greater coverage, better quality and resolution of affordable aerial platforms make the number of practical applications difficult to determine. UAVs increase speed and reduce the cost of remote data collection. Even a relatively small and cheap quadcopter, for example DJI Phantom, provides several minutes of effective working time/flight at working height up to 200–300 m above the ground. This allows measurements of small engineering objects in a limited terrain. A low-cost UAV equipped with remote sensors or camera allows frequent flights at low altitudes in almost any area (Huang et al. 2017).

The use of small unmanned measurement systems for acquiring geoscientific data has grown rapidly in recent years. Unmanned aerial vehicles (UAVs) are increasingly being used to monitor small areas, e.g., small water reservoirs (ponds, urban lakes) and can be a good alternative to satellites, due to better resolution and detail. UAVs have become significant tools to perform a photographic inventory of land and engineering structures. The most often used UAVs are low cost and not very sophisticated multicopters, which are becoming increasingly popular as a platform for photogrammetric cameras (Čermáková et al. 2016; Harvey et al. 2016; Yucel and Turan 2016).

In recent years, UAV platform has been developed to provide a solution for divergent water applications. Despite their increased capabilities, the use of drones in geophysical sciences is usually restricted to image acquisition for generating high-resolution maps (Tauro et al. 2016). Some authors offered a solution—the use of UAV systems to reconstruct topography in coastal environment (Mancini et al. 2013), shallow braided rivers (Javernick et al. 2014) and lakes. More sensors will be developed over time allowing, for example, in situ measurements (Kageyama et al. 2016; Koparan and Koc 2016), pollution meters (Zang et al. 2012) and so on.

Only few researchers have addressed the problem of using fixed-wing UAVs in water studies (Everaerts 2008). They mainly concentrate on an advanced platform dedicated to the largest water reservoirs (Zolich et al. 2015). Most of the publications have been limited to off-the-shelf multicopters (Aguirre-Gómez et al. 2016; Dietrich 2017). Despite the efficiency of such methods presented in the literature, their use is restricted to easy-to-access environments. They are limited in flight times and payload, with typical endurance of 15–20 min and payloads of up to 0.5 kg for consumer-level systems. UAV operation is also limited during windy weather conditions.

Due to their bigger size and more challenging service, fixed-wing platforms are considered as more complex, specialized tools. The main limitations are price, legal constraints and the need to ensure the technical expertise of the pilot. As a result, the UAV use usually requires cooperation with commercial companies (McEvoy et al. 2016). The application of the multi-rotor UAV in shoreline measurements is limited due to their range and working time. Therefore, the authors offer a much longer operation time of the measuring system by using the self-constructed and self-built fixed-wing unmanned aerial system (UAS), equipped with the autopilot and camera. A low-cost drone can operate for 35–40 min collecting photos for further elaboration. A properly configured autopilot can bring the sensor-equipped aircraft to the location where the measurements are to be made. After the surveys UAV can automatically return and perform a safe landing.

The study presents a reliable methodology for shallow lake shoreline investigation using a low-cost tailless fixed-wing UAV system. The research was implemented on a small, eutrophic urban inland reservoir located in the northern part of Poland— Lake Suskie. The geodetic TS, and RTK/GNSS measurements, hydroacoustic soundings and experimental aerial mapping were conducted by the authors in 2012–2015. The article specifically describes the UAV system used for experimental measurements, the obtained results and the accuracy analysis. The results show that even a low-cost fixed-wing UAV can provide an excellent tool for accurately surveying a shallow lake shoreline and providing valuable geoinformation data collected definitely faster than through traditional geodetic methods.

## 2. *Methodology*

The objective of any remote sensing (the process of measuring an object or phenomenon of interest from a distance), is to provide observation of some physical parameter in a mapping frame at a given time or time period (Toth and Jóźków 2016). Many studies have quantified analyzing the usefulness of UAVs for monitoring, inspection and surveys applications in water environment. Most of them are based on off-the-shelf constructions. Only few demonstrate advantages of an open, independent aerial platform (Sørensen et al. 2017).

A lot of research has been done using a different UAVs type platform. Using UAVs is now widespread across a range of disciplines (Anderson and Gaston 2013; Liu et al. 2014; Smith et al. 2016). The most popular environmental applications are: landslide monitoring (Lucieer et al. 2014a), measuring changes in coastal morphology (Casella et al. 2014; Gonçalves and Henriques 2015; Papakonstantinou et al. 2016), monitoring glacier movement (Immerzeel et al. 2014; Ryan et al. 2015), studying Antarctic moss beds (Lucieer et al. 2014b), soil erosion monitoring (d'Oleire-Oltmanns et al. 2012), fluvial

geomorphology (Mori et al. 2002; Tamminga et al. 2015; Woodget et al. 2015) and forest research (Tang and Shao 2015; Wallace et al. 2012).

These studies are conducted using different UAV platforms delivered by various manufacturers (i.e., hybrid, flapping-wing, fixed-wing, coaxial, duct-fan, single rotor, and multi-rotor). Numerous documents in the literature describe the state-of-the-art development of UAVs (Cai et al. 2014). These reports provide a brief overview of small-scale unmanned aerial vehicles (UAVs) based on the information summarized from 132 models available worldwide. Their research was primarily concerned with the UAV platforms developed by academic institutions. Nowadays, there are also many complex commercial aerial mapping systems on the geodetic market. The review of UAS technology for photogrammetry and remote sensing applications with emphasis on regulations, acquisition systems, navigation and orientation can be found in the literature (Colomina and Molina 2014; Liu et al. 2014).

Recently, the most common platforms are multirotor and fixed-wing. Both have advantages and disadvantages which define their potential applications. Each UAV has a unique take-off and landing system (McEvoy et al. 2016). The fixed-wing systems require a larger clear area for both take-off and landing. They also need launching systems such as bungee cords or rails along with landing airbags, parachutes or nets (Gülch 2012). The multi-rotor model could take-off and land vertically from almost any location. Flight time and flight speed also vary between UAV models. The flight time for the fixed-wing models was between 30 and 90 min per flight, whereas the flight time for multirotor models was usually less than 20 min per flight. It gives a wide area mission coverage within one basic package of battery. Fixed-wing UAVs move at a speed of approximately 15 m/s while multi-rotor UAVs move at approximately 3 m/s. A fixed-wing body has better weather resistance, can be operated with stronger wind (up to 15 m/s) and during little rain (Haala et al. 2012). In case of minor damage, it allows for stable, continuous flight and safe landing. The fixed-wing platform is cheaper to build, has a simpler construction and fewer electronic components. The basic disadvantages of the wing are the need for open space for takeoff and landing and the need to be controlled by an experienced operator.

Considering the arguments above, the authors propose to build an open, low-cost unmanned aerial system (UAS) based on a popular fixed-wing Skywalker X-5 platform and the widely used open source flight control Pixhawk system with a dual processor, equipped with a global positioning system, data teletransmission module, etc. (Meier et al. 2012).

For economic reasons and payload restriction, the low-cost UAV platforms utilized for mapping purpose are mainly based on inexpensive passive sensors. A typical UAV imaging system consists of low-cost consumer-grade cameras. To obtain true ortho-photo mosaic and extract the 3D structure from multiple overlapping photographs, the structure from motion (SfM) algorithm is used. In most cases it is also combined with multi-view stereo (MVS) to automatically produce high-resolution digital elevation models (DEMs) (AgiSoft 2010; Snavely et al. 2008). This methodology is now well described and has been used in many studies (Fonstad and Marcus 2005; Wu 2013).

The growing popularity of aerial solutions increases the amount of flexible and scalable UAV software. New applications can automate navigation tasks and analyze the collected data. The photogrammetric SFM/MVS software falls into two categories: commercial, proprietary software (easy to use, but closed like a black box) or open source software (more complex workflow that requires advance knowledge) (Shervais 2015).

To process data from a designed UAS platform different SFM/MVS software was considered. Open source applications like VisualSFM coupled with CMP-MVS (Jancosek and Pajdla 2011) or SFMToolkit (Johnson et al. 2014; Westoby et al. 2012) were tested. Due to relatively early stage of our platform and certain limitations of these applications (maximum dimension of threshold, and others), the authors have decided to use one of the well-known, proprietary solutions—AgiSoft PhotoScan Pro from Agisoft.

### 2.1. Study Area

The presented research was implemented on a shallow, eutrophic inland reservoir. Lake Suskie is a small, urban lake located in the northern part of Poland (Fig. 1). It is situated in Susz Town, in the south-western part of Iławskie Lakeland, in the drainage basin of the Liwa River, a second-rank, east tributary to the Nogat River (Lossow et al. 2004). The lake is shallow with little-diversified bottom with max depth of 5.3 m (Choiński 2006; Inland Fisheries Institute in Olsztyn 1963). The lake has no natural surface inflows. The drainage basin on the eastern side is dewatered by melioration ditches. Heavily polluted 62-ha reservoir, located in the Iława Lake District, plays a significant role both in the water management of the Susz municipality and recreation for the region's inhabitants. The existing analog bathymetric plan of Lake Suskie and morphometric card had been developed by the Inland Fisheries Institute in Olsztyn (IRŚ) on the basis of historical measurements (Inland Fisheries Institute in Olsztyn 1963). Unfortunately, the historical analog bathymetric plan of the bottom shape and shore line differs from reality. The old bathymetry and morphometry were developed by the IRŚ in the previous century (in 1963).

### 2.2. Work Purpose

The main purpose of presented work is to reveal the potential application of the proposed tailless fixed-wing UAV platform and the data-processing chain within the case study of a typical shallow water reservoir.

The authors commissioned by the city of Susz regional authority have begun a study of Lake Suskie in the frame of a reclamation project. During the preliminary inventory of the lake, it turned out that the shape of the bottom and the course of the coastline differ significantly from those presented in previous morphometric data, adversely affecting the current morphometric data of the lake. In the years of 2012–2013, the authors have conducted new bathymetric surveys using dual-frequency, single-beam hydroacoustic system. The coastline has also been updated with TS and RTK geodetic techniques. However, classical shoreline measurement techniques have proved to be time-consuming and inaccurate because of the emergence of vegetation, wetlands and trees that adversely affect RTK/GNSS measurements.

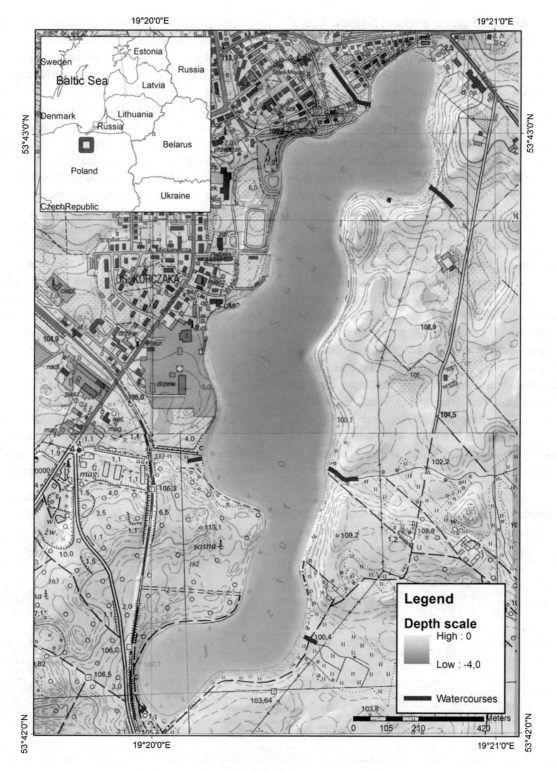

Figure 1
Location of Lake Suskie, the study area

For this reason, the authors proposed the use of UAVs to inventory the shape of the coastline.

The proposed methodology for shoreline investigation and finally for bathymetry and morphometry calculation requires the integration of several measurements systems. Direct geodetic, GNSS, hydroacoustic and finally proposed aerial measurements are used to quickly and reliably collect accurate data. The new challenge was to implement a modern, low-cost and efficient UAV system that supports an effective collection of shoreline data across large areas of the water environment.

Figure 2 shows the proposed procedure and lists the methods of coastline extracting used during the experimental study and further analysis. At first, potentially useful methodologies for shoreline acquisition were analyzed. The available data sources were discovered and used (the bathymetric plans and morphometric cards, aerial photos and satellite images, topographic maps and cadastral data from existing databases). Then direct geodetic, bathymetric and aerial measurements were conducted and the collected data were processed to calculate shoreline changes. The extracted shorelines were compared. Finally, the chosen UAV shoreline was applied to elaborate up-to-date bathymetry and morphometry of the lake.

The potential influence of data acquisition methodology and changes between different method results was presented on the most characteristic, east part of the lake shoreline (Figs. 2, 4). To highlight the advantages of the UAV system, the comparison between different measurement techniques was discussed. The results from the unmanned aerial platform were assigned and compared with complementary techniques and classical TS and GNSS methods of coastline detection on shallow reservoirs. The optimal solution was chosen for new bathymetric elaboration and new morphometric parameters calculation. A reliable Digital Elevation Model of the lake bottom was made and processed in the following steps. At the end, the results of the bathymetric campaign were discussed. The presented methodology was proposed to be applied during a sediment and morphometry analysis task on the typical small, eutrophic urban inland reservoir (Lake Suskie).

### 2.3. Geodetic/Bathymetric Data Collecting and Processing

According to the Lake Suskie reclamation project, new bathymetric surveys were conducted to elaborate up-to-date bottom elevation model and to calculate morphometric parameters. The integrated

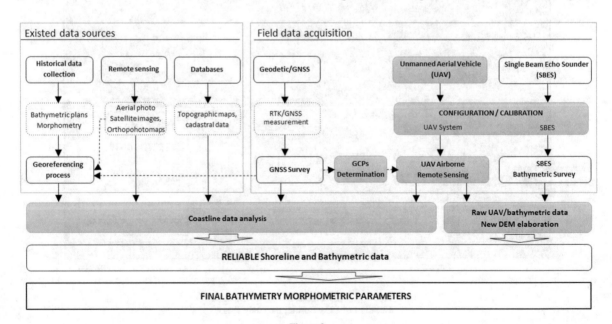

Figure 2
Flowchart of the adopted methodology

bathymetric system used during depth measurements basically consists of: RTK/GNSS positioning, an SBES hydroacoustic bottom detection system and special GNSS, hydrographic and GIS software. During the project, Topcon Hiper Pro receivers were used and RTK/GNSS positioning was completed based on the Ground-Based Augmentation System (GBAS) permanent reference station ASG-EUPOS (Active Geodetic Network—EUPOS). Additionally, the raw GNSS/static/data were used for accurate water level calculations. The hydrographic SBES equipment included two single-beam digital hydrographic Echo Sounders: a Simrad EA501 P and a Reson Navisound 515. Additionally, a YSI 600R sonde for SBES calibration was used. The EA 501 P system basically consists of a 200-kHz transducer, transceiver and personal computer. Dual channel Navisound 515 uses two-frequency transducers (38 and 200 kHz). The bathymetric measurements were carried out in two stages. The first part covered the central deepest region of Lake Suskie. The second was conducted in the northern and southern part of the lake. The basic measurement profiles were designed in the east–west direction at spatial sampling of 10 m. After conducting all the stages of field measurement campaign on Lake Suskie, the bathymetric raw data were processed.

reason why the authors have built their own prototype of unmanned aerial system (UAS) based on Skywalker X-5 fixed-wing platform.

The proposed UAS consists of five main components:

1. Unmanned aerial vehicle (UAV).
2. Sensor (camera).
3. Radio control system (RC).
4. Telemetry link.
5. Ground control station (GCS).

7A tailless fixed-wing aircraft is made of expanded polyolefin (EPO). The UAV is based on separate components available on the market, in hobby stores. All individual components have been purchased, configured and finally calibrated by the authors (Table 1). In addition to a simple based platform, it also contains radio control, telemetry and autopilot systems, GNSS positioning and a high torque brushless motor. The heart of the drone is an open source autopilot called Pixhawk and peripheral sensors, i.e., a GPS receiver with magnetometer, Pitots' tube, servo motors, an electronic speed controller with a brushless engine. Pixhawk works with Ardupilot firmware, which provides full automatic flights and camera triggering. Frsky's transmitter Taranis is a radio control (RC) system used as the safety background for steering of the drone. It makes it

## 3. UAV Measurements

The last element necessary for the development of the current bathymetry and morphometric data was the updating of the shoreline course. For this purpose, the authors propose a new methodology for the coastline shape inventory with the use of UAV. The whole area of Lake Suskie covered by scheduled and executed experimental aerial measurement was divided into four basic parts/flights. In the present work the most interesting middle/east part of the lake was presented and analyzed (Fig. 4).

### 3.1. UAV Fixed-Wing Prototype

Lake shoreline measurements and the inventory of underwater vegetation as well as adjacent areas cover large areas and take considerable time. That is one

Table 1

*Basic technical specifications of flying wing UAV system*

| Characteristics | UAS parameters |
| --- | --- |
| UAV type | Tailless fixed flying wing "Skywalker X-5" |
| Wingspan | 1180 mm |
| Total mass | 1.35 kg |
| Autopilot | PixHawk v2.45 |
| Radio control | FrSky Taranis X9D PLUS |
| Engine | High torque brushless motor AXI 2808/24 Gold |
| Camera | Sony RX100 20 MP, with Seagull controller |
| Battery life | 45 min (LiPo 3S, 11.1 V, 4600 mAh) |
| Max. flight time | 40 min |
| Max flight speed | 18 m/s |
| Max. wind speed | 15 m/s |
| Min. flight altitude | 70 m above ground level |
| Telemetry and RC range | Up to 2 km |

Figure 3
UAV system fitted with Sony RX100 camera

possible to operate a drone in four or more different flight modes: manual, assisted (FBWA—fly by wire alfa), full automatic (AUTO), and auto return to launch (RTL). Telemetry data are based on two 433-MHz radios. The radio link module is connected directly to the flight computer (the Pixhawk flight controller). The fully charged lithium polymer (LiPO) battery 4600 mAh provides a maximum flight time of 45 min. In addition, the built-in safety features in the flight computer manage potentially dangerous situations: UAV can land manually or automatically when the battery is running low. Our fixed-wing prototype with 1180 mm wingspan can lift a Sony RX100 20MP camera (weighing 300 g with Seagull controller). The total weight of the UAV system is 1.35 kg, including the battery and camera. Our UAV flight wing tolerates wind speed of up to 15 m/s (a threshold that was never exceeded during data collection). The fixed-wing system requires a relatively small, clear take-off and landing area. To launch the flying wing a bungee system is obligatory. This solution provides high reliability of each take-off. A belly landing can be done in the auto, assisted or manual mode. An open source software named Mission Planner (GCS) is responsible for planning and operating the photogrammetry missions, setting up the drone and calibrations.

To acquire spatial resolution photographs during the Lake Suskie experimental measurements, a fixed-wing UAS was assembled as follows (Fig. 3).

## 3.2. Data Acquisition/Image Acquisition

Our autonomous fixed-wing system can operate and take photographs automatically according to a flight plan (prepared in Mission Planner software). The drone is operational for approximately 45 min on a single battery. A Sony RX100 II camera with 20 effective megapixels was used to collect the photographs. The photo capture rate was controlled by the flight system control board, which was programmed to emit a trigger pulse at a desired frequency. The shutter of the camera was triggered by the Seagull steering controller, which was connected to the UAV control system. The stable flight of the UAV increases image quality and precision, while wind reduces it. The UAV was manually launched from a flat area near the lake. Manual control during the flight negatively affects image quality and resolution, especially in windy conditions. That is why images were acquired at a fixed height, under automatic flight control, while manual control was used for landing. Before each stage/flight

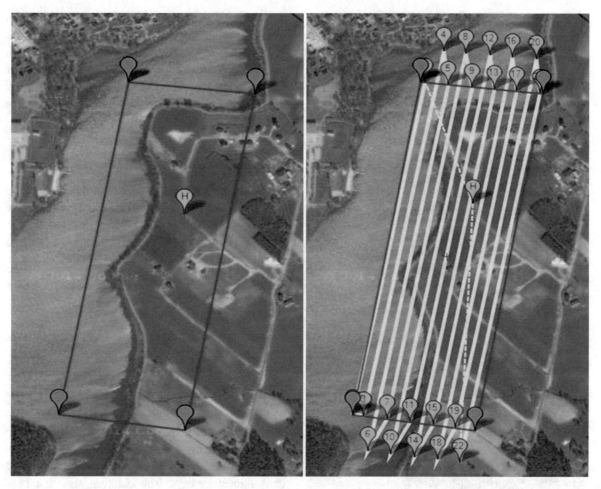

Figure 4
Investigated area (on the left). A flight path across the study area (on the right)

of the project, the front-lap image overlap was calculated to be 70%, with a side-lap image overlap of 50%, for a setting of one photograph per second.

The Lake Suskie is largely surrounded by forest or by single high trees. That is why flight altitudes ranged from 100 to 160 m, depending on the covered area and the height of surrounding trees. During experimental measurements the ground control system was used to set flight parameters and to monitor the flight status. A remote control unit (RC) was used to operate the tailless fixed-wing platform manually in case of emergency.

Before the mission the flight plan was designed. The whole measurement area of Lake Suskie was divided into four parts. One of the tested areas presented in the research is shown in Fig. 4. This

experimental flight area has 33 ha (including water) and covers 1600 m of shoreline. The take-off and landing site, marked as H in Fig. 4, was away from the edge of the lake. Because of high trees, hilly terrain and nearby construction works operating the UAV was neither easy nor comfortable.

### 3.3. GNSS Survey, GCPs Determination

To convert the image coordinates into geographical coordinates, ground control points (GCP) for the area of interest are required. Eight GPCs were used on the most interesting middle-east part of the lake described in the research (one flight from four covering the whole lake, Figs. 4, 5).

Figure 5
Ground control points location

The scale, resolution and precision of ground control points are the major factors affecting UAV image quality (Niethammer et al. 2012). GCPs were determined using GNSS Topcon HiperPro geodetic receivers. The raw GNSS/static/data were used for GCPs coordinates calculations. Static sessions were completed based on the four Ground-Based Augmentation System (GBAS) permanent reference stations ASG-EUPOS (Active Geodetic Network—EUPOS): ELBL, GRUD, ILAW, STRG. The BLH WGS'84 horizontal geodetic coordinates were transformed into projected coordinates (ETRS 1989 Poland CS 2000 Zone 7). Vertical measurements

were reduced to Kronsztad'86 (vertical map datum in Poland—normal heights above sea level). The RMS error of GCP measurements was less than 10 mm horizontally and 15 vertically. The distribution of GCPs across the region of one stage of all flights, used for the georectification procedure, is shown as yellow dots in Fig. 5.

### 3.4. Coastline Extraction and Comparison

To evaluate a possible impact of data acquisition on the coarse line location, all existing materials have been analyzed and prepared for further work. The following data were discovered and acquired from the succeeding sources:

- Historical bathymetric plan from Inland Fisheries Institute in Olsztyn,
- Orthophoto from a Polish geoportal (http://www.geoportal.gov.pl), topographic and cadastral maps from government and regional offices,
- RTK surveys from on-site measurement campaign,
- UAV orthomosaic from on-site UAV surveys.

All the data were collected, analyzed and processed using distinctive techniques. The bathymetric plan provided by the Inland Fisheries Institute in Olsztyn in 1963 had been prepared on the basis of old conventional surveys. The old plan was first converted into a digital format and next digitized for further work. The feature classes representing course line were extracted from adjusted historical bathymetric data.

The classical manned aerial photo was acquired as an image from web map service (WMS) from a Polish geoportal. The topographic maps and cadastral data were obtained as vector layers in ESRI shape formats. They were transformed into common database and used as supported layers for aerial maps. The coarse line was digitized and saved as polyline layers.

The direct RTK/GNNS shoreline measurements were made with Trimble R8 geodetic RTK/GNSS receiver used for collecting float/fixed field data based on the single GBAS ASG-EUPOS reference station in Iława (ILAW). During the RTK a survey of 608 fixed/float points was collected. Unfortunately, as many as 228 points were measured by the float solution because of the trees and the difficult satellite situation. That is why the inventory of the coastline was completed using the developed unmanned aerial system.

The typical workflow was used for UAV image processing. It was based on the SfM method used in the Agisoft PhotoScan software. The processing chain consisted of the following steps: image matching, georeferencing, digital elevation model creation, orthomosaics, point cloud generation and texture model creation.

All steps were performed with ESRI ArcMap, version 10.4.1. All of the GIS datasets were created in the Polish projected "ETRS 1989 Poland CS 2000 Zone 7" coordinate system.

## 4. Results

The proposed methodology for acquiring reliable coastline and bathymetric information was implemented during Lake Suskie project. After performing comprehensive GNSS, hydroacoustic and aerial measurements on the reservoir, the authors have compiled a new shoreline and up-to-date bathymetric and morphometric data.

### 4.1. Elaboration of Lake Suskie Boundaries

The collected data were processed and analyzed using ESRI ArcGIS version 10.4.1 with the 3D Analyst/Spatial Analyst extension. The data were firstly imported into common ESRI geodatabase structure and processed sequentially. As a result four different shorelines were prepared. The visualization of the first three (historical, orthophoto and RTK) are presented in Fig. 7.

The UAV data require an additional task. In the presented part of the area 310 photographs were captured on the average 156 m flying altitude, encompassing coverage area of 34,53 ha. Figure 6 presents orthophotos and the DEM of the lake shoreline and the adjacent natural topography created during image processing. On the basis of our UAV mission the resolution of elaborated orthomosaics was at the level of 1–6 cm per pixel. These high-resolution UAV images were next used to determine the lake borders. The on-screen digitization method was next used for shoreline extraction.

The calculated parameters of Lake Suskie are shown in Table 2. According to the UAV measurements conducted in 2015 the total area of the lake is 62.00 ha and the shoreline length is 5790 m. The calculated area is similar to the area obtained from satellite images and RTK surveys. The shoreline length differs by 81 and 33 m, respectively. According to the historical bathymetric plan the UAV area was reduced from 62.70 to 62.00 hectares. However, the shoreline increased from 5600 to 5790 m.

Figure 7 presents comparisons of coastline extraction methods. The top left—the historical morphometric plans from IRS, lower—the orthophotomap from Head Office of Land Surveying and Cartography in Warsaw, the bottom left—RTK survey on the topographic map, on the right, the comparison of coastlines (historical, orthophoto, RTK, UAV).

### 4.2. Lake Suskie DEM Generation and Morphometry Calculation

After conducting all the stages of field experimental measurements on Lake Suskie, the current coastline with border of reeds, 3D bathymetric model and morphometric parameters were elaborated. The raw data from hydroacoustic sounding were processed and bathymetric points were converted into an ESRI multipoint feature class. The inventory of the coastline included historical data and Web Map Service (WMS) analysis, aerial remote sensing data from UAV and RTK/GNSS direct measurements. Figure 8 presents the northern part of the lake, including the following layers: isobaths, the coastline shape with measured points, the vegetation area. The red line shows the old coastline extracted from archival bathymetric plan.

The quality of the DEM is basically a function of the accuracy of individual survey points, field survey strategy and the method of interpolation (Heritage et al. 2009). The most popular ways to create DEM models are the regular grid surface (Grid) and triangulated irregular network (TIN) (El-Sheimy et al. 2005). The Lake Suskie DEM was generated with the use of ESRI ArcGIS 10.4.1 (Fig. 9).

Figure 6
Study area orthophotos and DEM created in the Agisoft PhotoScan

Table 2

*Area of the lake and shoreline length*

| Method | Bathymetric plans | Orthophoto | RTK | UAV |
|---|---|---|---|---|
| Date | 1967 | 2014 | 2014 | 2015 |
| Resolution (m) | – | 0.50 | – | 0.06 |
| Area (ha) | 62.70 | 61.99 | 61.99 | 62.00 |
| Shoreline length (m) | 5 600 | 5709 | 5823 | 5790 |

Bathymetric parameters such as volumetric and area calculations were derived using the TIN model. Contours, depth ranges, and the shaded relief map were derived from a DEM grid. The TIN surface was created using the collected bathymetric data points and the lake boundary inputs. The TIN consists of connected data points that form a network of triangles representing the bottom surface of the lake. This grid was created using the ArcMap Topo to Raster Tool and had a spatial resolution of 5 m. Then the contours were generated and converted to polygon feature classes. They were attributed to show 0.5-m depth ranges across the lake. At the end, the contour lines were edited to improve accuracy and to smooth the lines.

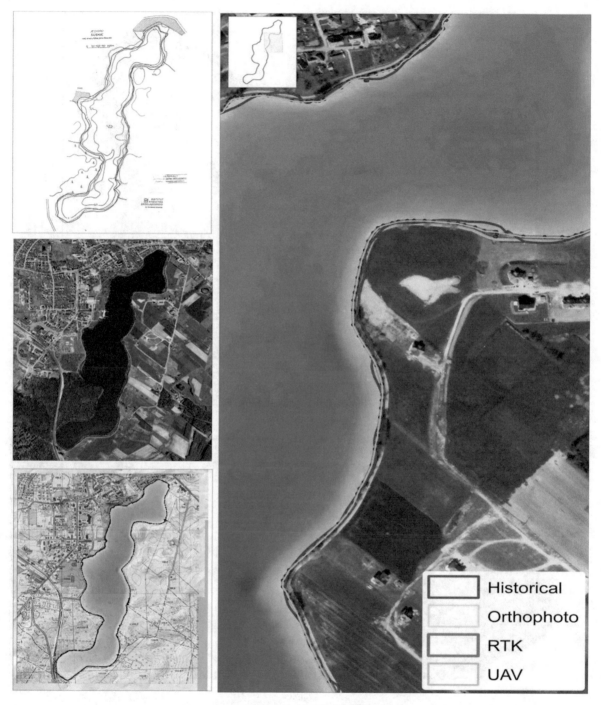

Figure 7
Coastline extraction methods comparisons

The new elaborated coastline and DEM is a source of quantitative information. Lake morphometry (e.g., depth, volume, size, etc.) facilitates the understanding of the physical and ecological dynamics of lakes (Hengl et al. 2009; Hutchinson 1957). The basic morphometric parameters of Lake Suskie

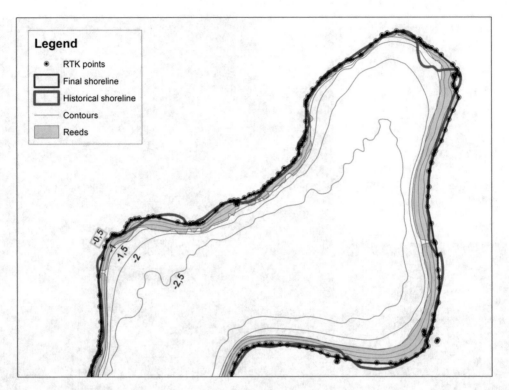

Figure 8
Measured points spatial distribution

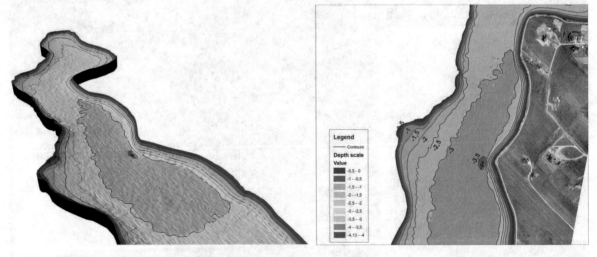

Figure 9
DEM of Lake Suskie and adjacent area

were calculated based on the methods and definitions described by Hutchinson, Wetzel, Choinski and Ławniczak (Choiński 2007; Hutchinson 1957; Wetzel 2001). The new Lake Suskie morphometry parameters on the basis of the last bathymetric, GNSS and UAV measurements were calculated for the entire lake. The following morphometric parameters describing Lake Suskie were elaborated: max. lake length (m), max. and mean width (m), area (ha), shoreline length (m), volume (m3), max. and mean

Table 3

*New morphometric parameters of Lake Suskie*

| | Lake Suskie | |
|---|---|---|
| Year | 1963 | 2015 |
| Source | Inland Fisheries Institute in Olsztyn | Popielarczyk, Templin |
| Maximum lake length ($L_{max}$)(m) | 2300.0 | 2251.9 |
| Maximum width ($B_{max}$) (m) | 475 | 449 |
| Mean width ($B_{min}$) (m) | 272 | 275 |
| Area (ha) | 62.7 | 62.0 |
| Shoreline length (m) | 5 600 | 5 790 |
| Shore Development Index $L_{dl}$ | 2.00 | 2.05 |
| Volume (m$^3$) | 1,491,4000 | 1,377,086 |
| Maximum depth ($D_{max}$) (m) | 5.3 | 4.1 |
| Mean depth ($D_{min}$) (m) | 2.4 | 2.2 |
| Relative depth ($Z_r$) (m) | 0.0067 | 0.0052 |

depth (m). Actual morphometry parameters are shown in Table 3.

It is important for these results to be considered in the future management of the shallow lake. Local authorities intended to use the new elaborated bathymetry and morphometry for the Lake Suskie reclamation project.

## 5. Discussion

Each of the presented methods of shorelines extraction is encumbered with errors, limiting its further use. The historical bathymetric plans, due to the paper form and lack of geodetic references, can only be used as information on the historical course of the coastline and contour lines to build the digital elevation model of the bottom. It can be then used for the analysis of spatio-temporal changes of the water reservoirs.

Photogrammetric data are a very good source to analyze the coastlines. However, the acquisition of classical manned aerial photos is still a very expensive and time-consuming process. The measurements are usually performed periodically. The orthophotomap used in the project is characterized by following parameters: the date of the picture taken for

orthophoto—2014, pixel size 0.5 m, RGB color. Due to the date, relatively low resolution (pixel size) and hard observation conditions on the shore, this method provides unreliable water environment information. The necessary geodata should be collected by direct measurements. RTK/GNSS measurements conducted under difficult satellite observation conditions have shown serious limitations during shoreline inventory. Nearly one-third of the measurements were characterized by a lack of a fixed solution. During the RTK measurements of the coastline or adjacent areas there are frequent problems with satellite signal loss or multipath effect. This is related to the fact that the banks of inland water reservoirs are often forested, or overgrown with single trees and brushes. This makes it much more difficult to perform GNSS geodetic measurements and sometimes even makes it impossible. Figure 10 shows some examples of the float RTK solutions in the immediate vicinity of the trees (Lake Suskie), even though the measurements were performed in early spring.

### 5.1. Accuracy Analysis

The inventory of the coastline and reed should be performed for further lake morphometry elaboration. So far, classical geodetic techniques or RTK/GNSS have traditionally been used. They are, however, time-consuming, and in some cases even impossible. Therefore, the authors decided to use UAV remote sensing for shoreline mapping. To check the coastline accuracy on the UAV-based orthophotomap, an accuracy analysis was performed on a fragment of the eastern part of the lake. The Trimble R8 geodetic RTK/GNSS receiver was used for collecting float/fixed field data based on the single GBAS ASG-EUPOS reference station in Iława (ILAW). To analyze the horizontal and vertical accuracy of the coordinates of terrain details as determined from the UAV, a fragment of a concrete bicycle path around the lake was used as a reference. At the same time the comparison of the coastline and reed line was performed. Figure 11 presents horizontal and vertical coordinate differences.

The horizontal differences of the coordinates range from − 0.033 to 0.020 m. The mean value is − 0.007 m. The results of the vertical surveys show

Figure 10
Float RTK/GNSS measurements problem

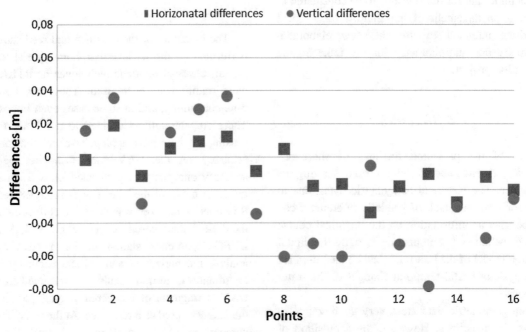

Figure 11
Accuracy of RTK/GNSS—UAV points analysis

that the differences between RTK and UAV heights range from − 0.078 to 0.037 m. The mean value of the vertical difference is − 0.021 m. This analysis confirms that the UAV technique can be successfully used for determining the position and height of the terrain points with an accuracy of a few centimeters.

## 5.2. Morphometry Analysis

In the presented research the authors describe geodetic, bathymetric and aerial measurements conducted on a small, urban lake, Lake Suskie, the final product of which is the development of a bathymetric map and morphometric parameters. The new integrated surveys show that the max depth of the lake is not 5.3 m (according to archival data from the Inland Fisheries Institute) but 4.1 m (Pawłowski 2014). The analysis of 1963 and 2013 survey suggests that over 50 years the surface area of Suskie Lake has changed from 62.7 to 62.0 ha and the lake volume has decreased from 1,491,400 to 1,377,086 m³. Moreover, meaningful changes in the shoreline were observed, which had a significant impact on the morphometric data of the lake. The coastline length has increased from 5600 to 5790 m. During the measurements and the raw data elaboration process all bathymetric data were referenced to the common, reference water level.

By analyzing archival maps, aerial photo and satellite images, it is evident on the archival bathymetric plan that the shape of the coastline is significantly different from archival geodesic materials (in some places the differences are several dozen meters). It can be concluded that the incompatibility of the coastline is due to the imperfection of the measurement methods used in the bathymetric process. The contemporary coastline only slightly differs from archival geodetic materials. On the other hand, it differs significantly from the old bathymetric plan.

As far as bathymetry is concerned, it is difficult to draw clear conclusions as to why the archival depth map differs from the current one. The old bathymetric plan indicates a maximum depth of 5.3 m centrally in the middle of the lake. Contemporary, professional hydroacoustic measurements have defined a maximum depth of 4.1 m in another part, 200 m northeast of the center of the lake. It should be emphasized that archival and current measurement results have been reduced to a common water level for analysis. The sonar echograms at two frequencies indicate a layer of loose organic sediment on the bottom of the entire lake, with a thickness of 20–40 cm. Ultimately, it can be concluded that differences in bathymetry and morphometric parameters are partly due to the

imperfection of old analog measurement methods (particularly positioning) and significant eutrophication, which happens in many lakes under anthropo pressure. Accuracy and reliability of raw measurement data significantly affect the bathymetry and can be a very important issue in the interpretation of the water environmental change. The presented results show that the bathymetric data and morphometric parameters of lakes, used for morphometric study and research in hydrobiology, limnology, fisheries and other environmental research, are highly dependent on the research methodology used, precise water level determination, reliable shoreline investigation, collected raw data, bathymetric and aerial maps elaboration processes.

## 5.3. Self-Built Fixed-Wing UAVs Future

Recent advances in drone technology, combined with low-weight sensors, provide a new method for mapping water areas at high resolution. A low-cost UAV system with a camera can quickly collect photographs of the shallow lake shoreline and inaccessible adjacent areas. Without human risk, it provides geoinformation with high accuracy. Therefore, the proposed and implemented low-cost UAV system can support the classical geodetic total station and RTK/GNSS technologies but, above all, it significantly supports the bathymetric measurements of inland, shallow lakes.

Nowadays, there are many complex aerial mapping systems on the geodetic market. The main advantages of a self-built fixed-wing platform over the commercial ones are undoubtedly a low price and a possibility of easy repair or improvement. Other advantages include own configuration of the UAS, a possibility to adapt various kinds of data collection and navigation sensors, and open source firmware being developed faster than commercial. Additionally, a mission planning tool is much more advanced. The main inconveniences of the self-built UAS are: much more time needed for building and testing and the lack of guaranty and technical support. The proposed fixed-wing UAS does not really compete with traditional photogrammetric flights, but offers interesting options for small water areas, where the cost of traditional flights is simply too high.

Our study confirms the relevance of UAVs platform in water applications and is consistent with previous results (Čermáková et al. 2016; Venturi et al. 2016; Yucel and Turan 2016). The solutions presented in those articles are based on off-the-shelf multi-rotor platforms. Their methodology and results validate the usefulness of UAVs platform for shoreline extraction and changes detection on small water reservoirs. These studies are preliminary and authors declare further study of UAVs, especially regarding the possibility to utilize other sensors on the platform (for example, for water pollution identification, chemical analyses, water samples). There are many platforms developed by academic institutions (Cai et al. 2014), but usually access to them is limited. There are some examples in literature describing the process of building UAVs platform (Anweiler and Piwowarski 2017; Sørensen et al. 2017). They are based on open standards and components but are usually constructed as a multi-rotor type of UAV. Our fixed-wing solution is similar, but offers a great opportunity to implement a new GNSS/IMU components and additional sensors. Both active and passive sensors can be used to improve the process of data acquisition (Dietrich 2017). The promising results suggest a further need for testing UAV platform to raise capacities and extend their applications.

## 6. Conclusions

The main objective of the present work was to describe an application of a low-cost tailless fixed-wing UAV for inland lake shoreline and adjacent area investigation. The proposed methodology uses an unmanned aerial system constructed and implemented specifically for small inland reservoirs (like Lake Suskie). A fixed-wing UAS, developed and implemented by the authors, can provide an excellent tool for accurate surveys of a shallow lake shoreline. It also provides valuable geoinformation data definitely faster than the traditional total station/RTK methods.

The constructed tailless UAV was made of widely available parts and open-source software. Images collected during the research were recorded using the Sony RX100 camera and elaborated with open source software. Precisely determined GCPs should be used to increase the accuracy of the final results. Static or RTK/GNSS methods need to be utilized for control points measured with 1 cm accuracy. The proposed solution improves the efficiency and makes it possible to get reliable results and accuracy comparable to the conventional techniques. Depending on the survey area and the UAV type, drones may offer a more economic survey platform than a crewed aircraft. They can also fly more slowly, at lower altitudes, allowing for safe use of high-resolution sensors. Just as aerial and satellite sensing have transformed scientific observations, allowing the resolution of large-scale physical processes, the pervasive use of drones is set to revolutionize geophysical sciences through the rapid and refined measurement of small-to-medium scale phenomena (Famiglietti et al. 2015; Tauro et al. 2016).

The proposed methodology shows that even a low-cost fixed-wing UAV can provide an excellent tool for accurately surveying a shallow lake shoreline and generate reliable geoinformation data collected faster than when traditional geodetic methods are employed. The presented study is only the first step to construct a universal, multifunctional platform for both passive and active sensors to acquire complex information about water environment.

Measurements made with the use of UAV methodology and orthomosaic seem to be one of the best options for water environment investigation. High-resolution (pixel at the level of centimeters) and up-to-date data provide reliable information on the coastline in a relatively short period of time. The proposed and implemented UAS platform can be easily extended by additional passive or active sensors. The raw data can be captured in a wider range of electromagnetic spectrum or by active sensors like laser scanners.

REFERENCES

AgiSoft. (2010). AgiSoft PhotoScan professional edition. http://www.agisoft.com. Accessed 27 Mar 2017.

Aguirre-Gómez, R., Salmerón-García, O., Gómez-Rodríguez, G., & Peralta-Higuera, A. (2016). Use of unmanned aerial vehicles and remote sensors in urban lakes studies in Mexico. *International Journal of Remote Sensing.* https://doi.org/10.1080/01431161.2016.1264031.

Anderson, K., & Gaston, K. J. (2013). Lightweight unmanned aerial vehicles will revolutionize spatial ecology. *Frontiers in Ecology and the Environment, 11*(3), 138–146. https://doi.org/10.1890/120150.

Anweiler, S., & Piwowarski, D. (2017). Multicopter platform prototype for environmental monitoring. *Journal of Cleaner Production, 155,* 204–211. https://doi.org/10.1016/j.jclepro.2016.10.132.

Cai, G., Dias, J., & Seneviratne, L. (2014). A survey of small-scale unmanned aerial vehicles: Recent advances and future development trends. *Unmanned Systems, 2*(2), 175–199. https://doi.org/10.1142/S2301385014300017.

Casella, E., Rovere, A., Pedroncini, A., Mucerino, L., Casella, M., Cusati, L. A., et al. (2014). Study of wave runup using numerical models and low-altitude aerial photogrammetry: A tool for coastal management. *Estuarine Coastal and Shelf Science, 149,* 160–167. https://doi.org/10.1016/j.ecss.2014.08.012.

Čermáková, I., Komárková, J., & Sedlák, P. (2016). Using UAV to detect shoreline changes: Case study—pohranov pond, Czech Republic. In: *International Archives of the Photogrammetry, Remote Sensing and Spatial Information Sciences—ISPRS Archives* (Vol. 2016–Jan, pp. 803–808). https://doi.org/10.5194/isprsarchives-XLI-B1-803-2016.

Choiński, A. (2006). *Polish Lake catalogue.* Poznań: Adam Mickiewicz University Publishing. (in Polish).

Choiński, A. (2007). *Physical limnology of Poland.* Poznań: Adam Mickiewicz University Publishing. (in Polish).

Colomina, I., & Molina, P. (2014). Unmanned aerial systems for photogrammetry and remote sensing: A review. *ISPRS Journal of Photogrammetry and Remote Sensing.* https://doi.org/10.1016/j.isprsjprs.2014.02.013.

d'Oleire-Oltmanns, S., Marzolff, I., Peter, K., & Ries, J. (2012). Unmanned aerial vehicle (UAV) for monitoring soil erosion in Morocco. *Remote Sensing, 4*(12), 3390–3416. https://doi.org/10.3390/rs4113390.

Dietrich, J. T. (2017). Bathymetric structure-from-motion: Extracting shallow stream bathymetry from multi-view stereo photogrammetry. *Earth Surface Processes and Landforms, 42*(2), 355–364. https://doi.org/10.1002/esp.4060.

El-Sheimy, N., Valeo, C., & Habib, A. (2005). *Digital terrain modeling: Acquisition, manipulation, and applications.* Norwood: Artech House.

Everaerts, J. (2008). The use of unmanned aerial vehicles (UAVs) for remote sensing and mapping. *The International Archives of the Photogrammetry Remote Sensing and Spatial Information Sciences, 37*(2008), 1187–1192.

Famiglietti, J. S., Cazenave, A., Eicker, A., Reager, J. T., Rodell, M., & Velicogna, I. (2015). Satellites provide the big picture. *Science, 349*(6249). http://science.sciencemag.org/content/349/6249/684.2. Accessed 21 Apr 2017.

Fonstad, M. A., & Marcus, W. A. (2005). Remote sensing of stream depths with hydraulically assisted bathymetry (HAB) models. *Geomorphology, 72*(1), 320–339. https://doi.org/10.1016/j.geomorph.2005.06.005.

Gonçalves, J. A., & Henriques, R. (2015). UAV photogrammetry for topographic monitoring of coastal areas. *ISPRS Journal of Photogrammetry and Remote Sensing, 104,* 101–111. https://doi.org/10.1016/j.isprsjprs.2015.02.009.

Gülch, E. (2012). Photogrammetric evaluation of multi-temporal fixed wing UAV imagery. *International Archives of the Photogrammetry Remote Sensing and Spatial Information Sciences, XXXVIII-1/(September),* 265–270. https://doi.org/10.5194/isprsarchives-XXXVIII-1-C22-265-2011.

Haala, N., Cramer, M., Weimer, F., & Trittler, M. (2012). Performance test on UAV-based data collection. *International Archives of the Photogrammetry Remote Sensing and Spatial Information Sciences, XXXVIII-1/,* 7–12. https://doi.org/10.5194/isprsarchives-XXXVIII-1-C22-7-2011.

Harvey, M. C., Rowland, J. V., & Luketina, K. M. (2016). Drone with thermal infrared camera provides high resolution georeferenced imagery of the Waikite geothermal area, New Zealand. *Journal of Volcanology and Geothermal Research, 325,* 61–69. https://doi.org/10.1016/j.jvolgeores.2016.06.014.

Heine, I., Stüve, P., Kleinschmit, B., & Itzerott, S. (2015). Reconstruction of lake level changes of groundwater-fed lakes in Northeastern Germany using rapideye time series. *Water, 7*(8), 4175–4199. https://doi.org/10.3390/w7084175.

Hengl, T., Reuter, H. I., & Institute for Environment and Sustainability (European Commission, Joint Research Centre). (2009). *Geomorphometry: Concepts software applications.* Oxford: Elsevier.

Heritage, G. L., Milan, D. J., Large, A. R. G., & Fuller, I. C. (2009). Influence of survey strategy and interpolation model on DEM quality. *Geomorphology, 112*(3–4), 334–344. https://doi.org/10.1016/j.geomorph.2009.06.024.

Huang, H., Long, J., Lin, H., Zhang, L., Yi, W., & Lei, B. (2017). Unmanned aerial vehicle based remote sensing method for monitoring a steep mountainous slope in the Three Gorges Reservoir, China. *Earth Science Informatics.* https://doi.org/10.1007/s12145-017-0291-9.

Hutchinson, G. E. (1957). *A treatise on limnology. Part 1: Geography and physics of lakes. Limnology* (Vol. 1). New York: Wiley.

Immerzeel, W. W., Kraaijenbrink, P. D. A., Shea, J. M., Shrestha, A. B., Pellicciotti, F., Bierkens, M. F. P., et al. (2014). High-resolution monitoring of Himalayan glacier dynamics using unmanned aerial vehicles. *Remote Sensing of Environment, 150,* 93–103. https://doi.org/10.1016/j.rse.2014.04.025.

Inland Fisheries Institute in Olsztyn. (1963). Bathymetric plan of Lake Suskie. Olsztyn.

Jancosek, M., & Pajdla, T. (2011). Multi-view reconstruction preserving weakly-supported surfaces. In: *CVPR 2011* (pp. 3121–3128). IEEE. https://doi.org/10.1109/CVPR.2011.5995693.

Javernick, L., Brasington, J., & Caruso, B. (2014). Modeling the topography of shallow braided rivers using structure-from-motion photogrammetry. *Geomorphology, 213,* 166–182. https://doi.org/10.1016/j.geomorph.2014.01.006.

Johnson, K., Nissen, E., Saripalli, S., Arrowsmith, J. R., McGarey, P., Scharer, K., et al. (2014). Rapid mapping of ultrafine fault zone topography with structure from motion. *Geosphere, 10*(5), 969–986. https://doi.org/10.1130/GES01017.1.

Kageyama, Y., Takahashi, J., Nishida, M., Kobori, B., & Nagamoto, D. (2016). Analysis of water quality in Miharu dam reservoir, Japan, using UAV data. *IEEJ Transactions on Electrical and Electronic Engineering, 11*(S1), S183–S185. https://doi.org/10.1002/tee.22253.

Koparan, C., & Koc, A. B. (2016). Unmanned Aerial Vehicle (UAV) assisted water sampling. In: 2016 ASABE international meeting (p. 1). American Society of Agricultural and Biological Engineers. https://doi.org/10.13031/aim.20162461157.

Liu, P., Chen, A. Y., Nan Huang, Y., Yu Han, J., Sung Lai, J., Chung Kang, S., et al. (2014). A review of rotorcraft unmanned aerial vehicle (UAV) developments and applications in civil engineering. *Smart Structures and Systems, 13*(6), 1065–1094. https://doi.org/10.12989/sss.2014.13.6.1065.

Łopata, M., Popielarczyk, D., Templin, T., Dunalska, J., Wiśniewski, G., Bigaj, I., et al. (2014). Spatial variability of nutrients (N, P) in a deep, temperate lake with a low trophic level supported by global navigation satellite systems, geographic information system and geostatistics. *Water Science and Technology, 69*(9), 1834–1845. https://doi.org/10.2166/wst.2014.084.

Lossow, K., Gawrońska, H., Łopata, M., & Jaworska, B. (2004). Selection criteria for restoration method on Lake Suskie. *Limnological Review, 4,* 143–152.

Lucieer, A., de Jong, S. M., & Turner, D. (2014a). Mapping landslide displacements using Structure from Motion (SfM) and image correlation of multi-temporal UAV photography. *Progress in Physical Geography, 38*(1), 97–116. https://doi.org/10.1177/0309133313515293.

Lucieer, A., Turner, D., King, D. H., & Robinson, S. A. (2014b). Using an unmanned aerial vehicle (UAV) to capture micro-topography of Antarctic moss beds. *International Journal of Applied Earth Observation and Geoinformation, 27,* 53–62. https://doi.org/10.1016/j.jag.2013.05.011.

Mancini, F., Dubbini, M., Gattelli, M., Stecchi, F., Fabbri, S., & Gabbianelli, G. (2013). Using unmanned aerial vehicles (UAV) for high-resolution reconstruction of topography: The structure from motion approach on coastal environments. *Remote Sensing, 5*(12), 6880–6898. https://doi.org/10.3390/rs5126880.

Marshall, D. M., Barnhart, R. K., Shappee, E., & Most, M. T. (2016). Introduction to unmanned aircraft systems, 2nd edn. Boca Raton, FL, USA: CRC Press.

McEvoy, J. F., Hall, G. P., & McDonald, P. G. (2016). Evaluation of unmanned aerial vehicle shape, flight path and camera type for waterfowl surveys: Disturbance effects and species recognition. *PeerJ, 4,* e1831. https://doi.org/10.7717/peerj.1831.

Meier, L., Tanskanen, P., Heng, L., Lee, G. H., Fraundorfer, F., & Pollefeys, M. (2012). PIXHAWK: A micro aerial vehicle design for autonomous flight using onboard computer vision. *Autonomous Robots, 33*(1–2), 21–39. https://doi.org/10.1007/s10514-012-9281-4.

Mori, Y., Takahashi, A., Mehlum, F., & Watanuki, Y. (2002). An application of optimal diving models to diving behaviour of Br{ü}nnich's guillemots. *Animal Behaviour, 64*(5), 739–745.

Nex, F., & Remondino, F. (2014). UAV for 3D mapping applications: A review. *Applied Geomatics, 6*(1), 1–15. https://doi.org/10.1007/s12518-013-0120-x.

Niedzielski, T., Witek, M., & Spallek, W. (2016). Observing river stages using unmanned aerial vehicles. *Hydrology and Earth System Sciences, 20*(8), 3193–3205. https://doi.org/10.5194/hess-20-3193-2016.

Niethammer, U., James, M. R., Rothmund, S., Travelletti, J., & Joswig, M. (2012). UAV-based remote sensing of the Super-Sauze landslide: Evaluation and results. *Engineering Geology, 128,* 2–11. https://doi.org/10.1016/j.enggeo.2011.03.012.

Papakonstantinou, A., Topouzelis, K., & Pavlogeorgatos, G. (2016). Coastline zones identification and 3D coastal mapping using UAV spatial data. *ISPRS International Journal of Geo-Information, 5*(6), 75. https://doi.org/10.3390/ijgi5060075.

Pawłowski, B. (2014). *Problems of reclamation of lakes with special emphasis on the Lake Suskie.* Toruń-Susz: Nicolaus Copernicus University in Toruń. **(in Polish)**.

Popielarczyk, D., & Templin, T. (2014). Application of integrated GNSS/hydroacoustic measurements and GIS geodatabase models for bottom analysis of Lake Hancza: The deepest inland reservoir in Poland. *Pure and Applied Geophysics, 171*(6), 997–1011. https://doi.org/10.1007/s00024-013-0683-9.

Popielarczyk, D., Templin, T., & Łopata, M. (2015). Using the geodetic and hydroacoustic measurements to investigate the bathymetric and morphometric parameters of Lake Hancza (Poland). *Open Geosciences.* https://doi.org/10.1515/geo-2015-0067.

Ryan, J. C., Hubbard, A. L., Box, J. E., Todd, J., Christoffersen, P., Carr, J. R., et al. (2015). UAV photogrammetry and structure from motion to assess calving dynamics at Store Glacier, a large outlet draining the Greenland ice sheet. *The Cryosphere, 9*(1), 1–11. https://doi.org/10.5194/tc-9-1-2015.

Shervais, K. (2015). Structure from Motion Introductory Guide. https://www.unavco.org/education/resources/modules-and-activities/field-geodesy/module-materials/sfm-intro-guide.pdf. Accessed 18 Apr 2017.

Shintani, C., & Fonstad, M. A. (2017). Comparing remote-sensing techniques collecting bathymetric data from a gravel-bed river. *International Journal of Remote Sensing.* https://doi.org/10.1080/01431161.2017.1280636.

Smith, M. W., Carrivick, J. L., & Quincey, D. J. (2016). Structure from motion photogrammetry in physical geography. *Progress in Physical Geography, 40*(2), 247–275. https://doi.org/10.1177/0309133315615805.

Snavely, N., Seitz, S. M., & Szeliski, R. (2008). Modeling the world from internet photo collections. *International Journal of Computer Vision, 80*(2), 189–210. https://doi.org/10.1007/s11263-007-0107-3.

Sørensen, L., Jacobsen, L., & Hansen, J. (2017). Low cost and flexible UAV deployment of sensors. *Sensors, 17*(1), 154. https://doi.org/10.3390/s17010154.

Szostak, M., Wezyk, P., & Tompalski, P. (2014). Aerial orthophoto and airborne laser scanning as monitoring tools for land cover dynamics: A case study from the Milicz forest district (Poland). *Pure and Applied Geophysics, 171*(6), 857–866. https://doi.org/10.1007/s00024-013-0668-8.

Szulwic, J., Burdziakowski, P., Janowski, A., Przyborski, M., Tysiąc, P., Wojtowicz, A., et al. (2015). Maritime laser scanning as the source for spatial data. *Polish Maritime Research, 22*(4), 9–14. https://doi.org/10.1515/pomr-2015-0064.

Tamminga, A. D., Eaton, B. C., & Hugenholtz, C. H. (2015). UAS-based remote sensing of fluvial change following an extreme flood event. *Earth Surface Processes and Landforms, 40*(11), 1464–1476. https://doi.org/10.1002/esp.3728.

Tang, L., & Shao, G. (2015). Drone remote sensing for forestry research and practices. *Journal of Forestry Research, 26*(4), 791–797. https://doi.org/10.1007/s11676-015-0088-y.

Tauro, F., Porfiri, M., & Grimaldi, S. (2016). Surface flow measurements from drones. *Journal of Hydrology, 540,* 240–245. https://doi.org/10.1016/j.jhydrol.2016.06.012.

Toth, C., & Jóźków, G. (2016). Remote sensing platforms and sensors: A survey. *ISPRS Journal of Photogrammetry and Remote Sensing, 115,* 22–36. https://doi.org/10.1016/j.isprsjprs.2015.10.004.

Venturi, S., Di Francesco, S., Materazzi, F., & Manciola, P. (2016). Unmanned aerial vehicles and Geographical Information System integrated analysis of vegetation in Trasimeno Lake, Italy. *Lakes and Reservoirs Research and Management, 21*(1), 5–19. https://doi.org/10.1111/lre.12117.

Wallace, L., Lucieer, A., Watson, C., & Turner, D. (2012). Development of a UAV-LiDAR system with application to forest inventory. *Remote Sensing, 4*(12), 1519–1543. https://doi.org/10.3390/rs4061519.

Westoby, M. J., Brasington, J., Glasser, N. F., Hambrey, M. J., & Reynolds, J. M. (2012). "Structure-from-Motion" photogrammetry: A low-cost, effective tool for geoscience applications. *Geomorphology, 179,* 300–314. https://doi.org/10.1016/j.geomorph.2012.08.021.

Wetzel, R. G. (2001). *Limnology: Lake and river ecosystems*. San Diego: Gulf Professional Publishing.

Woodget, A. S., Carbonneau, P. E., Visser, F., & Maddock, I. P. (2015). Quantifying submerged fluvial topography using hyperspatial resolution UAS imagery and structure from motion photogrammetry. *Earth Surface Processes and Landforms, 40*(1), 47–64. https://doi.org/10.1002/esp.3613.

Wu, C. (2013). Towards linear-time incremental structure from motion. In: *2013 international conference on 3D vision* (pp. 127–134). IEEE. https://doi.org/10.1109/3DV.2013.25.

Yucel, M. A., & Turan, R. Y. (2016). Areal change detection and 3D modeling of mine lakes using high-resolution unmanned aerial vehicle images. *Arabian Journal for Science and Engineering, 41*(12), 4867–4878. https://doi.org/10.1007/s13369-016-2182-7.

Zang, W., Lin, J., Wang, Y., & Tao, H. (2012). Investigating small-scale water pollution with UAV remote sensing technology. In: *World Automation Congress (WAC)*, 2012 (pp. 1–4).

Zolich, A., Johansen, T. A., Cisek, K., & Klausen, K. (2015). Unmanned aerial system architecture for maritime missions. Design and hardware description. In: *2015 Workshop on Research, Education and Development of Unmanned Aerial Systems (RED-UAS)* (pp. 342–350). IEEE. https://doi.org/10.1109/RED-UAS.2015.7441026.

(Received  April 24, 2017, revised  October 16, 2017, accepted  October 20, 2017, Published online  October 29, 2017)

Pure Appl. Geophys. 175 (2018), 3285–3302
© 2018 The Author(s)
https://doi.org/10.1007/s00024-018-1843-8

❙ **Pure and Applied Geophysics**

# Automated Snow Extent Mapping Based on Orthophoto Images from Unmanned Aerial Vehicles

Tomasz Niedzielski,[1] ⓘ Waldemar Spallek,[1] and Matylda Witek-Kasprzak[1]

*Abstract*—The paper presents the application of the *k*-means clustering in the process of automated snow extent mapping using orthophoto images generated using the Structure-from-Motion (SfM) algorithm from oblique aerial photographs taken by unmanned aerial vehicle (UAV). A simple classification approach has been implemented to discriminate between snow-free and snow-covered terrain. The procedure uses the *k*-means clustering and classifies orthophoto images based on the three-dimensional space of red–green–blue (RGB) or near-infrared–red–green (NIRRG) or near-infrared–green–blue (NIRGB) bands. To test the method, several field experiments have been carried out, both in situations when snow cover was continuous and when it was patchy. The experiments have been conducted using three fixed-wing UAVs (swinglet CAM by senseFly, eBee by senseFly, and Birdie by FlyTech UAV) on 10/04/2015, 23/03/2016, and 16/03/2017 within three test sites in the Izerskie Mountains in southwestern Poland. The resulting snow extent maps, produced automatically using the classification method, have been validated against real snow extents delineated through a visual analysis and interpretation offered by human analysts. For the simplest classification setup, which assumes two classes in the *k*-means clustering, the extent of snow patches was estimated accurately, with areal underestimation of 4.6% (RGB) and overestimation of 5.5% (NIRGB). For continuous snow cover with sparse discontinuities at places where trees or bushes protruded from snow, the agreement between automatically produced snow extent maps and observations was better, i.e. 1.5% (underestimation with RGB) and 0.7–0.9% (overestimation, either with RGB or with NIRRG). Shadows on snow were found to be mainly responsible for the misclassification.

**Key words:** Unmanned aerial vehicle, snow cover, snow extent mapping, *k*-means clustering, structure-from-motion.

## 1. Introduction

A key environmental variable that controls snowmelt peak discharges is snow cover (Tekeli et al. 2005), especially in the mountains (Hock et al. 2006). Snow water equivalent (SWE) for a given basin allows us to estimate the volume of water that may be mobilized during snowmelt episodes. Hence, the SWE estimation procedures have a practical potential as they may be used to forecast snowmelt peak flows. Although SWE can be estimated using several remote sensing techniques (Kunzi et al. 1982; Chang et al. 1987; Pulliainen and Hallikainen 2001; Tedesco et al. 2004; Pulliainen 2006; Takala et al. 2011), its direct calculation involves multiplication of snow depth (HS) measurements by snow density ($\rho$) estimates (Jonas et al. 2009).

Although it is rather difficult to get spatially continuous estimates of snow density, the HS raster maps can be produced using: interpolation from pointwise data (Erxleben et al. 2002; Dyer and Mote 2006; Pulliainen 2006), terrestrial light detection and ranging (LiDAR) and tachymetric measurements (Prokop 2008; Prokop et al. 2008; Grünewald et al. 2010; Prokop et al. 2015; Schön et al. 2015), airborne LiDAR (Deems et al. 2013) and satellite sensors (Hall et al. 2002; Romanov and Tarpley 2007). The in situ measurements in small basins have recently become substituted by high-resolution snow mapping offered by unmanned aerial vehicles (UAVs). The HS estimation using UAVs is based on applying the DoD procedure, abbreviated after DEM (digital elevation model) of differences, which allows for the subtraction of a snow-free digital surface model (DSM) from a DSM with snow cover (Vander Jagt et al. 2015; de Michele et al. 2016; Bühler et al. 2016; Harder et al.

Handling Editor: Dr. Ismail Gultepe.

[1] Department of Geoinformatics and Cartography, Faculty of Earth Science and Environmental Management, University of Wrocław, pl. Uniwersytecki 1, 50-137 Wrocław, Poland. E-mail: tomasz.niedzielski@uwr.edu.pl

2016; Bühler et al. 2017; Miziński and Niedzielski 2017). The latter dataset is produced using the Structure-from-Motion (SfM) algorithm run on oblique aerial images acquired by a UAV.

Not only HS, but also snow extent (SE) characterizes spatial distribution of snow. The large-scale estimation of SE is common and is carried out using satellite observations (e.g. Dozier 1989; Rosenthal and Dozier 1996; Robinson and Frei 2000; Molotch et al. 2004). However, the spatial resolution of satellite data constrains the identification of snow patches and is not suitable for evaluating discontinuous snow cover in small basins. The knowledge about SE in small basins is important when using UAV-based HS information from the vicinities of edges of snow-covered terrain. Namely, it may be useful in deciding if small HS values actually correspond to snow-covered terrain or if they are artefacts.

The problem of coarse spatial resolution of satellite-based SE reconstructions can be solved using oblique terrestrial or aerial high-resolution imagery. The UAV-acquired (Zhang et al. 2012a) and vessel-based (Zhang et al. 2012b) sea ice images were utilized to automatically make "ice" or "no-ice" mosaics, which is a similar task to the search for "snow" or "no-snow" grid cells. In contrast, Julitta et al. (2014) proposed a method for detecting snow-covered terrain on a basis of processing photographs taken by the Earth-fixed camera from the PhenoCam network (phenocam.sr.unh.edu, access date: 19/02/2018). The common idea of the latter three papers was the use of the $k$-means unsupervised classification to produce a dichotomous SE numerical map. This paper provides further evidences for the applicability of the $k$-means method in the automated SE reconstruction. Namely, our study complements the three papers by combining several methodical approaches therein contained. Firstly, we use the $k$-means clustering with more than two classes following the concept of Zhang et al. (2012a, b). We do it to test the potential of the method in detecting shadowed snow cover, thus our objective conceptually differs from the identification of ice type carried out by Zhang et al. (2012a, b). Secondly, we utilize UAV-acquired aerial images as inputs to the $k$-means-based snow detection algorithm; thus, we entirely modify the camera location proposed by Julitta et al. (2014) (a terrestrial one-view camera position was replaced by airborne moving camera which takes overlapping photos to generate the SfM-based orthophotomap which is subsequently processed) and adopt the camera location from Zhang et al. (2012a). Table 1 shows the differences between the approaches employed in this paper and those utilized in the said three articles. It is apparent from Table 1 that, apart form the above-mentioned differences, none of the three papers discussed in this paragraph uses near-infrared images, which were found useful in the UAV-based HS reconstructions (Bühler et al. 2017; Miziński and Niedzielski 2017).

The objective of this paper is therefore to check the usefulness of the $k$-means clustering, with two to four classes, for the unsupervised classification of the UAV-acquired visible-light and near-infrared images as well as for their incorporation into the production of numerical SE maps.

Table 1

*Comparison of the observed targets (sea ice/snow cover), camera locations (earth-fixed/shipborne/airborne), spectrum of images (visible light/near infrared) and number of clusters (two/more than two) between studies which make use of the k-means-based mapping in cold environments*

|  | Zhang et al. (2012a) | Zhang et al. (2012b) | Julitta et al. (2014) | This paper |
|---|---|---|---|---|
| Sea ice extent | + | + | − | − |
| Snow cover extent | − | − | + | + |
| Earth-fixed camera | − | − | + | − |
| Shipborne camera | − | + | − | − |
| Airborne camera | + | − | − | + |
| Visible-light images | + | + | + | + |
| Near-infrared images | − | − | − | + |
| Two-cluster analysis | + | + | + | + |
| More than two-cluster analysis | + | + | − | + |

## 2. Data

Numerous UAV flights targeted at a few study areas in the Kwisa River catchment in the Izerskie Mountains (part of the Sudetes, SW Poland) were performed. Two of them were: Rozdroże Izerskie (extensive mountain pass located at 767 m a.s.l., with nearby mountain meadow of size 100 × 110 m) and Polana Izerska (mountain meadow of size 250 × 170 m, with elevations ranging from 951 to 976 m a.s.l.). Aerial images of snow-covered terrain were acquired to cover three specific conditions: patchy snow cover (Rozdroże Izerskie on 10/04/2015), continuous snow cover (Polana Izerska W on 23/03/2016) and continuous snow cover with signatures of thawing in the vicinity of vegetation (Polana Izerska E on 16/03/2017). The study areas of Rozdroże Izerskie and Polana Izerska along with the selected three test sites are presented in Fig. 1.

Observations in Rozdroże Izerskie were carried out using the fixed-wing UAV named swinglet CAM (produced by senseFly, weight 0.5 kg, wingspan 80 cm), while fieldwork in Polana Izerska was conducted using the other fixed-wing UAVs, namely eBee (manufactured by senseFly, weight 0.7 kg, wingspan 96 cm) and Birdie (manufactured by Fly-Tech UAV, weight 1.0 kg, wingspan 98 cm). The swinglet CAM drone was equipped with a single camera bay, to which either Canon IXUS 220HS (red–green–blue = RGB) or Canon PowerShot ELPH 300HS (near-infrared–green–blue = NIRGB) cameras were mounted. Similarly, the one-bay eBee drone was equipped with removable cameras: Canon S110 RGB (RGB) or Canon S110 NIR (near-infrared–red–green = NIRRG). In Birdie's bay, Parrot Sequoia sensor (high-resolution RGB camera with low-resolution four individual bands: NIR, red-edge = RE, red = R, green = G) was installed. Wavelengths for which spectral responses reveal maximum values for a few cameras are juxtaposed in Table 2. Five UAV missions were completed: 1 × Rozdroże Izerskie RGB (swinglet CAM), 1 × Rozdroże Izerskie NIRGB (swinglet CAM), 1 × Polana Izerska W RGB (eBee), 1 × Polana Izerska W NIRRG (eBee), 1 × Polana Izerska E RGB (Birdie). From the available Parrot Sequoia bands, we used only the high-resolution three-band RGB camera to ensure the similar resolution of all sensors used. The altitudes above takeoff (ATO) of the flights were kept similar, namely 123–151 m, at which height the resolution of the ground surface in each image was approximately 4.1–4.5 cm/px. The UAVs and the data acquisition equipment used during the fieldwork are shown in Fig. 2. Table 3 juxtaposes basic UAV flight parameters and the number of images acquired in each flight.

## 3. Methods

The SfM algorithm, implemented in Agisoft Photoscan Professional version 1.2.5.2680, was used to produce orthophotomaps. Georeferencing was based on measurements carried out by standard onboard GPS receivers, and the geotagged images were processed in Agisoft Photoscan. We delineated three 100 × 100 m orthophoto image squares (Fig. 1). As a result, five fragments of orthophoto images were extracted (3× RGB, 1× NIRGB and 1× NIRRG). They became inputs to the analysis which aimed at the automated production of SE maps on the basis of the above-mentioned $k$-means clustering. In addition, they were used to produce reference SE maps, which were prepared by GIS experts who visually inspected the orthophotomaps and digitized terrain covered with snow. For a specific site and specific camera spectrum, the two SE spatial data, i.e. automatically and manually produced SE maps, were subsequently compared to validate the performance of the unsupervised classification.

In this section, we use a very simple approach to automatically estimate SE on the basis of orthophoto images produced from photographs taken by UAVs. The full automation is attained through the use of the unsupervised classification. Following the concept of Zhang et al. (2012a, b) and Julitta et al. (2014), the $k$-means clustering is utilized to discriminate between snow-covered and snow-free terrain.

Figure 3 presents the flowchart of the $k$-means-based production of SE numerical maps on the basis of UAV-based orthophoto images. The input raster is thus a fragment of the orthophotomap. It can be either RGB or NIRRG or NIRGB spatial data. Such an input raster is split into three 2D arrays (with spatial

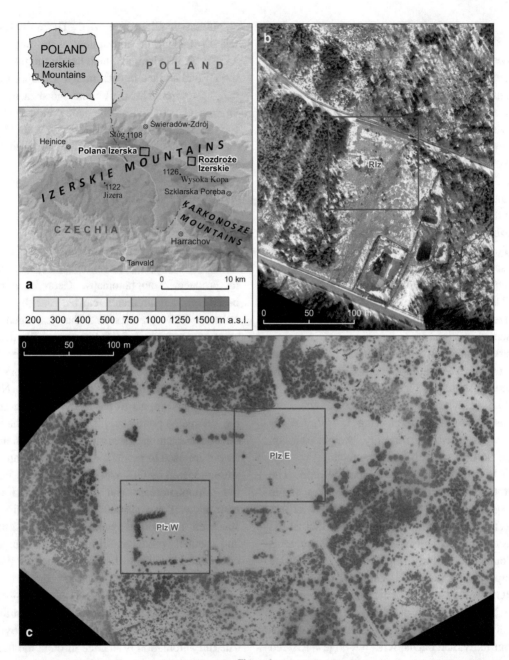

Figure 1
Locations of three test sites

relations kept), each corresponding to one of three bands. For instance, if RGB data are processed, the first 2D array includes $R$ values, the 2D second array stores $G$ values, while the third 2D array consists of $B$ values. Then, the three arrays are merged so that a single nonspatial array is composed of three rows

which are used to store band values produced through flattening of specific 2D arrays. For example, $R$ values in the first row are flattened from the 2D array for the $R$ band, $G$ values in the second row are flattened from the 2D array for the $G$ band, and $B$ values in the third row are flattened from the 2D array for the

Table 2

*Cameras used along with wavelengths for which spectral responses are maximum*

| Camera code | Camera | Wavelength with maximum spectral response (nm) | | | |
|---|---|---|---|---|---|
| | | NIR | *R* | *G* | *B* |
| RGB1 | Canon IXUS 220HS | – | Not specified | | |
| NIRGB | Canon PowerShot ELPH 300HS* | 700 | – | 520 | 480 |
| RGB2 | Canon S110 RGB | – | 660 | 520 | 450 |
| NIRRG | Canon S110 NIR | 850 | 625 | 550 | – |
| RGB3 | Parrot Sequoia RGB | – | Not specified | | |

*Modified

Figure 2
UAVs used to acquire aerial images, i.e. swinglet CAM (**a**), eBee (**b**), Birdie (**c**), along with the data acquisition equipment during fieldwork (**d**). The equipment visible in **d** comprises: eBee box (gray chassis), ground base station (grey rugged notebook), geodetic GPS receiver (red box), Parrot Sequoia box (small black chassis), Birdie box (big black chassis), UAV equipment box (medium black chassis), eBee drone (on the top of equipment box), UAV flight register (red folder), and 13-m telescope pole (black fishing rod)

*B* band. The nonspatial array becomes the input to the *k*-means clustering. According to Zhang et al. (2012a, b), we allow more than two classes, which is attained through the use of a loop (Fig. 3).

We use the *k*-means implementation available in the OpenCV (Open Source Computer Vision) library (opencv.org), in particular OpenCV-Python.

The cv2.kmeans function is utilized with the following setup of input parameters: data (RGB or NIRRG or NIRGB orthophoto images of size $n \times m$ flattened to three R/G/B or NIR/R/G or NIR/G/B vectors, each of length $nm$, forming the nonspatial array), number of clusters (integers ranging from 2 to 4), iteration termination criteria (maximum

Table 3

*Parameters of UAV flights*

| Site | Date | UAV model | Camera | Side overlap (%) | Desired resolution (cm/px) | Altitude ATO (m) | Begin of flight (UTC) | Temperature* (°C) | Number of images | Flight duration |
|------|------|-----------|--------|------------------|----------------------------|------------------|-----------------------|-------------------|------------------|-----------------|
| RIz | 10/04/2015 | swinglet CAM | RGB1 | 60 | 4.5 | 123 | 9:24 | + 11.6 | 71 | 19′05″ |
| RIz | 10/04/2015 | swinglet CAM | NIRGB | 60 | 4.5 | 123 | 9:53 | + 11.0 | 66 | 16′48″ |
| PIz W | 23/03/2016 | eBee | RGB2 | 60 | 4.5 | 123 | 13:24 | 0 | 88 | 11′50″ |
| PIz W | 23/03/2016 | eBee | NIRRG | 60 | 4.5 | 123 | 13:47 | − 0.2 | 83 | 8′53″ |
| PIz E | 16/03/2017 | Birdie | RGB3 | 60 | 4.1 | 151 | 11:57 | + 4.1 | 178** | ≈ 20′ |

RIz—Rozdroże Izerskie, PIz W—Polana Izerska W, PIz E—Polana Izerska E

RGB1—Canon IXUS 220HS, NIRGB—Canon PowerShot ELPH 300HS (modified)

RGB2—Canon S110 RGB, NIRRG—Canon S110 NIR

RGB3—RGB camera from Parrot Sequoia multispectral sensor

*Temperature measured in the nearest weather stations at 2 m (Kamienica for the site of Rozdroże Izerskie, Polana Izerska for the sites of Polana Izerska W and Polana Izerska E)

**178 Photographs were taken over a large terrain, from which 39 images were selected to produce the orthophotomap of the site of Polana Izerska E

number of iterations of 10, iterations break when threshold $\varepsilon$ attains 1), number of attempts with different initial labels (integer equals to 10), and method of choosing initial centres of classes (random centres are assumed). Although RGB images are the most common in our analysis, we also test the applicability of the $k$-means method in automated SE mapping using two different NIR cameras (Table 2). The choice of the number of clusters (2, 3, 4) is explained by the need to carry out two specific exercises: detection of "snow" and "no-snow" raster cells (2 classes), identification of "snow" and "no-snow" raster cells with possible "artefact" detection (3 or 4 classes). The artefacts may be of different origins, for instance may be caused by SfM-failures or shadows. The convergence criteria are selected to ensure a reasonable computational time to finish jobs. The random determination of centres enables us to avoid preferring particular band configurations to reduce potential bias. The output comprises: compactness statistics (sum of squared distances measured from each point to the corresponding centre), labels (vector of length $nm$ with labels to assign individual data to specific clusters coded as [0,1] for 2 classes, [0,1,2] for 3 classes etc.), centres (array of cluster centres, i.e. each centre is represented by three numbers R, G, B or NIR, R, G or NIR, G, B).

Subsequently, the classified arrays containing codes inherited from labels are reshaped into 2D matrices, the number of which is equal to $k_{max} - k_{min} + 1$ (e.g. if we allow $k = 2, 3, 4$, we produce $4 - 2 + 1 = 3$ reshaped 2D arrays). Then, spatial reference, which needs to be extracted from the input raster at the beginning of the entire procedure, is added to the reshaped arrays. The most straightforward case (2 classes) aims to classify the terrain into snow-free and snow-covered terrain, the latter being the estimate of SE. If orthophoto images include interfering elements, such as for instance SfM artefacts, more than two classes may be assumed to detect such features. The final result is the raster map, each cell of which represents one of $k$ possible values.

The motivation for the proposed method is that SE and HS may be jointly used to refine SWE estimates (Elder et al. 1998). The values of HS quantify snow cover thickness in three dimensions, and therefore they directly contribute to estimations of SWE. However, SE coveys a simple zero/one (no-snow/snow) message solely in planar view, and its role in SWE assessment is indirect and may be sought in refining HS estimates: (1) at edges of snow patches, (2) along lines where continuous snow cover becomes discontinuous and subsequently transits to snow-free terrain or (3) in the vicinity of land cover objects protruding from continuous snow cover.

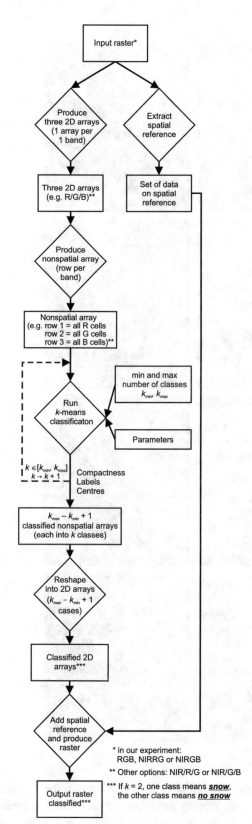

Namely, based on HS maps, snow cover exists if $HS > 0$. However, for thin snow cover (e.g. in one of the above-mentioned enumerated cases) HS estimation errors may be considerable, with potential insignificance of HS determinations. Having produced accurate SE numerical maps, it is thus possible to superimpose two extents ($HS > 0$ and $SE \neq 0$) and find a real number $h_0$ such that $HS > h_0$, leading to the reduction of HS reconstruction uncertainty. As a consequence, SWE estimates may be refined. It is worth noting here that the similar comparison of SE and HS data has recently been carried out by Wang et al. (2017) who assumed $HS \geq 4$ cm for the purpose of validating SE maps.

## 4. Results

To choose the number of classes ($k$) for further analysis, a simple exercise was carried out. The differences between the classifications into two, three and four clusters were analyzed against a background of the RGB orthophoto image (Fig. 4). The analysis concerned three test sites. In the first test site (Rozdroże Izerskie on 10/04/2015), snow cover was discontinuous and shadows were cast by trees onto terrain. In the second test site (Polana Izerska W on 23/03/2016), snow cover was continuous and shadows were not present. In the third site (Polana Izerska E on 16/03/2017), snow cover continuously covered the terrain (apart from close vicinities of trees) and evident shadows were cast onto terrain. It is apparent from Fig. 4 that the most natural results were obtained with $k = 2$ (snow/no-snow classes). Misfit was noticed for the two sites in places where shadows were clearly visible in images. Increasing the number of clusters to three produced an additional class which was difficult to interpret. Namely, in Rozdroże Izerskie, the intermediate class either corresponded to shadows cast onto the terrain or captured the ground not covered with snow (light green vegetation, mainly grass), while in Polana Izerska E the

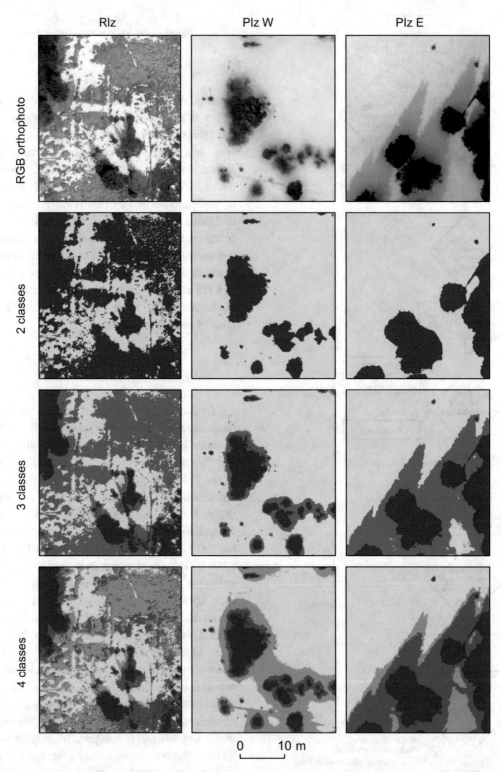

Rlz          Plz W          Plz E

RGB orthophoto

2 classes

3 classes

4 classes

0      10 m

intermediate class was found to work well in delineating SE and detecting shadows. The intermediate
class was very small for Polana Izerska W, captured

sparse tree branches and, importantly, the algorithm
did not identify shadows, which were not present
(Fig. 4). In the four-cluster case, the interpretation

◄Figure 4
RGB orthophotos zoomed in to trees selected from three test sites (codes explained in Fig. 1) with visible shadows (RIz, PIzE) and without shadows (PIzW), and classifications of these images into two, three and four classes

remained complex. While the findings for Rozdroże Izerskie and Polana Izerska E were similar to the three-cluster analysis (two intermediate classes, when combined, had the similar spatial extent to one intermediate class), the results for Polana Izerska W was different (one of two intermediate classes very significantly overrepresented tree branches, which in the 3-cluster case corresponded to spatially small class). Although the three- and four-cluster results reveal, in selected cases, some potential in identifying shadow-driven disturbances in SE maps, it is difficult to numerically identify the intermediate classes and their meaning. In particular, our exercises confirm the potential of the three-cluster classification to detect shadows on continuous snow cover, but fail to identify shadows when snow cover is discontinuous. Therefore, in the subsequent analyses the simple dichotomous setup is explored.

Figure 5a, b presents the input RGB and NIRGB orthophoto images of the partially snowed (snow patches covered approximately 24–25% of terrain—see Table 4) site of Rozdroże Izerskie along with the corresponding analyses of these orthophotomaps (rows 2–4 of the figure). The SE maps based on manual digitization and automated classification are presented in Fig. 5c–f, with the validation of the automated approach expressed as the difference between the two maps (G–H). The RGB- and NIRGB-based results are similar, with very comparable patterns of SE obtained from manual and automated approaches in the uncovered meadow. Considerable differences were identified along western and eastern edges of the study site of Rozdroże Izerskie, where the forests meets the meadow. Figure 6 presents this effect in a larger cartographic scale. It is apparent from the figure that shadows are responsible for the most evident mismatches; however, the shadow impact is different for RGB and NIRGB analyses. All in all, the k-means procedure applied to the RGB orthophoto image led to the underestimation of area of snow patches by 4.6%

with respect to the human-digitized layer (Table 4). In contrast, if the NIRGB orthophotomap was taken as input, the true area of snow patches was overestimated by the k-means clustering by 5.5% with respect to the digitized data (Table 4). It is worth noting that the sky was clear or only slightly overcast at the time of observations (in the nearby World Meteorological Organization, WMO station in Liberec, located approximately 32 km from Rozdroże Izerskie, 0 okta was observed at 9:00 UTC and 2 oktas were recorded at 10:00 UTC). Therefore, the effect of shadows was highly pronounced in the case study from the site of Rozdroże Izerskie on 10/04/2015.

Figure 7 shows the similar analysis for the site of Polana Izerska W on 23/03/2016, when snow cover was continuous (snow occupied approximately 91–92% of terrain—see Table 4). Both RGB- and NIRRG-based SE maps produced using the k-means clustering were found to agree well with their human-digitized analogues. Discrepancies occurred mainly at the contact between the trees and meadow. The overestimations cover only 0.7–0.9% of the human-digitized layer (Table 4). In the case study from the site of Polana Izerska on 23/03/2016, shadows were not present and thus they did not impact the accuracy of the automated mapping procedure. The sky was highly overcast at the time of observations. Indeed, in the nearby WMO station in Liberec, approximately 25 km from Polana Izerska, the following total cloud cover characteristics were recorded: 6 oktas at 13:00 UTC and 8 oktas at 14:00 UTC.

Figure 8 presents the results obtained for the other 100 × 100 m site in Polana Izerska E on 16/03/2017, with continuous snow cover which began to melt around vegetation (snow occupied approximately 95% of terrain—see Table 4). The SE map was generated by the k-means method on the basis of the RGB orthophotomap and was shown to be in agreement with its digitized version produced by GIS experts. The differences were as small as 1.5% of SE inferred by the experts. Although the shadows cast by trees were highly visible on the orthophoto image, their impact on the performance of the automated SE mapping accuracy was not very significant. The sunny spells were only intermittent as the sky was overcast. The total cloud cover in the WMO station in

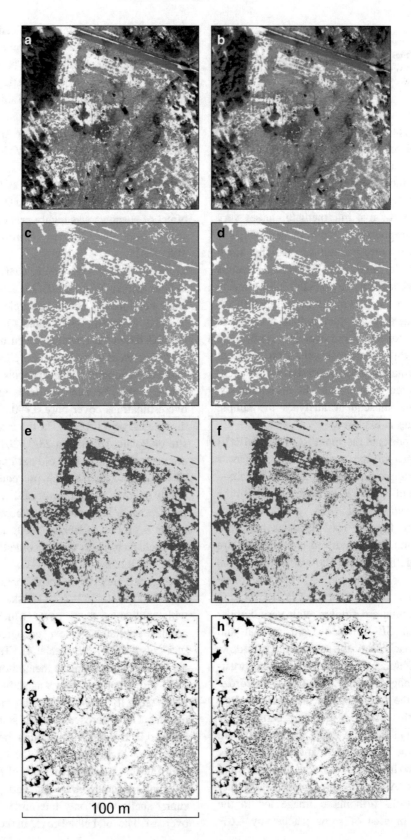

100 m

◀Figure 5
Orthophoto images based on RGB (left) or NIRGB (right)
photographs taken by swinglet CAM in the site of Rozdroże
Izerskie on 10/04/2015 (**a**, **b**), SE maps based on digitization of
RGB or NIRGB orthophotomap by GIS expert (**c**, **d**), automatically
produced SE maps based on running the *k*-means mapping
procedure on RGB or NIRGB orthophoto image (**e**, **f**), and
difference between manual and automated SE reconstructions
based on RGB or NIRGB data (**g**, **h**)

Liberec was 8 oktas at 12:00 UTC. The sunbeam
reflected from snow in the southeastern part of the
test site, but the scattered cloud cover cast shadow
onto the rest of the area, leading to a considerable
reduction of light delivery in the southwestern part of
the test site (Fig. 8a). The uneven lighting conditions
had no impact on snow detection (Fig. 8b).

Table 4

*Areas of snow-covered terrain computed on the basis of manual and automated procedures*

| Site | Total area (m²) | Bands to produce orthophoto | Area covered with snow | | Difference | |
|------|-------|------------|--------------------------------------------|----------------------------------------------|---------------|---------------------------|
| | | | Based on manual digitization (M) (m²) | Based on automatic classification (C) (m²) | M–C (m²) | M–C as percent of M (%) |
| RIz | 10,000 | RGB | 2482.9 | 2368.2 | 114.7 | 4.6 |
| RIz | 10,000 | NIRGB | 2392.1 | 2523.3 | – 131.2 | – 5.5 |
| PIz W | 10,000 | RGB | 9148.4 | 9229.1 | – 80.7 | – 0.9 |
| PIz W | 10,000 | NIRRG | 9159.3 | 9227.5 | – 68.2 | – 0.7 |
| PIz E | 10,000 | RGB | 9469.5 | 9358.8 | 140.3 | 1.5 |

RIz—Rozdroże Izerskie

PIz W—Polana Izerska W

PIz E—Polana Izerska E

RGB consists of bands: red, green, and blue

NIRGB consists of bands: near-infrared, green, and blue

NIRRG consists of bands: near-infrared, red, and green

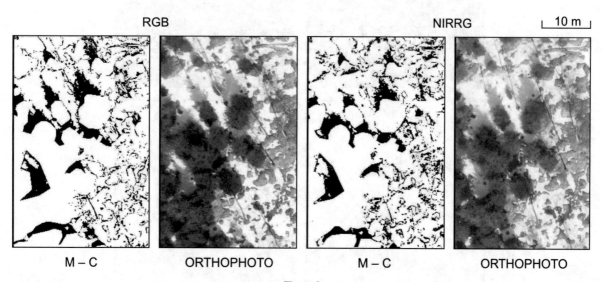

Figure 6
Differences (M–C) between manually digitized (M) and automatically classified (C) snow extent for RGB- and NIRGB-based analyses
(Rozdroże Izerskie)

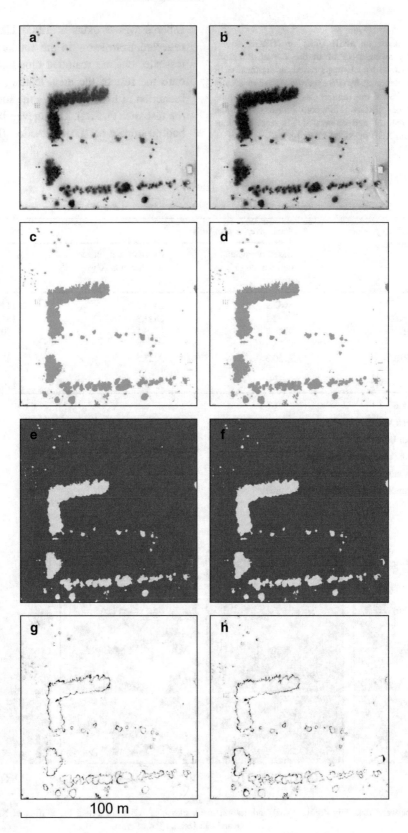

100 m

◄Figure 7
Orthophoto images based on RGB (left) or NIRRG (right) photographs taken by eBee in the site of Polana Izerska on 23/03/2016 (**a**, **b**), SE maps based on digitization of RGB or NIRRG orthophotomap by GIS expert (**c**, **d**), automatically produced SE map based on running the *k*-means mapping procedure on RGB or NIRRG orthophoto images (**e**, **f**), and difference between manual and automated SE reconstructions based on RGB or NIRRG data (**g**, **h**)

To check the performance of the snow mapping approach, we carried out a simple experiment, the aim of which was to evaluate the classification of features with similar spectral responses (lake ice vs. snow) and assess the skills of the investigated clustering method in detecting snow in images with spectrally different characteristics (very thin snow layer with protruding grass). Data for the ice/snow exercise were acquired by eBee UAV on 29/12/2016 in Polana Izerska, while aerial images for the snow/grass case were collected by eBee on 11/03/2016 in the district of Drożyna in Świeradów-Zdrój (approx. 2.5 km northeastward from Polana Izerska). Orthophotomaps of these areas were produced and squares of sizes $12 \times 12$ m were extracted so that mainly snow–ice and snow–grass boundaries occurred within each square, respectively. Figure 9 presents the two orthophotos with the corresponding SE maps produced automatically using the *k*-means method. In most places, the classifier correctly discriminated snow from ice (with one misclassification in the northern, i.e. top-centre, part of the image). In addition, water on ice and water flowing through dyke was also correctly classified as "no-snow". Interestingly, snow on ice was successfully detected. Figure 9 also shows good skills of the *k*-means-based SE mapping method in identifying a thin layer of snow overlaying the grass. The mosaic of "snow" and "no-snow" classes agrees well with the orthophoto.

## 5. Discussion and Conclusions

The tested method for automated SE mapping, based on processing high-resolution orthophoto images produced by the SfM algorithm from UAV-taken photographs, enables us to discriminate between snow-covered and snow-free terrain. The performance of the method is promising because, in the investigated case studies, the errors of estimating the area of snow-covered terrain are between $-6\%$ and $6\%$.

Indeed, the spatial extent of highly discontinuous snow cover was underestimated (overestimated) by 4.6% (5.5%) when RGB (NIRGB) orthophotomap served as input. Such an acceptable accuracy was improved when the method was used to estimate SE of continuous snow cover, with trees or bushes as the only protruding land cover elements. Namely, the areal underestimation (overestimation) of SE for continuous snow cover was of 1.5% (0.7–0.9%), with no clear picture whether the use of RGB or NIRRG camera influences the results.

The errors seem to be controlled either by the presence or nonexistence of shadows cast on the surface (driven by lighting conditions) or by the spatial pattern of snow (weather-driven variable that includes both patchy and continuous snow surfaces). The experiment with the use of the *k*-means method with more than two clusters, previously carried out by Zhang et al. (2012a) for detecting sea ice and ice type, enables us to identify shadows on continuous snow cover when three classes are allowed. Even though the finding has practical implications, further work is needed to judge which classes correspond to shadows.

The impact of shadows on the performance of SE mapping using the *k*-means method is uneven and depends on how intensive the shadows are with respect to the non-shadowed terrain. The SE maps based on the UAV observations carried out on 10/04/2015, when total cloud cover varied between 0 and 2 oktas, were found to be highly dependent on the shadows. However, the SE map produced from UAV-taken photographs on 16/03/2017, when total cloud cover was approximately equal to 8 oktas but simultaneously rare sunny spells occurred, was not impacted by shadows which were visible in the input orthophotomap. Therefore, it can be argued that the intensity of shadows matters, and only those which produce very high contrasts with respect to their surrounding deteriorate the skills of the automated mapping procedure. Figure 10 presents the responses of individual R/G/B bands for a fragment of the

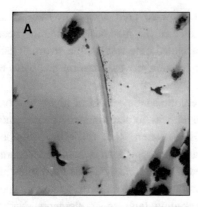

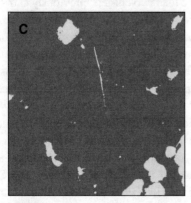

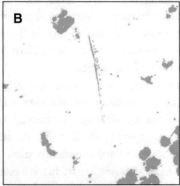

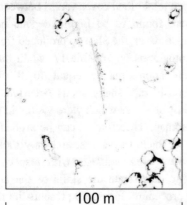

100 m

◀Figure 8
Orthophoto image based on RGB photographs taken by Birdie in the site of Polana Izerska on 16/03/2017 (**a**), SE map based on digitization of RGB orthophotomap by GIS expert (**b**), automatically produced SE map based on running the *k*-means mapping procedure on RGB orthophoto image (**c**), difference between manual and automated SE reconstructions based on RGB data (**d**)

orthophoto image for the test site of Rozdroże Izerskie. The signature of the shadows is particularly well seen for the R band, while the lowest response is noticed for the B band.

The pattern of snow, i.e. continuous or discontinuous snow surface, was also found to impact the performance of the proposed SE mapping method. The approach was the most skillful in the process of mapping SE on 23/03/2016 when snow cover was continuous and stable, with air temperatures around 0 °C (Table 3). A slightly worse performance was reported for mapping SE on 16/03/2017 when snow cover was continuous with significant signatures of thawing in the vicinity of vegetation, with air temperatures around 4 °C (Table 3). The worst results, but still within a reasonable ±6% error, were obtained for mapping SE during a snowmelt episode at the beginning of spring on 10/04/2015 when snow cover was highly patchy, with air temperatures around 11–12 °C (Table 3).

It is known that snow distribution is controlled by different processes (or dissimilar intensities of these processes) in open areas and in forests, and this impacts the performance of snow cover mapping methods (e.g. Vikhamar and Solberg 2002). The *k*-means classifier used in this paper was found to have uneven skills of snow detection, depending on the observed surface. If snow distribution observed from top view is discontinuous (either due to thin layer of snow or due to the presence of trees) or if snow signal is interfered by shadows, the errors of SE mapping are higher. Although our UAV-based snow cover mapping is incomparable with satellite products due to differences in spatial resolution and coverage, there exist certain similarities in the performance of UAV- and satellite-based SE estimates. For instance, the deterioration of snow detection skills was also noticed between forests and open areas when MODIS

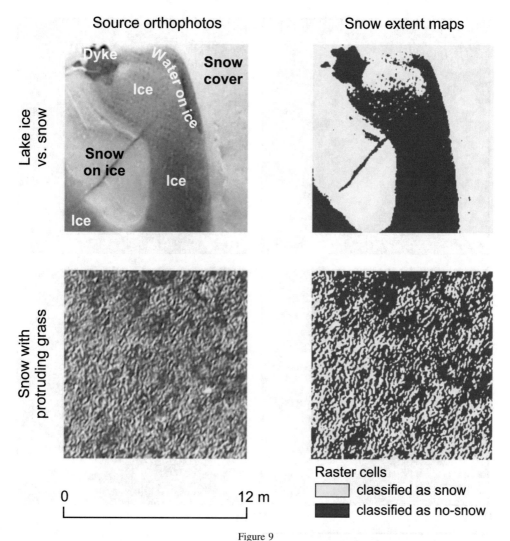

Figure 9
Results of the experiment aimed at verifying the automated SE mapping method for: objects with similar spectral responses (lake ice vs. snow) and meadow covered with thin snow cover (snow with protruding grass)

products are used (Wang et al. 2017). Limitations of optical sensors in forested areas were described by Rittger et al. (2013) who, in the context of assessing MODIS snow cover products, emphasized that some portion of snow under thick canopies cannot be observed, but for moderate canopies the penetration may be possible for some optical sensors.

Although SE layers may be determined from HS reconstructions (SE raster cells for which HS > 0), which may now be produced using the SfM processing of UAV-taken photographs (e.g. Bühler et al. 2016), our approach is very simple and thus computationally undemanding. We omit HS reconstructions

and consider the SE field in two dimensions (a planar view) which ends up with the dichotomous output (1—snow vs. 0—no-snow).

The method outlined in this paper was tested on data collected in three sites, with different UAVs operating over each location. This confirms that the approach is independent of the data acquisition platform. Collecting spatial data using UAVs along with their processing in near-real time is known as "rapid mapping" (Tampubolon and Reinhardt 2015). Our platform-independent solution fits this idea as SE maps can be produced in the field, immediately after spatial data have been collected by a UAV.

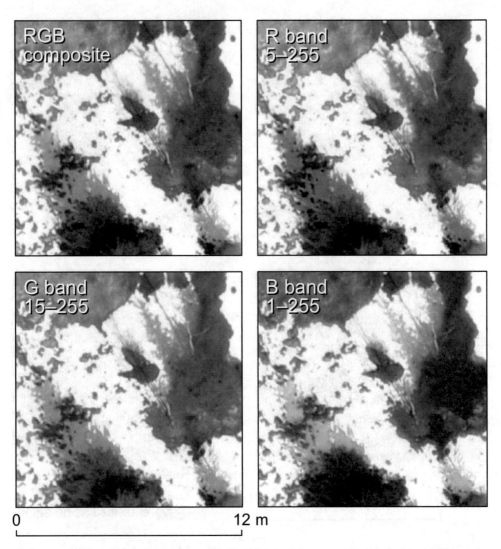

Figure 10
Fragment of orthophoto image of patchy snow cover in the test site of Rozdroże Izerskie, zoomed in to trees that cast shadow on terrain, along with responses of individual R/G/B bands

Information on SE, especially when combined with HS to estimate SWE, is important for water management purposes, since it enables basin managers to assess the risk of snowmelt floods.

## Acknowledgements

The research was financed by the National Centre for Research and Development, Poland, through Grant No. LIDER/012/223/L-5/13/NCBR/2014. The mobile Laboratory for Unmanned Aerial Observations of Earth of the University of Wrocław, including UAVs and the associated equipment, was financed by several Polish institutions: the National Centre for Research and Development (LIDER/012/223/L-5/13/NCBR/2014), Ministry of Science and Higher Education (IP2014 032773), National Science Centre (2011/01/D/ST10/04171), the University of Wrocław (statutory funds). The authors thank Joanna Remisz and Jacek Ślopek for technical help in operating UAVs. The authors kindly acknowledge the authorities of Świeradów Forest Inspectorate, Poland, for productive partnership and support. The authors are particularly grateful to Mr. Lubomir Leszczyński and Ms. Katarzyna Męcina for their support in fieldwork management. We used ArcGIS with Python, equipped with the libraries: arcpy, sys, numpy and cv2. The SYNOP data, collected by the Institute of Meteorology and Water Management–National Research Institute (Instytut Meteorologii i Gospodarki Wodnej—Państwowy Instytut Badawczy; IMGW–PIB), were acquired from http://www.ogimet.com.

**Open Access** This article is distributed under the terms of the Creative Commons Attribution 4.0 International License (http://creativecommons.org/licenses/by/4.0/), which permits unrestricted use, distribution, and reproduction in any medium, provided you give appropriate credit to the original author(s) and the source, provide a link to the Creative Commons license, and indicate if changes were made.

## References

Bühler, Y., Adams, M. S., Bösch, R., & Stoffel, A. (2016). Mapping snow depth in alpine terrain with unmanned aerial systems (UASs): Potential and limitations. *The Cryosphere, 10,* 1075–1088.

Bühler, Y., Adams, M. S., Stoffel, A., & Bösch, R. (2017). Photogrammetric reconstruction of homogenous snow surfaces in alpine terrain applying near-infrared UAS imagery. *International Journal of Remote Sensing.* https://doi.org/10.1080/01431161.2016.1275060.

Chang, A. T. C., Foster, J. L., & Hall, D. K. (1987). Nimbus-7 derived global snow cover parameters. *Annals of Glaciology, 9,* 39–45.

de Michele, C., Avanzi, F., Passoni, D., Barzaghi, R., Pinto, L., Dosso, P., et al. (2016). Using a fixed-wing UAS to map snow depth distribution: An evaluation at peak accumulation. *The Cryosphere, 10,* 511–522.

Deems, J. S., Painter, T. H., & Finnegan, D. C. (2013). Lidar measurement of snow depth: A review. *Journal of Glaciology, 59,* 467–479.

Dozier, J. (1989). Spectral signature of alpine snow cover from the landsat thematic mapper. *Remote Sensing of Environment, 28,* 9–22.

Dyer, J., & Mote, T. (2006). Spatial variability and trends in observed snow depth over North America. *Geophysical Research Letters.* https://doi.org/10.1029/2006GL027258.

Elder, K., Rosenthal, W., & Davis, R. E. (1998). Estimating the spatial distribution of snow water equivalence in a montane watershed. *Hydrological Processes, 12,* 1793–1808.

Erxleben, J., Elder, K., & Davis, R. E. (2002). Comparison of spatial interpolation methods for es-timating snow distribution in the Colorado Rocky Mountains. *Hydrological Processes, 16,* 3627–3649.

Grünewald, T., Schirmer, M., Mott, R., & Lehning, M. (2010). Spatial and temporal variability of snow depth and ablation rates in a small mountain catchment. *The Cryosphere, 4,* 215–225.

Hall, D. K., Riggs, G. A., Salomonson, V. V., DiGirolamo, N. E., & Bayr, K. J. (2002). MODIS snow-cover products. *Remote Sensing of Environment, 83,* 181–194.

Harder, P., Schirmer, M., Pomeroy, J., & Helgason, W. (2016). Accuracy of snow depth estimation in mountain and prairie environments by an unmanned aerial vehicle. *The Cryosphere, 10,* 2559–2571.

Hock, R., Rees, G., Williams, M. W., & Ramirez, E. (2006). Preface contribution from glaciers and snow cover to runoff from mountains in different climates. *Hydrological Processes, 20,* 2089–2090.

Jonas, T., Marty, C., & Magnusson, J. (2009). Estimating the snow water equivalent from snow depth measurements in the Swiss Alps. *Journal of Hydrology, 378,* 161–167.

Julitta, T., Cremonese, E., Migliavacca, M., Colombo, R., Galvagno, M., Siniscalco, C., et al. (2014). Using digital camera images to analyse snowmelt and phenology of a subalpine grassland. *Agricultural and Forest Meteorology, 198–199,* 116–125.

Kunzi, K.F., Patil, S., & Rott H. (1982). Snow-covered parameters retrieved from NIMBUS-7 SMMR data. *IEEE Transactions on Geoscience and Remote SensingGE-20,* 452–467.

Miziński, B., & Niedzielski, T. (2017). Fully-automated estimation of snow depth in near real time with the use of unmanned aerial vehicles without utilizing ground control points. *Cold Regions Science and Technology, 138,* 63–72.

Molotch, N. P., Fassnacht, S. R., Bales, R. C., & Helfrich, S. R. (2004). Estimating the distribution of snow water equivalent and snow extent beneath cloud cover in the Salt Verde River basin, Arizona. *Hydrological Processes, 18,* 1595–1611.

Prokop, A. (2008). Assessing the applicability of terrestrial laser scanning for spatial snow depth measurements. *Cold Regions Science and Technology, 54,* 155–163.

Prokop, A., Schirmer, M., Rub, M., Lehning, M., & Stocker, M. (2008). A comparison of measurement methods: Terrestrial laser scanning, tachymetry and snow probing for the determination of the spatial snow-depth distribution on slopes. *Annals of Glaciology, 49,* 210–216.

Prokop, A., Schön, P., Singer, F., Pulfer, G., Naaim, M., Thibert, E., et al. (2015). Merging terrestrial laser scanning technology with photogrammetric and total station data for the determination of avalanche modeling parameters. *Cold Regions Science and Technology, 110,* 223–230.

Pulliainen, J., & Hallikainen, M. (2001). Retrieval of regional snow water equivalent from space-borne passive microwave observations. *Remote Sensing of Environment, 75,* 76–85.

Pulliainen, J. (2006). Mapping of snow water equivalent and snow depth in boreal and sub-arctic zones by assimilating space-borne microwave radiometer data and ground-based observations. *Remote Sensing of Environment, 101,* 257–269.

Rittger, K., Painter, T. H., & Dozier, J. (2013). Assessment of methods for mapping snow cover from MODIS. *Advances in Water Resources, 51,* 367–380.

Robinson, D. A., & Frei, A. (2000). Seasonal variability of northern hemisphere snow extent using visible satellite data. *The Professional Geographer, 52,* 307–315.

Romanov, P., & Tarpley, D. (2007). Enhanced algorithm for estimating snow depth from geostationary satellites. *Remote Sensing of Environment, 108,* 97–110.

Rosenthal, W., & Dozier, J. (1996). Automated mapping of montane snow cover at subpixel resolution from the landsat thematic mapper. *Water Resources Research, 32,* 115–130.

Schön, P., Prokop, A., Vionnet, V., Guyomarc'h, G., Naaim-Bouvet, F., & Heiser, M. (2015). Improving a terrain-based parameter for the assessment of snow depths with TLS data in the Col du Lac Blanc area. *Cold Regions Science and Technology, 114*, 15–26.

Takala, M., Luojus, K., Pulliainen, J., Derksen, C., Lemmetyinen, J., Krn, J.-P., et al. (2011). Estimating northern hemisphere snow water equivalent for climate research through assimilation of space-borne radiometer data and ground-based meas-urements. *Remote Sensing of Environment, 115*, 3517–3529.

Tampubolon, W., & Reinhardt, W. (2015). UAV data processing for rapid mapping activities. *ISPRS Archives, XL-3/W3*, 371–377.

Tedesco, M., Pulliainen, J., Takala, M., Hallikainen, M., & Pampaloni, P. (2004). Artificial neural network-based techniques for the retrieval of SWE and snow depth from SSM/I data. *Remote Sensing of Environment, 90*, 76–85.

Tekeli, A. E., Akyürek, Z., Şorman, A. A., Şensoy, A., & Şorman, A. Ü. (2005). Using MODIS snow cover maps in modeling snowmelt runoff process in the eastern part of Turkey. *Remote Sensing of Environment, 97*, 216–230.

Vander, J. B., Lucieer, A., Wallace, L., Turner, D., & Durand, M. (2015). Snow depth retrieval with UAS using photogrammetric techniques. *Geosciences, 5*, 264–285.

Vikhamar, D., & Solberg, R. (2002). Subpixel mapping of snow cover in forests by optical remote sensing. *Remote Sensing of Environment, 84*, 69–82.

Wang, X., Zhu, Y., Chen, Y., Zheng, H., Liu, H., Huang, H., et al. (2017). Influences of forest on MODIS snow cover mapping and snow variations in the Amur River basin in Northeast Asia during 2000–2014. *Hydrological Processes, 31*, 3225–3241.

Zhang, Q., Skjetne, R., Lset, S., & Marchenko, A. (2012a). Digital image processing for sea ice observations in support to Arctic DP operations. In: ASME 2012 31st International Conference on Ocean, Offshore and Arctic Engineering, American Society of Mechanical Engineers, pp. 555–561.

Zhang, Q., van der Werff, S., Metrikin, I., Lset, S., & Skjetne, R. (2012b). Image processing for the analysis of an evolving broken-ice field in model testing. In: ASME 2012 31st International Conference on Ocean, Offshore and Arctic Engineering, American Society of Mechanical Engineers, pp. 597–605.

(Received  December 14, 2017, revised  March 13, 2018, accepted  March 17, 2018, Published online  April 26, 2018)

Pure Appl. Geophys. 175 (2018), 3303–3324
© 2017 The Author(s)
This article is an open access publication
https://doi.org/10.1007/s00024-017-1748-y

# Multitemporal Accuracy and Precision Assessment of Unmanned Aerial System Photogrammetry for Slope-Scale Snow Depth Maps in Alpine Terrain

MARC S. ADAMS,[1] Yves BÜHLER,[2] and REINHARD FROMM[1]

*Abstract*—Reliable and timely information on the spatio-temporal distribution of snow in alpine terrain plays an important role for a wide range of applications. Unmanned aerial system (UAS) photogrammetry is increasingly applied to cost-efficiently map the snow depth at very high resolution with flexible applicability. However, crucial questions regarding quality and repeatability of this technique are still under discussion. Here we present a multitemporal accuracy and precision assessment of UAS photogrammetry for snow depth mapping on the slope-scale. We mapped a 0.12 km² large snow-covered study site, located in a high-alpine valley in Western Austria. 12 UAS flights were performed to acquire imagery at 0.05 m ground sampling distance in visible (VIS) and near-infrared (NIR) wavelengths with a modified commercial, off-the-shelf sensor mounted on a custom-built fixed-wing UAS. The imagery was processed with structure-from-motion photogrammetry software to generate orthophotos, digital surface models (DSMs) and snow depth maps (SDMs). Accuracy of DSMs and SDMs were assessed with terrestrial laser scanning and manual snow depth probing, respectively. The results show that under good illumination conditions (study site in full sunlight), the DSMs and SDMs were acquired with an accuracy of $\leq 0.25$ and $\leq 0.29$ m (both at $1\sigma$), respectively. In case of poorly illuminated snow surfaces (study site shadowed), the NIR imagery provided higher accuracy (0.19 m; 0.23 m) than VIS imagery (0.49 m; 0.37 m). The precision of the $UAS_{SDMs}$ was 0.04 m for a small, stable area and below 0.33 m for the whole study site (both at $1\sigma$).

**Key words:** Unmanned aerial vehicles, terrestrial laser scanning, manual snow depth probing, digital surface models, validation, error.

## 1. Introduction

The spatial distribution of snow depth in alpine environments is highly heterogeneous (Elder et al.

1998). This is mainly owed to the complex interaction between alpine terrain and meteorological factors, such as precipitation and surface energy fluxes, as well as the redistribution of snow by wind, sloughing or avalanche activity (Cline et al. 1998; Elder et al. 1991). Area-wide approaches to determine snow depth [e.g., based on automatic weather station (AWS) data combined with medium-resolution satellite imagery (Foppa et al. 2007)] are not able to capture its high local variability (Ginzler et al. 2013). However, detailed information on slope-scale snow depth distribution plays an important role for many applications in snow science and practice, including numerical modelling of snow drift (Durand et al. 2005; Beyers et al. 2004), ecological studies on alpine flora and fauna (Bilodeau et al. 2013; Peng et al. 2010), planning avalanche hazard mitigation measures (Margreth and Romang 2010; Fuchs et al. 2007), avalanche forecasting and warning (Helbig et al. 2015; Vernay et al. 2015), avalanche event documentation, e.g., for hazard zone mapping (Holub and Fuchs 2009; Decaulne 2007), prediction and assessment of flood hazard resulting from snow melt (Painter et al. 2016; Schöber et al. 2014) or as an input for the optimisation of numerical simulation models in avalanche dynamics research (Fischer et al. 2015; Teich et al. 2014). Manually measuring this information in situ is labour-intensive, potentially hazardous or even impossible (Nolin 2010). Therefore, a wide range of terrestrial, airborne and spaceborne remote and close-range sensing techniques have been applied to retrieve digital surface models (DSMs)/snow depth maps (SDMs) at the slope-scale (Deems et al. 2013; Dietz et al. 2012; Rees 2006). One of the most recent techniques is unmanned aerial system (UAS) photogrammetry,

[1] Department of Natural Hazards, Austrian Research Centre for Forests (BFW), Hofburg Rennweg 1, 6020 Innsbruck, Austria. E-mail: marc.adams@bfw.gv.at

[2] WSL Institute for Snow and Avalanche Research SLF, Flüelastrasse 11, 7260 Davos Dorf, Switzerland.

which has quickly become a wide-spread method for geodata collection in different fields of earth science (Colomina and Molina 2014; Nex and Remondino 2013). This development has been fostered by the proliferation of easy-to-use UAS platforms and sensors, as well as recent progress in the field of computer vision [structure-from-motion (Koenderink and van Doorn 1991) and multi-view stereopsis (Furukawa and Ponce 2009)], considerably reducing requirements for photogrammetric processing of aerial imagery (Mancini et al. 2013). Despite some drawbacks (e.g., range limited to slope-scale, legal regulations, necessity for stable flight weather conditions), UAS photogrammetry offers many advantages over established techniques for snow depth mapping: compared to manned aircraft campaigns, UAS can acquire imagery at a much lower cost (e.g., for equipment, training, maintenance, operation) (Harder et al. 2016), higher operational flexibility (Vander Jagt et al. 2015), as well as higher flexibility and choice regarding the sensors' spatial and radiometric resolution, including an option for UAS-based laser scanning (Whitehead and Hugenholtz 2014); compared to terrestrial laser scanning (TLS), UAS photogrammetry is more flexible regarding deployment in alpine terrain [high-accuracy UAS positioning or point cloud registration routines as presented by Miziński and Niedzielski (2017) make georeferencing targets obsolete] and it does not suffer the limitations of the line-of-sight due to acute viewing angles or occlusions (Marti et al. 2016; Harder et al. 2016). However, while the above-mentioned techniques are well-established, their quality and repeatability well-known (Hartzell et al. 2015; Müller et al. 2014), crucial questions regarding the accuracy and precision of UAS-based snow depth mapping are still under discussion (Avanzi et al. 2017). Several contributions have recently been published, reporting on the application of UAS photogrammetry to snow depth mapping, using both multicopter and fixed-wing UAS. In all of these studies, the UAS results were validated with reference data including:

i. Global navigation satellite system (GNSS) measurements of the snow surface and/or manual snow depth probing (MP) (Miziński and Niedzielski 2017; De Michele et al. 2016; Harder et al. 2016; Lendzioch et al. 2016; Bühler et al. 2016; Vander Jagt et al. 2015).

ii. Very high resolution optical satellite imagery (Marti et al. 2016).

iii. A large-frame aerial camera mounted on a manned aircraft (Boesch et al. 2016).

iv. A multi station in scanning mode (Avanzi et al. 2017).

However, all these assessments were made based on a comparatively small number of UAS measurements (1–3 flights), except for Harder et al. (2016); the majority used small amounts of discrete samples (GNSS and MP measurements); most studies evaluated the use of imagery collected in the visible part of the spectrum (VIS) (except for Miziński and Niedzielski 2017; Bühler et al. 2016; Boesch et al. 2016), however, several authors have pointed to the benefits of using near-infrared (NIR) imagery for snow mapping (Bühler et al. 2015; Nolin and Dozier 2000). Nolan et al. (2015) performed a large-scale accuracy and precision assessment of imagery collected with a consumer-grade digital camera mounted on a manned aircraft over large areas, with GNSS and airborne laser scanning data. However, since the employed methodology differs substantially from the presented study (size of target area, georeferencing routine, employed platform), results were not directly compared.

In this contribution, we present a multitemporal assessment (12 UAS flights) of the accuracy and precision of UAS photogrammetry for snow depth mapping. Adding to findings from the above-mentioned studies, we used TLS data to assess the accuracy of $UAS_{DSM}$, MP as reference data for $UAS_{SDM}$ accuracy and calculated precision by inter-comparison of UAS results. VIS and NIR imagery was used to map snow depth with a fixed-wing UAS.

## 2. Materials and Methods

### 2.1. Study Site

The study site is located in the Tuxer Alps of North Tyrol, Austria (47°10′N; 11°38′E), between the Northern Calcareous Alps and the Main Alpine

Ridge. It lies at approximately 2020 m a.s.l., near the head of a north–south running valley. The area features a typical inner-alpine climate, with annual precipitation between 1200 and 1700 mm (period 1983–2003) and snow depths of 1–2 m (Schaffhauser and Fromm 2008). The land cover of the site is mainly characterised by (partially boggy) alpine grasslands, mixed with various types of small scrubs (height < 1 m). In the west and north, large clusters of dwarf pine (*Pinus mugo*, height 1–3 m) and singular or groups of stone pine (*Pinus cembra*) are present. Several small streams run parallel to the valley axis, some of which drain into a pond (approximate size 0.006 km$^2$) in the north. The topography of the site (mean slope angle 6°) is dominated by the flat valley bottom; the steepest areas lie in the east and west, where the lower sections of the adjoining slopes reach into the study site. Large boulders (max. width < 30 m, max. height < 5 m) are scattered in the centre of the site. Multiple small buildings are situated in the north and east, connected by a network of gravel roads, which are partially cleared in winter. An overview of the site is provided in Fig. 1; it highlights where TLS, UAS and MP data were collected, as well as the location of the AWS and reference points (RPs). The site was chosen on account of its good accessibility, even during periods with high avalanche danger, and well-established infrastructure (power supply and network connection) (Adams et al. 2016).

## 2.2. Data Acquisition and Processing

We collected data at the study site during four measurement campaigns in 'snow-on' conditions in February and March 2015 (Fig. 2, upper image). This allowed us to take different snow pack properties (snow depth, snow type at surface) and illumination conditions at the study site into account. For example, the snow depth measured at the AWS, ranged between 0.68 m (13 February, 1 p.m.) and 1.01 m (3 March, 2 a.m.).

Each campaign consisted of:

- Two to four UAS flights.
- One to two TLS scans.
- 149 MP measurements (February campaigns only).

The UAS and TLS data were acquired over an area of interest (AOI) of 0.12 km$^2$ (Fig. 1). It was located in the centre of the valley floor, where MP measurements were performed, too. Due to the geometric properties of the measurement setup, TLS data was only collected on 70% of the area of the core AOI (not considering occlusions). Reference 'snow-off' UAS imagery was acquired on 21 August 2015.

### 2.2.1 Unmanned Aerial Systems

The aerial imagery was collected with a Multiplex Mentor Elapor fixed-wing UAS (Fig. 2—lower image, Table 1) at different times of the day. The original Mentor model was modified to add UAS capabilities, it was fitted with navigation sensors to determine its absolute position (GNSS) and orientation (inertial measurement unit); this data was managed by the on-board autopilot for autonomous flight (3DR ArduPilotMega); pre-flight mission planning to define the flight path, height and speed was performed in the open-source software Mission Planner (Table 2); an additional on-board GNSS unit (SM GPS-Logger 2) recorded 10 Hz positional data ($x$, $y$ and $z$). After completing each flight, the on-board GNSS data was synchronised with the recorded imagery (geotagging), to facilitate the image processing (Adams et al. 2016). The UAS had a maximum flight time of 40 min, during which it could map up to 0.6 km$^2$ at 2,000 m a.s.l. in wintry conditions.

A Sony NEX5R digital camera was installed in the UAS fuselage to record the imagery on all the flights. It weighed 0.4 kg and was fitted with a 50 mm Sony prime lens (0.2 kg). The camera's 16-megapixel APS-C sensor was modified by removing the built-in short-pass filter, increasing its sensitivity in the near-infrared from 700 to 1100 nm. This allowed us to mount the lens with different notch filters to record data in various parts of the electromagnetic spectrum: VIS ($\lambda = 350$–680 nm), NIR700 ($\lambda > 700$ nm) and NIR830 ($\lambda > 830$ nm). Each flight was carried out with a single camera on-board the UAS, set to record imagery at a defined wavelength, and the filters changed between retrievals. The camera was

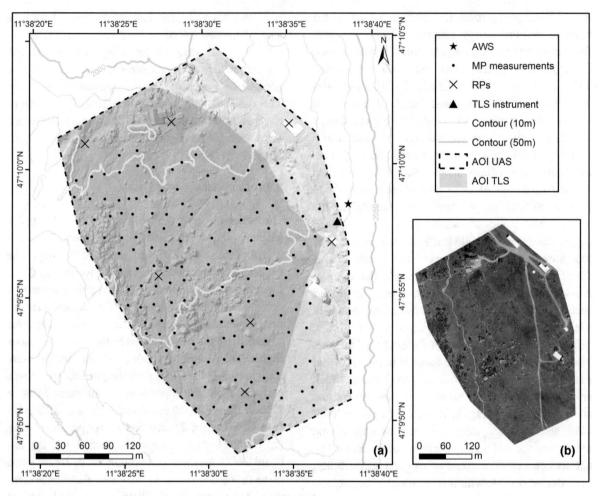

Figure 1
Study site overview; **a** outline of areas of interest (AOI) for UAS and TLS data acquisition, as well as points where MP measurements were performed; positions of instruments (TLS and AWS) and RPs; 10 and 50 m contour lines were derived from airborne laser scanning data; 'snow-off' reference data included as hillshaded $UAS_{DSM}$ [background of **a**] and orthophoto (**b**)

triggered via infrared signal, recording images at 1.25 Hz. Basic camera settings were fixed pre-flight, as no telemetry was available (Table 2); imagery was recorded with manual focus. The high image overlap (80% along- and 90% cross-track) was chosen based on the authors' own experience, as well as recommendations from authors of similar studies dealing with UAS-based mapping of low contrast surfaces (Harder et al. 2016; Klemas 2015). We performed no internal camera calibration.

The study site was surrounded by high peaks, which cast a shadow on the valley floor from 1 p.m. onwards. This allowed a direct comparison of imagery collected in good (full sunlight) and in poor illumination conditions (shadow) on the same day. During all the campaigns, the sky was clear or only partially cloudy, with no precipitation; the nearby AWS (located at 2041 m a.s.l.) recorded the air temperatures between $-8°$ and $5°$C, at only very light winds ($V_{max} < 3$ m s$^{-1}$) 7 m above ground level. These can be considered good weather conditions for our UAS flights, especially considering the alpine environment. However, higher wind speeds and lower air temperatures can be expected at our typical flight height (400 m above ground level).

In lieu of survey-grade GNSS sensors on-board the UAS, indirect georeferencing had to be used (Harwin et al. 2015; Vander Jagt et al. 2015).

Figure 2
Central part of study site on 11 February 2015 (upper image);
launching Mentor UAS (lower image)

Table 1

*Technical specifications of the Mentor UAS (Adams et al. 2016)*

| UAS type | Fixed-wing (custom-built) |
|---|---|
| Dimensions | 1.6 m (wing span) |
| | 1.2 m (fuselage) |
| Engine | 1 electrical, brushless motor |
| Flight time | 30–40 min |
| Max. range/coverage | 1500 m/0.6 km$^2$ |
| Empty weight | 2.3 kg |
| Max. take-off weight | 2.8 kg |
| Max. payload weight | 0.5 kg |
| Navigation | 3DR APM 2.6 (IMU, barometer) |
| | 3DR uBlox GNSS with Compass Kit uBlox LEA-6H module |
| Wireless communication | Graupner MX-20 HOTT 2,4 GHz (sender) |
| | Frequency 2400 ... 2484.5 MHz |
| | Graupner GR-16 HOTT 2.4 GHz (receiver) |
| LiPo battery | LiPolice GreenLine Light Edition 5s 4900 mAh (0.6 kg) |

Therefore, prior to each campaign, we distributed 10–20 RPs, consisting of 0.4 × 0.4 m black-and-white checkered wooden boards, within the AOI. We

Table 2

*Typical UAS flight and camera settings*

| Overlap (along-/cross-track) | 80/90% |
|---|---|
| Flight height | 400 m above ground level |
| Flight speed | 12–14 m s$^{-1}$ |
| Image format | JPEG (high quality) |
| Brightness compensation (fixed) | 0 |
| Exposure (fixed) | 1/320–1/800 |
| ISO (automatic) | 100–400 |
| Aperture (automatic) | f/2.5–18 |

surveyed the location of each RP using a Trimble GEO-XT 2008, with an expected accuracy in the decimetre range (Adams et al. 2016). The data was corrected real time in the field and differentially during post-processing. Final RP coordinates were averaged from more than 200 point measurements made at each RP location. However, the resulting overall georeferencing errors (especially in $z$-direction) proved too high and resulted in implausible SDMs. Therefore, the $z$ values used for georeferencing the 'snow-off' UAS$_{DSM}$ were extracted from an airborne laser scanning DSM from 2009, while retaining the $x$- and $y$-coordinates surveyed with the GNSS. To georeference the 'snow-on' UAS$_{DSMs}$, seven natural or man-made RPs were chosen. These RPs had remained snow-free throughout the winter (e.g., centre of flat stones in the river bed, corner of wooden patio outside hut) (Fig. 1). Their coordinates were extracted from the 'snow-off' UAS data. Thus, the 'snow-on' data could be referenced using a stable set of RPs, ensuring minimal systematic error introduced by the georeferencing procedure (Adams et al. 2016). However, this resulted in a comparatively small amount of seven RPs.

All the UAS imagery was processed with Agisoft's PhotoScan Pro (version 1.2.3), a commercially available photogrammetric software suite, that is widely used in the UAS community (Tonkin et al. 2014). It is credited to be among the most reliable (Sona et al. 2014) and accurate (Gini et al. 2013) software packages available. PhotoScan is based on a structure-from-motion algorithm (Verhoeven 2011) and provides a complete, photogrammetric workflow, with particular emphasis on multi-view stereopsis (Harwin et al. 2015). This workflow consists of the following principal steps (Vander Jagt et al. 2015):

i. Tie-point matching.

ii. Bundle adjustment (constrained by assigning high weights to the RP coordinates).

iii. Linear seven-parameter conversion; removal of nonlinear deformations.

iv. Dense point cloud generation with multi-view stereo reconstruction.

v. Triangulation of dense point cloud into mesh, subsequently generating DSM and orthophotos.

In a related study, Boesch et al. (2015) analysed PhotoScan's suitability for snow depth mapping and the best combination of processing parameters. Therefore, all the imagery was processed with the following alignment parameters: accuracy—highest, pair selection—reference, key point limit—40,000, tie-point limit—10,000. The dense point cloud was generated with the settings: quality—medium, depth filtering—moderate. One of the main reasons for corrupt UAS imagery is motion blur, which results from shutter speeds that are too slow in relation to the movement of the UAS (Turner et al. 2015; Immerzeel et al. 2014). This applies in particular to motion in direction of the UAS' roll-axis, resulting from crosswinds, and increases with the length of the camera lens (Morgenthal and Hallermann 2016). As reported in Bühler et al. (2016), in our experience, fixed-wing UAS are generally more susceptible to crosswinds and thus less stable in windy conditions, than some multicopters. The sensor on-board our fixed-wing UAS was not stabilised by a gimbal. To systematically evaluate our imagery, we routinely calculated the 'quality index' (QI) during pre-processing in PhotoScan (Adams et al. 2016). As reported in the PhotoScan documentation (Agisoft 2016), it provides a normalised value for the sharpness of the imagery; images with $QI < 0.5$ are recommended to be excluded from photogrammetric processing. Orthophotos and DSM were exported from PhotoScan in 0.05 and 0.2 m resolution, respectively. We calculated snow depth for each pixel by subtracting the 'snow-off' DSM from the 'snow-on' DSM. This follows the definition by Fierz et al. (2009), where snow depth is the vertical distance from the base to the surface of the snow pack.

### 2.2.2 Terrestrial Laser Scanning

We used two Riegl long-range TLS instruments to collect the validation data: a LPM-321 (Fig. 3—left image) and a LPM 98-2k (Fig. 3—right image). Both the instruments operate at 905 nm wavelength, therefore, the penetration depths into the snow surface are only a few millimetres (Dozier and Painter 2004). They were positioned in a purpose-built shed, overlooking the valley (Fig. 1), and set to map the valley floor in a single scan window. The LPM-321 was used for the first campaign; for the subsequent campaigns, the LPM 98-2k was installed in a fixed, weatherproof transparent glass fibre enclosure. We set up the LPM 98-2k to continuously and automatically acquire scans from the study site approximately every 6 h, and a datalink allowed remote access [detailed setup description and technical specifications of the TLS instruments are provided in Adams et al. (2016)].

To georeference the TLS data, five RPs, consisting of 0.3–0.5 m rectangular aluminium plates, coated with highly reflective material, were installed in the target area prior to the UAS campaigns. Their positions were surveyed with a Trimble M3 total station [expected accuracy $\pm 0.002$ m ($1\sigma$), plus 2 ppm distance dependent error]. The RPs were scanned by the TLS instrument before and after mapping the valley floor. Point clouds from both the TLS instruments were processed in RiPROFILE (version 1.5.7). Here the locations of the RPs in the global coordinate system and the scanner-own coordinate system were linked by minimising the standard deviation of the residues. An unfavourable geometry of the measurement setup and the inherent scanning routine of the instruments caused inhomogeneous point distances. To counter this distant dependent point density, mean $z$ values were calculated within a 0.2 m raster (corresponding to $DSM_{UAS}$ resolution) and the raster centre location plotted for validation. The accuracy assessment of the $UAS_{DSMs}$ was performed with $TLS_{DSMs}$, not the calculated snow depth values. No additional co-registration of these DSMs was performed.

Figure 3
TLS instruments Riegl LPM-321 (left image) and LPM 98-2k (right image) in operation at the study site on 11 February 2015 and 13 February 2015, respectively (Adams et al. 2016)

### 2.2.3 Manual Snow Depth Probing

MP measurements were performed during both the February campaigns. The snowpack was sounded at each checkpoint with an avalanche probe. The checkpoints were distributed randomly within the AOI, roughly following a grid pattern to avoid spatial bias. At each checkpoint, five measurements were performed by probing all the four corners and the centre of a $2 \times 2$ m$^2$. The snow depths were recorded to the nearest centimetre. Additionally, a GNSS (Garmin GPSMap 64s) was used to record the geographic coordinates of the square's centre. The data was collected after completing campaign one, but is assumed to also be valid for campaign two, as the AWS recorded no intervening snowfall, and snow melt/settling was minimal (0.03 m). For validating the UAS-based snow depth maps, the centre location of the MP checkpoints was corrected by plotting them on the UAS orthophotos and manually adjusting their

position. This was necessary as the Garmin GNSS has a nominal accuracy of only $\pm 3$ m [1 standard deviation ($\sigma$)]. Additionally, it ensured the correct position of the checkpoints relative to the UAS results. To minimise the effect of the micro-topography below the snowpack on the results, the mean value of the five measurements was calculated. For accuracy assessment, the UAS$_{SDM}$ values of all the pixels within a 2 m radius around a checkpoint were averaged (Adams et al. 2016).

### 2.3. Accuracy and Precision Assessment

To evaluate the performance of the UAS for slope-scale snow depth mapping in alpine terrain, we need to answer the following questions: (i) How well do the UAS-based DSMs and SDMs correspond to measurements taken with established, state-of-the-art techniques? (ii) How reliable are the UAS results in terms of their reproducibility? These questions correspond to determining the accuracy and precision of the UAS results, respectively (Nolan et al. 2015).

### 2.3.1 Accuracy

Two reference data sets were used for accuracy assessment:

1. The TLS measurements allowed an assessment of UAS$_{DSM}$ accuracy at high spatial resolution (mean point distance: 0.2 m). The TLS instruments effectively surveyed a very large number of (pseudo-) checkpoints within the AOI, at high accuracy; the LPM-321 operates at a nominal accuracy of $\pm$ 0.025 m (1$\sigma$) plus a distance dependent error of $\leq$ 20 ppm (Grünewald et al. 2010; Riegl 2010); the LPM 98-2k at $\pm$ 0.05 m (1$\sigma$), plus a distance dependent error of $\leq$ 20 ppm (Schaffhauser et al. 2008; Riegl 2006). Considering all the areas surveyed by the TLS are within 300 m range of the instruments, the nominal TLS accuracy is between $\pm$ 0.031 and $\pm$ 0.056 m (1$\sigma$) for LPM-321 and LPM 98-2k, respectively. However, these values assume that the area illuminated by the laser beam (the footprint) is circular; this implies an incidence angle $\theta = 0°$ on a planar surface (Prokop 2008). $\theta$ is defined as the angle between the vector normal to the measured surface and the incoming laser beam (Jörg et al. 2006). The size of the footprint ($\delta$) generally increases with an increase of distance from the instrument, beam divergence and $\theta$ (when only considering planar surfaces) (Prokop et al. 2008). According to Jörg et al. (2006), $\delta$ remains below 1 m in diameter for close range ($<$ 500 m) TLS measurements, even at unfavourable scanning angles (i.e., $\theta < 75°$). In the present case, the TLS instrument surveys the valley floor from a small mound at the base of the slope east of the AOI, resulting in high $\theta$ ($>$ 75°) and thus large $\delta$ values ($>$ 1 m diameter). Therefore, we calculated $\theta$ and the resulting $\delta$ for the 11 February 2015 TLS data set (change in snow depth and the position of the TLS to the following campaigns were considered negligible). Subsequently, the correlation between $\delta$ and UAS$_{DSM}$ error was determined. Following the general practice in statistics, the Bravais-Pearson correlation coefficient $r$ was calculated for normally distributed data and the Spearman's rank correlation coefficient $r_{SP}$ for non-normally distributed data (Fahrmeir et al. 2011).

2. The MP measurements were the basis for the assessment of the UAS$_{SDM}$ accuracy. Snow depth values were surveyed at comparatively low spatial resolution (mean distance between checkpoints: 18 m). However, this data has a high vertical accuracy, as a majority of the checkpoints were located above the frozen ground; the penetration depth of the probe is therefore considered to be within $\pm$ 0.02 m. Similar values are reported in comparative studies (e.g., Harder et al. 2016; Nolan et al. 2015). As this area is an unmanaged (high-) alpine grassland, it features a jagged micro-topography with local terrain height variations in the decimetre range. However, the high ground sampling distance of the UAS$_{DSMs}$ (0.2 m) and the MP sampling routine (Sect. 2.2.3) are considered to be able to account for these variations.

Authors of the comparable studies (e.g., Fras et al. 2016; Hugenholtz et al. 2013; Harder et al. 2016) used checkpoints surveyed with high-accuracy GNSS as reference data. Such data are not included in the presented study, as it focusses on the area-wide, multitemporal evaluation of UAS-based photogrammetry of snow-covered surfaces with TLS. Such a comparison has only been marginally covered in the literature published to date (Sect. 1). Additionally, MP data was included for a direct assessment of the UAS' snow depth mapping accuracy. This study focusses on vertical accuracy assessment, as no planimetric offset could be derived from MP or TLS data. Thus, the error of the UAS results was calculated as a difference in $z$ value between UAS and the reference data sets (Müller et al. 2014).

We followed the accuracy assessment procedure outlined in Höhle and Höhle (2009), which was also adapted in similar studies [e.g., Fras et al. (2016) and Müller et al. (2014)]. Thus, the normality of UAS$_{DSM}$ and UAS$_{SDM}$ error distributions was checked by visually interpreting their histograms and quantile–quantile (Q–Q) plots. Q–Q plots juxtapose theoretical quantiles of a normal distribution with the quantiles

Table 3

*Accuracy measures applied to normally (A) and non-normally (B) distributed errors; n is the number of tested points, and $\Delta h_i$ denotes the difference from reference data for a point i (Höhle and Höhle 2009)*

| Accuracy measure | Notational expression |
|---|---|
| **A** | |
| Mean error | $ME(\mu) = \frac{1}{n}\sum_{i=1}^{n}\Delta h_i$ |
| Standard deviation | $SD = \sqrt{\frac{1}{(n-1)}\sum_{i=1}^{n}(\Delta h_i - \mu)^2}$ |
| Mean absolute error | $MAE = \frac{\sum_{i=1}^{n}|\Delta h_i|}{n}$ |
| Root mean square error | $RMSE = \sqrt{\frac{1}{n}\sum_{i=1}^{n}\Delta h_i^2}$ |
| **B** | |
| 50% quantile (median) | $Q_{\Delta h}(0.5) = m_{\Delta h}$ |
| 68.3% quantile | $Q_{|\Delta h|}(0.683)$ |
| 95% quantile | $Q_{|\Delta h|}(0.95)$ |
| Normalised median absolute deviation | $NMAD = 1.4826 \; \text{median}_i(|\Delta h_i - m_{\Delta h}|)$ |

of the empirical distribution function. If the latter is normally distributed, the Q–Q plot will result in a straight line; strong deviation indicates non-normal distribution (Höhle and Höhle 2009). Additionally, skewness and kurtosis were calculated. Based on recommendations from Höhle and Höhle (2009) and Willmott and Matsuura (2006), different accuracy measures were applied to normally and non-normally distributed errors (Table 3).

All the data sets were referenced to and compared within common global projected planimetric (MGI Austria GK West; EPSG Code 31254) and vertical coordinate systems (Gebrauchshöhen Adria; WKID 5778).

### 2.3.2 Precision

Performing several UAS flights per campaign day allowed assessing the precision, i.e., the reproducibility of the UAS results. As argued by Fabris and Pesci (2005) and Nolan et al. (2015), precision assessment of photogrammetric DSMs by intercomparison generally provides the basis for two different assumptions: (i) if the intervening changes of the reference surfaces between flights are negligible, it yields an empirical estimate of the internal precision of the UAS data acquisition and processing setup; (ii) in case real height changes of the snow surface (e.g., due to wind drift, snow fall, snow melt/settling) occur between two flights, it allows one to track the magnitude of these changes.

On all the campaign days, the AWS recorded air temperatures below 5 °C, snow temperatures below

Table 4

*Overview of UAS campaigns, details of data acquisitions (columns one through three), camera settings and output ('imagery' columns), (pre-)processing results as reported in PhotoScan ('photogrammetric processing' columns) (Adams et al. 2016)*

| Date | Time | Campaign/flight number | Imagery | | | | | Photogrammetric processing | | | |
|---|---|---|---|---|---|---|---|---|---|---|---|
| | | | Sensor | #Photos | Aperture | Exposure | ISO | Mean QI | Overlap | Marker error (XYZ) [m] | Reprojection error (RMS/max.) [m] |
| 11.02.2015 | 11:00 | 1/1 | NIR830 | 1064 | f/4.5–8 | 1/320 | 100 | 0 | 10 | 0.14 | 0.2/4.3 |
| | 12:00 | 1/2 | VIS[a] | 527 | f/10–18 | 1/320 | 100 | 0.62 | 9 | 0.09 | 0.3/1.7 |
| | 13:00 | 1/3 | VIS[a] | 897 | f/8–18 | 1/400 | 100 | 0.52 | 11 | 0.08 | 0.3/1.4 |
| 13.02.2015 | 10:30 | 2/1 | VIS | 884 | f/7.1–14 | 1/500 | 100 | 0.57 | 16 | 0.08 | 0.3/1.4 |
| | 13:30 | 2/2 | NIR830 | 973 | f/4–9 | 1/320 | 100 | 0.72 | 6 | 0.22 | 0.9/3.7 |
| | 14:00 | 2/3 | NIR700 | 863 | f/4–16 | 1/400 | 100 | 0.66 | 10 | 0.1 | 0.3/4.2 |
| | 15:00 | 2/4 | VIS | 680 | f/5–18 | 1/320 | 100 | 0.65 | 4 | 0.03 | 0.3/1.4 |
| 03.03.2015 | 10:30 | 3/1 | VIS | 652 | f/8–18 | 1/500 | 100 | 0.71 | 7 | 0.16 | 0.3/1.7 |
| | 13:00 | 3/2 | NIR830 | 965 | f/4–7.1 | 1/500 | 100 | 0.40 | 7 | 0.19 | 0.3/1.6 |
| 13.03.2015 | 13:30 | 4/1 | VIS[a] | 920 | f/7.1–13 | 1/500 | 100 | 0.81 | 11 | 0.11 | 0.4/4.6 |
| | 14:30 | 4/2 | VIS[a] | 500 | f/4–11 | 1/800 | 100 | 0 | 7 | 0.2 | 0.3/1.2 |
| | 15:30 | 4/3 | NIR830 | 544 | f/2.5–14 | 1/400 | 400 | 0.88 | 9 | 0.12 | 0.5/1.9 |
| 21.08.2015 | 13:00 | – | VIS | 1371 | f/7.1 | 1/320–1/2000 | 400 | 0.81 | 36 | 0.4 | 0.4/2.8 |

[a]Indicates data sets used for precision assessment in Sect. 2.3.2

− 3 °C, calm winds ($< 3$ m s$^{-1}$), no precipitation and a snow settling of less than 0.03 m. We, therefore, follow assumption (i) in this paper when interpreting the precision assessment results. Precision of the UAS$_{SDMs}$ was determined following Fabris and Pesci (2005), by calculating $\Delta h_i$ residuals for each pixel of two UAS flights performed on the same day (Table 4). Standard deviation (SD) of the $\Delta h_i$ residuals distribution was reported as precision value. No separate precision calculations were performed for UAS$_{DSMs}$, as the reference 'snow-off' UAS$_{DSM}$ was the same for all the campaigns. We performed the assessment for:

i. A small area we considered to be the best-case scenario (snow heavily compacted, therefore, intermittent snow depth change was zero; high-contrast snow surface, therefore, high-point density expected in photogrammetric processing; planar surface with little elevation change, therefore, there is no influence of topography on the result).
ii. The whole AOI.

## 3. Results and Discussion

### 3.1. Unmanned Aerial System

Four 'snow-on' UAS campaigns were conducted between 11 February and 13 March 2015; details on data acquisition, camera settings and quality reports from photogrammetric (pre-) processing are presented in Table 4. 12 UAS flights were performed to record approximately 11,000 images in total, of which 9500 were used in photogrammetric image processing. Seven VIS, one NIR700 and four NIR830 data sets were acquired between 10.30 a.m. and 4 p.m. Each flight took approximately 35 min. Camera settings were chosen according to the illumination conditions prior to UAS launch; priority was given to exposure (1/320–1/500), ISO was set at 100 for most flights, while aperture was adapted dynamically by the camera for each image (typically between f/4 and f/18).

Results from the QI calculation showed that two-thirds of the UAS imagery have a satisfactory average QI above 0.62. Low average QI values were reported for imagery recorded on flights 1/3 (0.52), 2/1 (0.57) and 3/2 (0.40); QI calculation failed for imagery from flights 1/1 and 4/2 (QI = 0) (Table 4). A visual check of the data sets confirmed a large amount of blurry imagery on flight 1/1, possibly due to an error in data acquisition; no apparent deficiencies with regard to image sharpness were detected in the other imagery. As the calculation of the QI is poorly documented and therefore essentially black-box, no details on the impact of other deficiencies in UAS imagery are available. Therefore, the reason for the low QI of some UAS imagery is unknown. Overlap was at 'nine', indicating that, on average, each point within the AOI was visible in the nine UAS images. The lowest overlap was calculated for flight 2/2, which, in turn, also features the highest marker (0.22 m) and reprojection error (0.9 m/3.7 − RMS/maximum error) of the 'snow-on' flights; all the marker errors reported in this section are mean values of all the seven RPs. The highest overlap by far (36) was achieved for the 'snow-off' UAS campaign. This was owed to the flight path design, which consisted of overcrossing flight lines parallel and orthogonal to the valley axis, as opposed to the winter flight lines, which were always orthogonal. The marker error for the summer reference flight was remarkably higher (0.4 m) than the average winter marker error (0.13 m). This may be due to the fact that the GNSS instrument used to survey the GCPs has a low accuracy (Sect. 2.2.1). Reprojection errors for all the data sets were below 0.4 m (RMSE) and below 1.9 m (maximum error). To calculate a statistically significant correlation between overlap, quality index and marker/reprojection error, the sample size ($n = 13$) is too small in the presented case; a visual interpretation of the results points to high overlap ($> 8.9$, when excluding outlier 36) leading to low marker error ($< 0.15$ m) and vice versa (valid for all flights except 2/4); little or no connection was found between the other parameters. This confirms results from previous studies, which have shown that high image overlap generally leads to a high signal-to-noise ratio in photogrammetric outputs and therefore low error at the GCPs (Zongjian et al. 2012; Haala 2011). This holds true especially when mapping low contrast surfaces, such as snow (Vander Jagt et al. 2015;

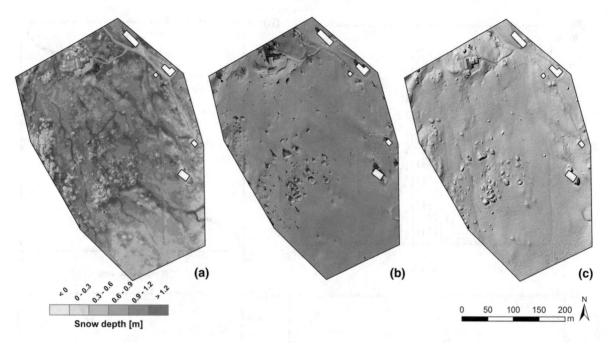

Figure 4
Results from photogrammetric processing of UAS imagery, generated on flight 3/1; SDM (**a**), orthophoto (**b**), hillshaded DSM (**c**)

Table 5

*Details of collected TLS data (Adams et al. 2016)*

| Date | Scanner settings | | | Number of points | | Point distances [m] | | | | Standard deviation residues |
|---|---|---|---|---|---|---|---|---|---|---|
| | Instrument | Res. *X* | Res. *Y* | AOI | Filtered | Mean | Min. | Max. | $1\sigma$ | |
| 11.02.2015 | LPM-321 | 0.063° | 0.063° | 271,511 | 171,983 | 0.13 | 0.001 | 5.46 | 0.10 | 0.07 |
| 14.02.2015 | LPM 98-2K | 0.054° | 0.054° | 268,155 | 183,064 | 0.16 | 0.001 | 8.77 | 0.12 | 0.11 |
| 03.03.2015 | LPM 98-2K | 0.107° | 0.108° | 63,077 | 56,766 | 0.29 | 0.001 | 9.68 | 0.21 | 0.07 |
| 11.03.2015 | LPM 98-2K | 0.107° | 0.108° | 69,371 | 57,298 | 0.26 | 0.001 | 20.64 | 0.22 | 0.11 |

Harder et al. 2016) or sand (Klemas 2015; Mancini et al. 2013).

In total, 12 'snow-on' and one 'snow off' orthophoto and DSM, as well as 12 SDMs for all the 'snow-on' flights were calculated. An example for an SDM (a), orthophoto (b) and hillshaded DSM (c) of flight 3/1, are shown in Fig. 4.

### 3.2. Terrestrial Laser Scanning and Manual Snow Depth Probing

Four TLS scans were selected for accuracy assessment of UAS$_{DSMs}$, based on their temporal proximity to UAS flights, quality and completeness.

One scan was performed with the LPM-321 (11 February 2015), three with the LPM 98-2K (14 February, 3 and 11 March 2015). The details of these scans are provided in Table 5. As described above, the number of measured TLS points (column 'AOI') was subsequently reduced to mitigate range bias (column 'Filtered'). In the section 'Point distances', descriptive statistics of the unfiltered point clouds are reported: The mean distance between points and consequently $1\sigma$ is lowest for the LPM-321 measurement (0.13 and 0.1 m, respectively). Average values for mean and $1\sigma$ of point distances for all the campaigns are at 0.21 and 0.16 m, respectively. Both measurements performed in March were recorded at

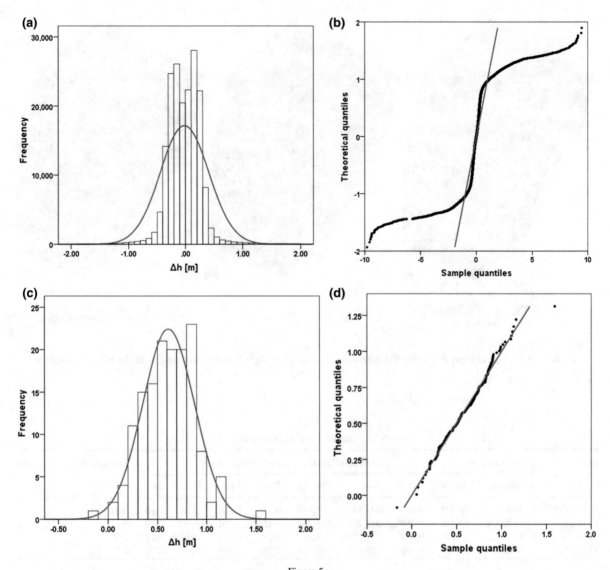

Figure 5

Visualisation of the error distributions of UAS$_{DSMs}$ (**a, b**) and UAS$_{SDMs}$ (**c, d**) for flight 2/1; histograms (**a, c**) of $\Delta h_i$ in [m] with superimposed normal distribution, frequency corresponds to the number of measurements; Q–Q plots of $\Delta h_i$ (**b, d**) (Höhle and Höhle 2009)

lower resolution than the February data sets (0.107° and 0.054° azimuth resolution, respectively) and thus show larger point distances. Average geolocation residues (column 'Standard deviation residues') were 0.09 m and show no connection with the type of instrument used.

Results from the analysis of $\theta$ for the TLS measurements conducted on 11 February 2015 show a normal distribution around a median of 82° ($1\sigma = 3.9°$). The corresponding $\delta$ values are left-skewed (skewness = 230) and comparatively large

(median = 0.47 m$^2$, 68.3% quantile = 1.24 m$^2$), considering the close range (< 300 m) and small beam divergence of the LPM-321 (typically 0.8 mrad). This confirms the assumption of an unfavourable measurement setup for TLS validation. An analysis of the spatial distribution of $\delta$ values shows that it is dominated by range. An analysis of the spatial distribution of $\delta$ values shows they are dominated by range, due to increasingly acute viewing angles ($r_{SP} = 0.94$ between $\delta$ and $\theta$). By comparison, increasing divergence of the laser beam

Table 6

*Overview of statistical analysis of the UAS$_{DSM}$ and UAS$_{SDM}$ errors for all flights*

| Accuracy measure | Campaign/flight number | | | | | | | | | | | |
|---|---|---|---|---|---|---|---|---|---|---|---|---|
| | 1/1 | 1/2 | 1/3 | 2/1 | 2/2 | 2/3 | 2/4 | 3/1 | 3/2 | 4/1 | 4/2 | 4/3 |
| UAS$_{DSM}$ error | | | | | | | | | | | | |
| Skewness | − 2 | − 2 | − 2 | 0 | 1 | − 1 | − 23 | 3 | 3 | 1 | 2 | 0 |
| Kurtosis | 57 | 101 | 128 | 114 | 146 | 143 | 1842 | 123 | 108 | 108 | 74 | 34 |
| Median | − 0.20 | 0.17 | 0.10 | − 0.02 | − 0.06 | 0.09 | − 0.19 | − 0.08 | 0.09 | − 0.12 | − 0.19 | − 0.34 |
| NMAD | 0.32 | 0.27 | 0.17 | 0.28 | 0.16 | 0.15 | 0.43 | 0.12 | 0.14 | 0.17 | 0.28 | 0.51 |
| 68.3% quantile | 0.41 | 0.34 | 0.23 | 0.25 | 0.19 | 0.19 | 0.49 | 0.19 | 0.21 | 0.26 | 0.35 | 0.55 |
| 95% quantile | 0.96 | 0.57 | 0.55 | 0.52 | 0.47 | 0.45 | 1.33 | 0.63 | 0.50 | 0.69 | 0.81 | 1.70 |
| UAS$_{SDM}$ error | | | | | | | | | | | | |
| Skewness | 0 | − 1 | 0 | 0 | − 1 | 0 | 0 | – | – | – | – | – |
| Kurtosis | 0 | 0 | 2 | 0 | 1 | 1 | 0 | – | – | – | – | – |
| Median | − 0.29 | − 0.06 | − 0.07 | − 0.24 | − 0.21 | − 0.15 | − 0.30 | – | – | – | – | – |
| SD | 0.36 | 0.26 | 0.21 | 0.21 | 0.14 | 0.19 | 0.35 | – | – | – | – | – |
| ME[a] | − 0.16 | 0.10 | 0.13 | − 0.03 | − 0.03 | 0.01 | − 0.12 | – | – | – | – | – |
| MAE[a] | 0.29 | 0.23 | 0.20 | 0.18 | 0.11 | 0.15 | 0.29 | – | – | – | – | – |
| RMSE[a] | 0.39 | 0.28 | 0.26 | 0.22 | 0.14 | 0.19 | 0.37 | – | – | – | – | – |

[a]Calculated without offset (Sect. 3.3.3)

with range (diameter increases by 0.24 m between 0 and 300 m), or local variations of the terrain slope angle have less influence on the result ($r_{SP} = − 0.3$ between $\delta$ and slope angle). We also checked correlation between $\delta$ and TLS error values for all the flights and found none ($r_{SP}$ between − 0.07 and 0.04). To sum up, although the calculated $\delta$ values are relatively large, compared to the size of the UAS$_{DSM}$ pixels, $\delta$ is independent from the magnitude of error in the UAS$_{DSMs}$.

149 MP checkpoints were measured in the late afternoon of 11 February 2015. The grid pattern of the data collection routine had an average spacing of 18 m. Average snow depth was 0.83 m, with a maximum of 1.4 m. Snow depth differences within the 2 × 2 m plots at each checkpoint were as high as 0.8 m, with an average of 0.23 m.

### 3.3. Accuracy and Precision Assessment

#### 3.3.1 Error Distribution

The histograms and Q–Q plots showed that distributions of UAS$_{DSM}$ and UAS$_{SDM}$ error followed a characteristic pattern. Examples of each type of distribution and plot are shown in Fig. 5.

The UAS$_{DSM}$ error distributions (Fig. 5a, b), show a high amount of values around the median (± 0.5 m) and clear deviation from the superimposed normal distribution in the histogram. The Q–Q plot confirms the impression of a non-normal error distribution; the deviation of the plotted values from the straight line indicates a large amount of outliers and therefore heavy tails of the error distribution (Höhle and Höhle 2009). These observations agree with the general notion that non-normal error distribution is very common in photogrammetric DSMs, as stated in textbooks and related studies (e.g., Müller et al. 2014; Maune 2007). The UAS$_{SDM}$ error distribution, on the other hand (Fig. 5c, d), shows good agreement with the normal distribution in the histogram. This observation is confirmed in the Q–Q plot; the plotted values are mostly located close to the line, indicating close resemblance between the empirical quantiles distribution and the theoretical quantiles of a normal distribution (Höhle and Höhle 2009). Skewness and kurtosis are shown in Table 6; skewness remains within ± 3 (except for flight 2/4) for UAS$_{DSM}$ and UAS$_{SDM}$ errors; however, kurtosis is high for UAS$_{DSM}$ errors (mean = 100, if excluding the outlier flight 2/4) and very low (mean = 0.6) in the UAS$_{SDM}$ errors (also apparent in histograms Fig. 5a, c). The general difference in error

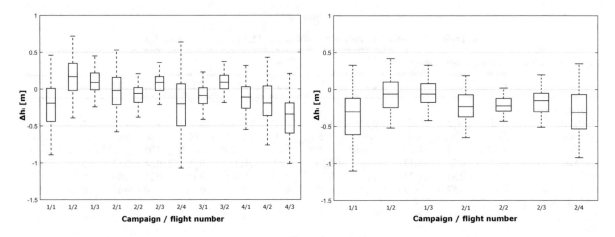

Figure 6
Boxplots of UAS$_{DSM}$ error (left image) and UAS$_{SDM}$ error (right image); $y$-axis shows $\Delta h_i$ [m], ID of UAS flight (campaign/flight number) are plotted on $x$-axis; whiskers in boxplot correspond to $1\sigma$, outliers not shown; boxplots truncated at 1/− 1.5 m for better visualisation

distributions and kurtosis values between UAS$_{DSM}$ and UAS$_{SDM}$ could be explained by the fact, that the majority of outliers in laser scanning and photogrammetric DSMs are caused by objects with high vertical offset from the terrain (Höhle and Höhle 2009). In the presented case, the TLS surveys the height of the snow surface and objects with high vertical offset (i.e., buildings, boulders, trees), thus potentially generating outliers (and therefore high kurtosis values); the MP routine, however, only samples the snow surface, therefore the accuracy assessment of UAS$_{SDM}$ is less prone to outliers and shows very low kurtosis values. Based on these observations, a normal distribution is assumed for the UAS$_{SDM}$ errors, and a non-normal distribution for the UAS$_{DSM}$ errors.

### 3.3.2 Accuracy Assessment of UAS$_{DSMs}$

Results of the UAS$_{DSM}$ accuracy assessment are provided in Table 6 and visualised in boxplots (Fig. 6). The results show a high error variability between the UAS flights. Low values for NMAD (< 0.2 m), 68.3% quantile (< 0.25 m) and 95% quantile (< 0.55 m) were determined for approximately half the flights (e.g. 1/3, 2/2 or 3/2). This translates to 68.3 and 95% of the UAS$_{DSM}$ errors of these flights being within a magnitude of ± 0.25 and ± 0.55 m, respectively (Höhle and Höhle 2009). The assessment of at least three flights (i.e., 1/1, 2/4

and 4/3), however, shows comparatively high values for the above-mentioned robust accuracy measures (> 0.3, > 0.4 and ≥ 1 m, respectively). The error medians are all within ± 0.2 m (except for flight 4/3). We analysed the correlation between indicators from the photogrammetric processing report (i.e., QI, overlap, marker error and reprojection error) and results from the UAS$_{DSM}$ accuracy assessment (i.e., median, NMAD, 68.3 and 95% quantiles). The highest correlations were found between marker error and NMAD ($r = -0.39$), and marker error and 68.3% quantile ($r = -0.35$); all the other pairings were $r < 0.3$. It therefore seems that from a statistical view-point, the photogrammetric processing indicators have little explanatory power regarding the accuracy of the UAS$_{DSM}$ when assessed with TLS. This could either be due to the fact that the chosen indicators are not statistically significant, or that the chosen sample size is too small to correctly show correlation between these indicators.

Exemplary results of the spatial distribution of errors from flights 1/1 through 1/3 are provided in Fig. 7. They show that the high (mostly negative) DSM$_{UAS}$ errors of flight 1/1 reported in Table 6 occur mainly in the east (near the TLS instrument) and far west of the AOI (close to a large boulder and stone pine cluster—inset in Fig. 1). The UAS$_{DSM}$ errors of flight 1/2 are mostly positive and are located in the AOI centre. While the overall error in UAS$_{DSM}$ of flight 1/3 is low (i.e., 68.3% quantile = 0.23 m, most

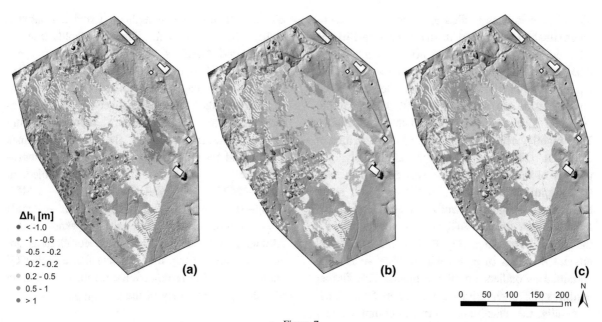

**Δh$_i$ [m]**
- $< -1.0$
- $-1 - -0.5$
- $-0.5 - -0.2$
- $-0.2 - 0.2$
- $0.2 - 0.5$
- $0.5 - 1$
- $> 1$

(a)        (b)        (c)

0  50  100  150  200
                      m

N

Figure 7

Results from accuracy assessment of UAS$_{DSMs}$ recorded on 11 February 2015 [flights 1/1 (**a**), 1/2 (**b**) and 1/3 (**c**)]; reds indicate negative, blues positive Δh$_i$ values (Adams et al. 2016)

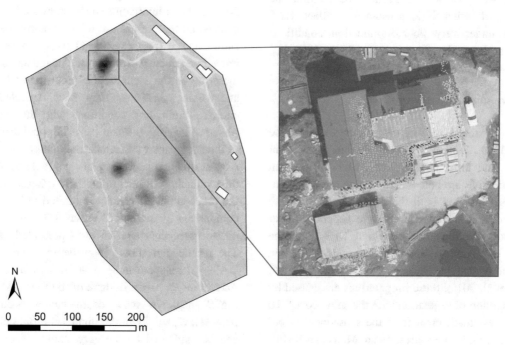

N

0  50  100  150  200
                      m

Figure 8

Location and occurrence of outliers; outlier heatmap of AOI—the darker the red, the more outliers in this area (left image); number of outlier occurrences at largest hotspot (red rectangle), coloured blue = 2 through red = 7 or more, per 0.2 m grid cell (right image); UAS orthophoto from 21 August 2015 in background of both figures plotted checkpoints within $\pm$ 0.2 m), an area of approximately 0.002 km$^2$ in the central part of the AOI shows high errors (0.5–1 m), surrounded by a 0.012 km$^2$ large area with errors in the 0.2–0.5 m range. Additionally, outliers were identified (i.e., errors outside the above-mentioned 95% quantile) and their location mapped for all the UAS flights. They mainly occur in the central and northern area of the AOI (Fig. 8, left image). Additionally, the number of outlier occurrences was counted in each 0.2 m grid cell (equivalent to UAS$_{DSM}$ cell size); high values are predominant on steep rock faces of large boulders (central AOI section, not shown) and the façades of buildings (Fig. 8, right image). A small amount of pixels ($n = 46$) within these areas were classified as outliers on all the flights. This further confirms the observations described in Sect. 3.3.1. Generally, the magnitude of error is complimentary in both statistical and spatial representations. However, the latter allows a more goal-oriented analysis of factors influencing UAS$_{DSM}$/UAS$_{SDM}$ error. An example was presented in a related publication, the potential of using NIR sensors to collect UAS imagery under very poor illumination conditions and its impact on the accuracy of UAS$_{DSMs}$/UAS$_{SDMs}$ (Bühler et al. 2017).

### 3.3.3 Accuracy Assessment of UAS$_{SDMs}$

Results of the UAS$_{SDM}$ error analysis are also included in Table 6 and Fig. 6 (boxplots—right image). The error median and five out of seven upper quantiles are below zero, indicating a systematic offset between both the data sets: snow depth values mapped with the UAS were generally lower than snow depth values measured with MP. The average of this offset was $- 0.19$ m. As described in detail in related publications (Adams et al. 2016; Bühler et al. 2016), these irregularities are caused by the interaction of vegetation with the snow cover. To exclude systematic error from the subsequent calculation of accuracy measures, mean, MAE and RMSE were determined after subtracting the offset from the measured values (Table 6). Compared to the UAS$_{DSM}$ errors described above, the boxplots show a lower spread of the UAS$_{SDM}$ errors. The UAS$_{SDM}$ RMSE of the middle five UAS flights (1/2 through

2/3) is $\leq$ 0.28 m, while flights 1/1 and 2/4 show a higher RMSE ($>$ 0.37 m). The same holds true for the reported MAE ($\leq$ 0.23 and $>$ 0.29 m, respectively) and SD ($\leq$ 0.26 and $>$ 0.35 m, respectively) (Fig. 6—right image). As with the UAS$_{DSM}$ errors, we analysed the correlation between UAS$_{SDM}$ accuracy measures and indicators reported during photogrammetric processing; the highest $r$ values were found for QI $-$ SD ($r = - 0.65$), RMS reprojection error $-$ SD ($r = - 0.64$), RMS reprojection error $-$ MAE ($r = - 0.68$), QI $-$ RMSE ($r = - 0.67$) and RMS reprojection error $-$ RMSE ($r = - 0.65$); all the other pairings were $r < 0.6$. Although, in absolute terms, these correlations are again not very strong, they indicate that a higher QI and/or lower RMS reprojection error may result in an overall higher accuracy of the UAS$_{SDMs}$.

### 3.3.4 Precision

As highlighted in related publications concerning UAS-based snow depth mapping (Bühler et al. 2016, 2017), illumination of the snow surface, sensor choice and the presence of minor disturbances of the snow surface (e.g., due to ski tracks) have a large impact on structure-from-motion image matching and therefore on the accuracy and precision of UAS$_{DSMs}$ and UAS$_{SDMs}$. Thus, only data collected under similar illumination conditions with the same VIS sensor setup from the same day (minimal disturbance of the snowpack by MP measurements) were considered in this precision assessment (Table 5). None of the available NIR data fitted these criteria; therefore, the precision assessment was limited to VIS data. A comparison of flights 1/2 versus 1/3 and 4/1 versus 4/2 for precision assessment is presented in Fig. 9. The profile plot shows snow depth values for four UAS$_{SDMs}$ mapped along a 90 m stretch of cleared road (transect A–B). In three of the four UAS$_{SDMs}$ in Fig. 9, negative snow depth values occur (grey patches), especially in the flight 1/3 profile plot. This can be explained by the very shallow snow depths along the cleared road (Sect. 2.3.2) and the slightly negative bias of the UAS$_{SDM}$ error (Fig. 6, Sect. 3.3.3). As described above, the precision is reported as the SD of the residues between both the acquisitions. For the profile line in Fig. 9 this was 0.04 m for

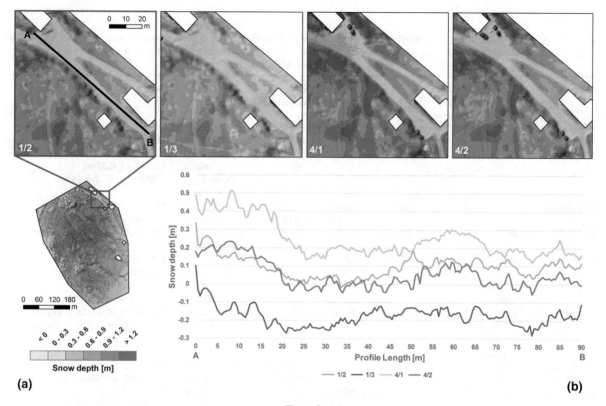

Figure 9

Precision assessment of UAS$_{SDMs}$; overview AOI snow depth values for flight 1/2 (**a**); detail-view of area marked with red rectangle for flights 1/2, 1/3, 4/1 and 4/2 (top row); snow depth values along profile (A–B) for all flights (**b**)

the comparison between both flights. For the entire AOI (approximately 6 million pixels), analysis of the residuals showed they were non-normally distributed; following the procedure applied to the accuracy assessment above, the 68.3 and 95% quantiles are reported: 0.25 and 0.55 m for 11 February, 0.33 and 0.99 m for 13 March, respectively.

### 3.4. Comparable Studies

To relate these results to the existing literature, Table 7 provides an overview of recent, comparable studies, dealing with accuracy assessment of UAS photogrammetry for snow depth mapping. A direct relation of findings from the presented study with comparable studies is limited, because:

i. Most studies are based on one to three UAS winter flights (all except Harder et al. 2016), thus temporal accuracy change cannot be investigated.

ii. Different kinds of reference data, with varying nominal accuracies were used.

iii. The employed methodologies for data processing and error analysis varied considerably and different accuracy measures were reported.

iv. The size and topography of the AOI varied considerably (0.007–0.65 km$^2$; steep mountain slopes vs. flat areas), incurring size- and terrain-effects on the results.

When putting aside these difficulties, it seems the accuracy reported here is within a similar range to the related studies. However, especially the implementation of recently available high-accuracy geolocation routines (e.g., real time kinematic GNSS) is able to substantially increase accuracy (Harder et al. 2016).

Of the studies presented in Table 7, only four reports on the precision of UAS$_{DSMs}$/UAS$_{SDMs}$: De Michele et al. (2016), Lendzioch et al. (2016) and Vander Jagt et al. (2015) report precision as the SD of

Table 7

*Overview of results from accuracy assessments in comparable studies for UAS-based snow depth mapping; geolocation errors reported in some listed studies are not included in this overview; multiple rows per study correspond to different sites and/or different reference measurements and related results*

| Author(s) | Size of AOI [km$^2$] | # UAS flights (winter) | UAS type | Sensor(s) | Reference measurement(s) | Evaluated parameters | Results |
|---|---|---|---|---|---|---|---|
| Avanzi et al. (2017) | 0.007 | 1 | Multicopter | RGB | Multistation[a] MP | DSM/point cloud (z values) Snow depth (z values) | SD = 0.056/0.025 m; RMSE = 0.069/0.036 m SD = 0.14–0.27 m; RMSE = 0.17–0.45 m |
| Boesch et al. (2016) | 0.35 | 2 | Multicopter | RGB & NIR | ADS100[b] GNSS | DSM (z values) | Median = 0.06–0.08 m Median = 0.1–0.17 m |
| De Michele et al. (2016) | 0.03 | 1 | Fixed-wing | RGB | MP | Snow depth (z values) | SD = 0.13 m; RMSE = 0.14 m |
| Harder et al. (2016) | 0.65 (site A) 0.32 (site B) | 22 18 | Fixed-wing | RGB | GNSS | DSM (z values) | SD = 0.06–0.08 m; RMSE = 0.08–0.012 m SD = 0.05–0.06 m RMSE = 0.08–0.09 m |
| Lendzioch et al. (2016) | 0.26[c] (site A) 0.005 (site B) | 2 | Multicopter | RGB | MP | Snow depth (z values) | SD = 0.22 m; RMSE = 0.22 m SD = 0.36 m; RMSE = 0.42 m |
| Marti et al. (2016) | 0.31 | 1 | Fixed-wing | RGB | Pléiades[d] GNSS | Snow depth (z values) | SD = 1.47 m; NMAD = 0.78 m SD < 0.63 m; NMAD < 0.38 m |
| Miziński and Niedzielski (2017) | 0.04 | 2 | Fixed-wing | RGB and NIR | MP | Snow depth (z values) | MAE = 0.33–0.43 m; RMSE = 0.41–0.58 m; NMAD = 0.37–0.55 m |
| Vander Jagt et al. (2015) | 0.007 | 1 | Multicopter | RGB | GNSS | Snow depth (z values) | RMSE = 0.9 m |
| Bühler et al. (2016)[e] | 0.12 (site A) 0.35 (site B) | 1 3 | Multicopter | RGB and NIR | MP | Snow depth (z values) | RMSE = 0.07–0.3 m RMSE = 0.15 m |
| Bühler et al. (2017)[e] | 0.12 (site A) 0.12 (site B) | 3 4 | Multicopter Fixed-wing | RGB and NIR | GNSS TLS | DSM (z values) | SD = 0.11–0.19 m; RMSE = 0.17–0.23 m RMSE = 0.18–0.77 m |

[a] Data collected in scanning mode

[b] Large-frame aerial camera on-board manned aircraft

[c] Total size of area – AOI-size not provided

[d] Very high resolution satellite imagery (two imagery triplets with 0.7 m ground sampling distance used to generate 1, 2 and 4 m DSMs)

[e] Studies also published within the project RPAS4SNOW

UAS$_{SDM}$ values of a single flight (0.1, 0.22–0.45, 0.21 m, respectively); Bühler et al. (2016) determined the precision by calculating the SD of residues for several UAS acquisitions along a limited, stable area (SD < 0.1 m), similar to the method used here. Precision results from the presented study are, therefore, in a similar range as in comparable studies, both with regard to small, stable areas and area-wide estimates.

## 4. Conclusions

In this work, we present a multitemporal assessment of the accuracy and precision of fixed-wing UAS photogrammetry for slope-scale snow depth mapping in alpine terrain. VIS and NIR imagery was collected with a modified off-the-shelf digital camera, mounted on a custom-built fixed-wing UAS. We performed 12 UAS flights during four campaigns in February and March 2015 in 'snow-on' and one in August 2015 in 'snow-off' conditions. The data were collected at a flat 0.12 $km^2$ study site, located at approximately 2000 m a.s.l. While all the UAS imagery was processed with the same parameter settings in structure-from-motion photogrammetry software, it was collected under different site- and UAS-specific settings. This allowed testing the setup under different conditions and investigating factors influencing the quality of the $UAS_{DSMs}$ and $UAS_{SDMs}$. Our assessment followed a threefold approach:

1. Accuracy assessment of $UAS_{DSMs}$ with TLS reference data.
2. Accuracy assessment of $UAS_{SDMs}$ with MP reference data.
3. Precision assessment of $UAS_{SDMs}$ by intercomparison of multiple UAS results.

Ad (1) To determine how well UAS photogrammetry was able to map the absolute height of the snow surface, an accuracy assessment of the $UAS_{DSMs}$ was performed with high-resolution, high-accuracy TLS reference data. While the choice of the study site location in a flat valley floor benefitted the UAS data acquisition (e.g., easy accessibility for MP and RP measurements), it resulted in an unfavourable TLS setup. The only slightly elevated location of the instrument over the valley floor resulted in a large amount of observations with high incidence angles, and caused large TLS footprints and occlusions (45% of AOI beyond TLS line-of-sight). However, the results showed no correlation between $UAS_{DSM}$ error magnitude and footprint size. Therefore, we considered the TLS data valid for $UAS_{DSM}$ accuracy assessment. $UAS_{DSM}$ error distribution was non-normal, but without systematic bias. The skewed error distribution resulted from outliers, which typically occurred at steep rock faces, trees or building façades. Robust accuracy measures were calculated for $UAS_{DSM}$ error and error maps interpreted visually. The results showed:

i. Low errors were mostly observed for VIS or NIR UAS imagery acquired with the AOI in full sunlight (68.3% quantile $\leq 0.35$ m; 95% quantile $\leq 0.57$ m).

ii. High errors were determined for VIS UAS imagery acquired with the AOI shadowed or with high amount of blurry imagery ($> 0.41$ m; $> 0.96$ m, respectively).

iii. For NIR imagery collected with the AOI shadowed, one flight showed errors in the same range as for imagery collected in full sunlight; a similar scenario in a different flight, however, showed very high errors. Visual interpretation suggested an underlying systematic error (e.g., poor image alignment due to changing illumination caused by shadow moving across AOI during image acquisition).

Ad (2) If UAS imagery was available for two or more points in time, height differences were calculated between the $UAS_{DSMs}$. In this case, one 'snow-off' $UAS_{DSM}$ was compared to several 'snow-on' $UAS_{DSMs}$ to determine relative surface height change, i.e., $UAS_{SDMs}$. The accuracy of $UAS_{SDMs}$ was assessed with MP data, collected for the February campaigns. The $UAS_{SDM}$ error was distributed normally, thus standard accuracy measures were applied. The results show a negative bias in the data, caused by the interaction of vegetation and snow cover (Adams et al. 2016; Bühler et al. 2016). This problem has also been described in comparable studies by other authors (e.g., Marti et al. 2016; Vander Jagt et al. 2015; Nolan et al. 2015). After correction of this bias, $UAS_{SDM}$ errors follow a pattern similar to the $UAS_{DSM}$ errors (RMSE $< 0.31$ and $> 0.39$ m for flights recorded under good and poor illumination conditions, respectively). The results also showed some indication of high QI and low RMS reprojection error being correlated to low $UAS_{SDM}$ error.

Ad (3) The precision of snow depth mapping with UAS photogrammetry was assessed by intercomparison of two sets of $UAS_{SDMs}$ recorded under similar site-specific settings (VIS imagery acquired with AOI in full sunlight). The results showed that over a small area with negligible intermittent height change, the normally distributed residues were within 0.04 m (SD) for both the comparisons; over the whole AOI, the 68.3% quantile of the non-normally distributed residues where within ± 0.25 to ± 0.33 m.

UAS-based snow depth mapping provides a reliable source of snow depth information at an unprecedented level of detail (decimetres to centimetres). On-demand UAS surveys at the slope-scale can be performed cost-efficiently to provide details on snow depth distribution and allow visual interpretation of the snow surface with orthophotos. In the presented case, testing different setups and sensors, and evaluating the spatial and temporal variations of accuracy and precision, provides further insight into UAS-based snow depth mapping. However, the employed indirect georeferencing technique proved to be a considerable drawback, because it was very time-consuming, limited the achievable accuracy and reduced the benefits of close-range sensing (Adams et al. 2016). Additionally, the TLS setup showed some weaknesses due to high incidence angles. Further developments could include applying the presented technique to an AOI with a geometric setup more suited to TLS measurements or the use of additional sensors for snow quality, rather than only snow quantity mapping.

## Acknowledgements

This research was funded within the Austrian Academy of Sciences (ÖAW) research programme Earth System Sciences (ESS)—International Geoscience Programme (IGCP)/Federal Ministry of Science, Research and Economy (BMWFW); project RPAS4SNOW. The authors would like to sincerely thank: Johann Zagajsek and his team for support at the test site; Andreas Huber and Armin Graf for supporting the field work; Julia Adams for proofreading the manuscript; both the reviewers and the editor for valuable comments, which helped improve the clarity and quality of this paper.

## REFERENCES

Adams, M. S., Bühler, Y., Boesch, R., Fromm, R., Stoffel, A. & Ginzler, C. (2016). Investigating the potential of low-cost remotely piloted aerial systems for monitoring the Alpine snow cover (RPAS4SNOW). Final Project Report, ÖAW—Austrian Academy of Sciences, Innsbruck (Austria), pp. 82.

Agisoft LLC. (2016). Agisoft PhotoScan user manual: Professional edition, Version 1.2.

Avanzi, F., Bianchi, A., Cina, A., De Michele, C., Maschio, P., Pagliari, D., Passoni, D. Pinto, L., Piras, M. & Rossi, L. (2017). Measuring the snowpack depth with unmanned aerial system photogrammetry: Comparison with manual probing and a 3D laser scanning over a sample plot. *The Cryosphere Discussion, in review.*

Beyers, J. H. M., Sundsbø, P. A., & Harms, T. M. (2004). Numerical simulation of three-dimensional, transient snow drifting around a cube. *Journal of Wind Engineering and Industrial Aerodynamics, 92*(9), 725–747.

Bilodeau, F., Gauthier, G., & Berteaux, D. (2013). The effect of snow cover on lemming population cycles in the Canadian high arctic. *Oecologia, 172,* 1007–1016.

Boesch, R., Bühler, Y., Ginzler, C., Adams, M. S., Fromm, R. & Graf, A. (2015). Optimizing channel weights for digital surface models with snow coverage. *ISPRS Archives, XL–3/W3.*

Boesch, R., Bühler, Y., Marty, M., & Ginzler, C. (2016). Comparison of digital surface models for snow depth mapping with UAV and aerial cameras. *The International Archives of the Photogrammetry, Remote Sensing and Spatial Information Sciences, XLI–B8,* 453–458.

Bühler, Y., Adams, M. S., Boesch, R., & Stoffel, A. (2016). Mapping snow depth in alpine terrain with unmanned aerial systems (UAS): Potential and limitations. *The Cryosphere, 10,* 1075–1088.

Bühler, Y., Adams, M., Stoffel, A., & Boesch, R. (2017). Photogrammetric reconstruction of homogenous snow surfaces in alpine terrain applying near infrared UAS imagery. *International Journal of Remote Sensing, 38,* 8–10.

Bühler, Y., Meier, L., & Ginzler, C. (2015). Potential of operational high spatial resolution near-infrared remote sensing instruments for snow surface type mapping. *IEEE Geoscience and Remote Sensing Letters, 12*(4), 821–825.

Cline, D. W., Bales, R. C., & Dozier, J. (1998). Estimating the spatial distribution of snow in mountain basins using remote sensing and energy balance modeling. *Water Resources Research, 34*(5), 1275–1285.

Colomina, I., & Molina, P. (2014). Unmanned aerial systems for photogrammetry and remote sensing: A review. *ISPRS Journal of Photogrammetry and Remote Sensing, 92,* 79–97.

De Michele, C., Avanzi, F., Passoni, D., Barzaghi, R., Pinto, L., Dosso, P., et al. (2016). Using a fixed-wing UAS to map snow depth distribution: An evaluation at peak accumulation. *The Cyrosphere, 10,* 511–522.

Decaulne, A. (2007). Snow-avalanche and debris-flow hazards in the fjords of north-western Iceland, mitigation and prevention. *Natural Hazards, 41*(1), 81–98.

Deems, J. S., Painter, T. H., & Finnegan, D. C. (2013). Lidar measurement of snow depth: A review. *Journal of Glaciology, 59,* 467–479.

Dietz, A. J., Kuenzer, C., Gassner, U., & Dech, S. (2012). Remote sensing of snow—A review of available methods. *International Journal of Remote Sensing, 33*(13), 4094–4134.

Dozier, J., & Painter, T. H. (2004). Multispectral and hyperspectral remote sensing of alpine snow properties. *Annual Review of Earth and Planetary Sciences, 32,* 465–494.

Durand, Y., Guyomarc'h, G., Mérindol, L., & Corripio, J. G. (2005). Improvement of a numerical snow drift model and field validation. *Cold Regions Science and Technology, 43*(1–2), 93–103.

Elder, K., Dozier, J., & Michaelsen, J. (1991). Snow accumulation and distribution in an alpine watershed. *Water Resources Research, 27*(7), 1541–1552.

Elder, K., Rosenthal, W., & Davis, R. E. (1998). Estimating the spatial distribution of snow water equivalence in a montane watershed. *Hydrological Processes, 12*(10–11), 1793–1808.

Fabris, M., & Pesci, A. (2005). Automated DEM extraction in digital aerial photogrammetry: Precisions and validation for mass movement monitoring. *Annales Geophysicae, 48,* 973–988.

Fahrmeir, L., Künstler, R., Pigeot, I., & Tutz, G. (2011). *Statistik: Der Weg zur Datenanalyse* (7th ed.). Heidelberg: Springer.

Fierz, C., Armstrong, R. L., Durand, Y., Etchevers, P., Greene, E., McClung, D. M., et al. (2009). *The International classification for seasonal snow on the ground.* Paris, France: IACS, UNESCO.

Fischer, J.-T., Kofler, A., Fellin, W., Granig, M., & Kleemayr, K. (2015). Multivariate parameter optimization for computational snow avalanche simulation. *Journal of Glaciology, 61*(229), 875–888.

Foppa, N., Stoffel, A., & Meister, R. (2007). Synergy of in situ and space borne observation for snow depth mapping in the Swiss Alps. *International Journal of Applied Earth Observation and Geoinformation, 9,* 294–310.

Fras, K. M., Kerin, A., Mesarič, M., Peterman, V & Grigillo, D. (2016). Assessment of the quality of digital terrain model produced from unmanned aerial system imagery. In *The international archives of the photogrammetry, remote sensing and spatial information sciences, XLI–B1,* XXIII ISPRS Congress, 12–19 July 2016, Prague, Czech Republic.

Fuchs, S., Thöni, M., McAlpin, M. C., Gruber, U., & Bründl, M. (2007). Avalanche hazard mitigation strategies assessed by cost effectiveness analyses and cost benefit analyses—Evidence from Davos, Switzerland. *Natural Hazards, 41*(1), 113–129.

Furukawa, Y. & Ponce, J. (2009). Dense 3D motion capture for human faces. In *Proceedings/CVPR, IEEE computer society conference on computer vision and pattern recognition.*

Gini, R., Pagliari, D., Passoni, D., Pinto, L., Sona, G., Dosso, P. (2013). UAV photogrammetry: block triangulation comparisons. International Archives of the Photogrammetry, Remote Sensing and Spatial Information Sciences, XL-1/W2, 157–162.

Ginzler, C., Marty, M., & Bühler, Y. (2013). Grossflächige hochaufgelöste Schneehöhenkarten aus digitalen Stereoluftbildern. *DGPF Tagungsband, 22,* 71–78.

Grünewald, T., Schirmer, M., Mott, R., & Lehning, M. (2010). Spatial and temporal variability of snow depth and ablation rates in a small mountain catchment. *The Cryosphere, 4,* 215–225.

Haala, N. (2011). Multiray photogrammetry and dense image matching. In D. Fritsch (Ed.), *Proceedings of photogrammetric week 2011, Wichmann/VDE Verlag, Berlin & Offenbach* (pp. 185–195).

Harder, P., Schirmer, M., Pomeroy, J., & Helgason, W. (2016). Accuracy of snow depth estimation in mountain and prairie environments by an unmanned aerial vehicle. *The Cryosphere, 10,* 2559–2571.

Hartzell, P. J., Gadomski, P. J., Glennie, C. L., Finnegan, D. C., & Deems, J. S. (2015). Rigorous error propagation for terrestrial laser scanning with application to snow volume uncertainty. *Journal of Glaciology, 61*(230), 1147–1158.

Harwin, S., Lucieer, A., & Osborn, J. (2015). The impact of the calibration method on the accuracy of point clouds derived using unmanned aerial vehicle multi-view stereopsis. *Remote Sensing, 7*(9), 11933–11953.

Helbig, N., van Herwijnen, A., & Jonas, T. (2015). Forecasting wet-snow avalanche probability in mountainous terrain. *Cold Regions Science and Technology, 120,* 219–226.

Höhle, J., & Höhle, M. (2009). Accuracy assessment of digital elevation models by means of robust statistical methods. *ISPRS Journal of Photogrammetry and Remote Sensing, 64*(4), 398–406.

Holub, M., & Fuchs, S. (2009). Mitigating mountain hazards in Austria—Legislation, risk transfer and awareness building. *Natural Hazards and Earth System Sciences, 9,* 523–537.

Hugenholtz, C. H., Whitehead, K., Brown, O., Barchyn, T. E., Moorman, B. J., LeClair, A., et al. (2013). Geomorphological mapping with a small unmanned aircraft system (sUAS): Feature detection and accuracy assessment of a photogrammetrically derived digital terrain model. *Geomorphology, 194,* 16–24.

Immerzeel, W. W., Kraaijenbrink, P. D. A., Shea, J. M., Shrestha, A. B., Pellicciotti, F., Bierkens, M. F. P., et al. (2014). High-resolution monitoring of Himalayan glacier dynamics using unmanned aerial vehicles. *Remote Sensing of Environment, 150,* 93–103.

Jörg, P., Fromm, R., Sailer, R. & Schaffhauser, A. (2006). Measuring snow depth with a terrestrial laser ranging system. In *ISSW international snow science workshop 2006, Telluride, Colorado, Proceedings* (pp. 452–460).

Klemas, V. V. (2015). Coastal and environmental remote sensing from unmanned aerial vehicles: An overview. *Journal of Coastal Research, 31*(5), 1260–1267.

Koenderink, J. J., & van Doorn, A. J. (1991). Affine structure from motion. *Journal of the Optical Society of America, 8*(2), 377–385.

Lendzioch, T., Langhammer, J. & Jenicek, M. (2016). Tracking forest and open area effects on snow accumulation by unmanned aerial vehicle photogrammetry. In *The international archives of the photogrammetry, remote sensing and spatial information sciences, XLIB1,* XXIII ISPRS Congress, 12–19 July 2016, Prague, Czech Republic, 2016.

Mancini, F., Dubbini, M., Gattelli, M., Stecchi, F., Fabbri, S., & Gabbianelli, G. (2013). Unmanned aerial vehicles (UAV) for high-resolution reconstruction of topography: The structure from motion approach on coastal environments. *Remote Sensing, 5,* 6880–6898.

Margreth, S., & Romang, S. (2010). Effectiveness of mitigation measures against natural hazards. *Cold Regions Science and Technology, 64*(2), 199–207.

Marti, R., Gascoin, S., Berthier, E., de Pinel, M., Houet, T., & Laffly, D. (2016). Mapping snow depth in open alpine terrain from stereo satellite imagery. *The Cryosphere, 10,* 1361–1380.

Maune, D. F. (2007). *Digital elevation model technologies and applications: The DEM user manual* (2nd ed.). Bethesda: ASPRS Publications.

Miziński, B., & Niedzielski, T. (2017). Fully-automated estimation of snow depth in near real time with the use of unmanned aerial vehicles without utilizing ground control points. *Cold Regions Science and Technology, 138*, 63–72.

Morgenthal, G., & Hallermann, N. (2016). Quality assessment of unmanned aerial vehicle (UAV) based visual inspection of structures. *Advances in Structural Engineering, 17*(3), 289–302.

Müller, J., Gärtner-Roer, I., Thee, P., & Ginzler, C. (2014). Accuracy assessment of airborne photogrammetrically derived high resolution digital elevation models in a high mountain environment. *ISPRS Journal of Photogrammetry and Remote Sensing, 98*, 58–69.

Nex, F., & Remondino, F. (2013). UAV for 3D mapping applications: A review. *Applied Geomatics, 6*(1), 1–15.

Nolan, M., Larsen, C., & Sturm, M. (2015). Mapping snow depth from manned aircraft on landscape scales at centimeter resolution using structure-from-motion photogrammetry. *Cryosphere, 9*, 1445–1463.

Nolin, A. W. (2010). Recent advances in remote sensing of seasonal snow. *Journal of Glaciology, 56*(200), 1141–1150.

Nolin, A. W., & Dozier, J. (2000). A Hyperspectral method for remotely sensing the grain size of snow. *Remote Sensing of Environment, 74*(2), 207–216.

Painter, T. H., Berisford, D. F., Boardman, J. W., Bormann, K. J., Deems, J. S., Gehrke, F., et al. (2016). The airborne snow observatory: Fusion of scanning lidar, imaging spectrometer, and physically-based modeling for mapping snow water equivalent and snow albedo. *Remote Sensing of the Environment, 184*, 139–152.

Peng, S., Piao, S., Ciais, P., & Fang, J. (2010). Change in winter snow depth and its impacts on vegetation in China. *Global Change Biology, 16*, 3004–3013.

Prokop, A. (2008). Assessing the applicability of terrestrial laser scanning for spatial snow depth measurements. *Cold Regions Science and Technology, 54*(3), 155–163.

Prokop, A., Schirmer, M., Rub, M., Lehning, M., & Stocker, M. (2008). A comparison of measurement methods: Terrestrial laser scanning, tachymetry and snow probing, for the determination of the spatial snow depth distribution on slopes. *Annals of Glaciology, 49*, 210–216.

Rees, W. G. (2006). *Remote sensing of snow and ice.* Boca Raton, FL: CRC Press.

Riegl. (2006). *Long-range laser profile measuring system LPM 98-2K—Technical documentation & user's instructions.*

Riegl. (2010). *Long-range laser profile measuring system LPM-321—Technical documentation & user's instructions.*

Schaffhauser, A., Adams, M., Fromm, R., Joerg, P., Luzi, G., Noferini, L., et al. (2008). Remote sensing based retrieval of snow cover properties. *Cold Regions Science and Technology, 54*, 164–175.

Schaffhauser, A. & Fromm, R. (2008). General statements on the temporal and spatial development of snow cover during snow fall periods particularly with consideration of snow drift. Deliverable D 7.5 in the project GALAHAD Advanced Remote Monitoring Techniques for Glaciers, Avalanches and Landslides Hazard Mitigation.

Schöber, J., Schneider, K., Helfricht, K., Schattan, P., Achleitner, S., Schöberl, F., et al. (2014). Snow cover characteristics in a glacierized catchment in the Tyrolean Alps-improved spatially distributed modelling by usage of Lidar data. *Journal of Hydrology, 519*, 3492–3510.

Sona, G., Pinto, L., & Pagliari, D. (2014). Experimental analysis of different software packages for orientation and digital surface modelling from UAV images. *Earth Science Informatics, 7*(2), 97–107.

Teich, M., Fischer, J.-T., Feistl, T., Bebi, P., Christen, M., & Grêt-Regamey, A. (2014). Computational snow avalanche simulation in forested terrain. *Natural Hazards and Earth System Sciences Discussions, 14*, 2233–2248.

Tonkin, T., Midgley, N. G., Graham, D. J., & Ladadz, J. (2014). The potential of small unmanned aircraft systems and structure-from-motion for topographic surveys: A test of emerging integrated approaches at Cwm Idwal, North Wales. *Geomorphology, 226*, 35–43.

Turner, D., Lucieer, A., & de Jong, S. M. (2015). Time series analysis of landslide dynamics using an unmanned aerial vehicle (UAV). *Remote Sensing, 7*(2), 1736–1757.

Vander Jagt, B., Lucieer, A., Wallace, L., Turner, D., & Durand, M. (2015). Snow depth retrieval with UAS using photogrammetric techniques. *Geosciences, 5*, 264–285.

Verhoeven, G. (2011). Taking computer vision aloft—Archaeological three-dimensional reconstructions from aerial photographs with photoscan. *Archaeological Prospection, 18*, 67–73.

Vernay, M., Lafaysse, M., Mérindol, L., Giraud, G., & Morin, S. (2015). Ensemble forecasting of snowpack conditions and avalanche hazard. *Cold Regions Science and Technology, 120*, 251–262.

Whitehead, K., & Hugenholtz, C. H. (2014). Remote sensing of the environment with small unmanned aircraft systems (UASs), part 1: A review of progress and challenges. *Journal of Unmanned Vehicle Systems, 2*(3), 69–85.

Willmott, C. J., & Matsuura, K. (2006). On the use of dimensioned measures of error to evaluate the performance of spatial interpolators. *International Journal of Geographical Information Science, 20*(1), 89–102.

Zongjian, L., Guozhong, S. & Feifei, X. (2012). UAV borne low altitude photogrammetry system. In *International archives of the photogrammetry, remote sensing and spatial information sciences, XXXIX-B1*, 2012 XXII ISPRS Congress, 25 August–01 September 2012, Melbourne, Australia.

(Received May 31, 2017, revised November 30, 2017, accepted December 4, 2017, Published online December 14, 2017)

Pure Appl. Geophys. 175 (2018), 3325–3342
© 2018 The Author(s)
This article is an open access publication
https://doi.org/10.1007/s00024-018-1767-3

| Pure and Applied Geophysics

# UAS as a Support for Atmospheric Aerosols Research: Case Study

Michał T. Chiliński,[1] Krzysztof M. Markowicz,[1] and Marek Kubicki[2]

*Abstract*—Small drones (multi-copters) have the potential to deliver valuable data for atmospheric research. They are especially useful for collecting vertical profiles of optical and microphysical properties of atmospheric aerosols. Miniaturization of sensors, such as aethalometers and particle counters, allows for collecting profiles of black carbon concentration, absorption coefficient, and particle size distribution. Vertical variability of single-scattering properties has a significant impact on radiative transfer and Earth's climate, but the base of global measurements is very limited. This results in high uncertainties of climate/radiation models. Vertical range of modern multi-copters is up to 2000 m, which is usually enough to study aerosols up to the top of planetary boundary layer on middle latitudes. In this study, we present the benefits coming from usage of small drones in atmospheric research. The experiment, described as a case study, was conducted at two stations (Swider and Warsaw) in Poland, from October 2014 to March 2015. For over 6 months, photoacoustic extinctiometers collected data at both stations. This enabled us to compare the stations and to establish ground reference of black carbon concentrations for vertical profiles collected by ceilometer and drone. At Swider station, we used Vaisala CL-31 ceilometer. It delivered vertical profiles of range corrected signal, which were analysed together with profiles acquired by micro-aethalometer AE-51 and Vaisala RS92-SGP radiosonde carried by a hexacopter drone. Near to the surface, black carbon gradient of $\approx 400$ ($\mu g/m^3$)/100 m was detected, which was below the ceilometer minimal altitude of detection. This confirmed the usefulness of drones and potential of their support for remote sensing techniques.

**Key words:** Atmospheric aerosols, black carbon, UAS, drone, unmanned aerial vehicle, vertical profiles.

## 1. Introduction

In the last few years, the development of technologies for small unmanned aerial systems (sUAS) made them affordable and easy to use as tools in research. Elevating small sensors above ground level has significant benefits for atmosphere physics, especially when it comes to aerosol research. Complex interactions between aerosols and solar radiation, together with very limited measurement options of vertical profiles of aerosols' optical and microphysical properties, make understanding and modeling Earth's climate difficult (Bond et al. 2013; IPCC 2013; Koch and Del Genio 2010; Myhre and Samset 2015). Columnar integrated data are available for many properties, but it is important to remember that for some aerosols, such as absorption of black carbon (BC), vertical distribution is even more important than total columnar values (Samset and Myhre 2011; Samset et al. 2013; Zarzycki and Bond 2010; Cook and Highwood 2004). Aerosols play an important role in radiative transfer and Earth's energy budget, both in terms of direct and indirect effects. Recently, we observe an increase of interest in air quality and its influence on human health. Owing to monitoring of large air polluters, local emissions are more strictly controlled. Air-quality monitoring and modeling focus on particle size distribution and their horizontal and vertical variability (Morawska et al. 1999; Vardoulakis et al. 2003; Chan et al. 2005). New ways of acquiring vertical profiles of aerosols near to the ground could significantly improve measurements and, later on, lead to constructing better models and more advanced data processing. Apart from aerosols, drones equipped with proper sensors are able to collect information on trace gases (Brady et al. 2016). Due to their limited endurance, caused by short-lasting electrical power supply, sUAS serve best in experiments where synergy of ground-based soundings and remote sensing data is used. Profiles delivered by tropospheric lidars tend to overlap, and hence, they cannot be used to detect aerosols in the layers close to the surface (Wandinger and Ansmann

---

[1] Institute of Geophysics, Faculty of Physics, University of Warsaw, Warsaw, Poland. E-mail: mich@igf.fuw.edu.pl
[2] Institute of Geophysics, Polish Academy of Sciences, Warsaw, Poland.

2002), which means that a researcher is forced to use near-field or dual field of view lidars (Lv et al. 2015).

Multi-rotor electric-powered drones perform very well as tools for capturing vertical profiles of quantities that are important for the purpose of atmospheric science. What makes them particularly useful is their vertical take-off and landing, they are also easy to operate, as they are equipped with sophisticated autopilot systems. Moreover, drones do not produce pollution, typical for combustion engines. Before multi-copters became popular, low-altitude vertical profiles of aerosols were often collected with helium-tethered balloons (Ferrero et al. 2014), but their usage is more limited, due to wind and space requirements. The balloons do not need electrical energy for lifting payload and are able to descend/ascend very slowly, reaching high spatial resolutions. On the other hand, drones can collect vertical profiles much faster than any balloon, reaching time resolution of $\approx 15$ min per profile. It is worth to mention that each individual profile could be acquired without any additional costs apart from the initial cost of an sUAS ($\approx 3000$ to 13000 USD) and its maintenance parts. This means that this method is very cost-efficient, especially during long field campaigns or measurements in remote locations. Recently many researchers are focused on vertical profiling over remote, fragile locations as Arctic. Many different platforms were used for such measurements as aircraft (Schwarz et al. 2010; Spackman et al. 2010), helicopter (Kupiszewski et al. 2013), or tethered balloon (Ferrero et al. 2016; Markowicz et al. 2017), but there is still need for sUAS dedicated to measurements in Arctic.

The aim of this work is to present the benefits of sUAS usage in atmospheric aerosols research, as proven by the described case study. Our field experiment was conducted in the last quarter of 2014 and first quarter of 2015 in Swider and Warsaw (Poland). The analysed cases contain ground-based measurements with photoacoustic devices, backscatter vertical profiles from ceilometer, and vertical profiles of BC concentration collected by a detector mounted on a detector installed on a drone. Among the presented profiles captured by sUAS, there are a few unique ones, such as severe smog conditions, untraceable by most of lidars.

Thermodynamical profiles acquired together with BC concentration were useful for the analysis of aerosols vertical distribution and relation between temperature inversion and accumulation of aerosols close to the surface.

## 2. Unmanned Aerial Systems in Atmospheric Aerosols Research

Modern science uses many different methods to acquire scientific data related to physics of atmosphere. One of them is data dedicated to optical and microphysical properties of atmospheric aerosols and their interactions with radiation. Most common division of methods discriminate in situ measurement and remote sensing retrieval of quantities as aerosol optical depth (AOD), extinction coefficient, scattering coefficient, absorption coefficient, mass, and particle number concentrations (Hess et al. 1998; Horvath 1993; McMurry 2000; Arnott et al. 2005; Bond et al. 1999; Wiedensohler et al. 2012). In situ measurements are limited spatially, but usually thanks to enough space, power supply and lack of mass limits, they deliver high time-resolution and low-noise results. In case of remote sensing methods, three basic types could be defined: ground-based, aerial, and satellite (Loeb et al. 2007). On ground level, AOD is measured by photometers (Holben et al. 1998), while vertical profiles of extinction/backscattering coefficient or depolarization ratio could be retrieved from tropospheric lidars. Satellites carry radiometers for columnar measurements and lidar for vertical profiles measured from top of atmosphere, down to the ground. Measurements done from an orbit extend to large spaces, covering the whole globe, but with low time-resolution, due to differences in revisit time and changes in cloud cover. During aerial campaigns, both types of measurement could be conducted. Remote sensing and in situ equipment are mounted on aircrafts during dedicated flight-campaigns. This type of campaigns delivers unique data, but incurs significant costs and requires long planning and preparation phase. This makes them episodic, so to speak, and dedicated to special events and basic research rather than monitoring of atmosphere and long-term measurements (Ramanathan et al. 2001; Welton et al.

2002). Small unmanned aerial systems, mostly multi-rotors, could help to fill the gap between in situ and aerial measurements in the aspect of temporal and spatial resolution of atmospheric aerosols measurements. Miniaturized sensors, based on full-scale sensors used in on-ground measurements, could be carried on altitudes up 2000 m above ground level and deliver vertical profiles of atmospheric aerosol selected properties in the same way as radio-sounding, but in repeatable way, without any need to use new equipment during each flight (Chilinski et al. 2016). Thanks to easy operation technique (professional training takes up to 2 weeks), small size, and almost no cost of a single flight, drones could significantly extend data set of local measurements and improve understanding of lidar vertical profiles together with ground data.

### 2.1. Miniature Sensors

The potential success of small UAS in atmospheric aerosols research lies in miniaturization of sensors, which are light enough to fit in payload limits. Mass is the crucial factor here, and in most cases, 1000–1500 g is the maximum payload; otherwise, it could significantly influence range and flight time. Moreover, 'flying sensors' meet the requirements of power efficiency and data storage/transmission. Along with the sensor, a power source must be carried. Some sensors are equipped with internal batteries and convenient when no additional power sources are needed, but usually, it is impossible to change such batteries, so after a flight, the vessel requires downtime for recharging. If we consider using external power supply, depending on drone design, two options are possible. In the first one, the power is supplied by drone's main batteries, through universal battery elimination circuit (UBEC) with right voltage range. This solution is the best in terms of mass, but could decrease flight time due to additional power consumption, although in comparison with main engines, using up $\approx 60A@22.2$ V (Chang et al. 2016), consumption of $\approx 1A@5$ V is minimal. Users do not have to remember about recharging batteries for sensors, but always have access to the main power source. The second option bases on dedicated external battery for powering payload. It adds some mass to the whole payload, but thanks to small batteries ($< 100$ g, up to 2 h of operation) it is possible to rotate batteries between flight and operate without downtimes, and with no significant reduction of flight time. In real life, all the above ideas are often applied; however, the external payload battery option is probably the most popular.

Apart from supplying power to a sensor, a method for data recording must be secured. There is an approach based on integrated solutions, e.g., internal micro-controller and embedded memory or external data-logger. Another possibility involves devices which transmit data to be stored on-ground, similar to common meteorological radiosondes. External data-logger with a micro-controller is especially efficient when multiple sensors are used during one flight. In such cases, data integration is easier when the entire data stream is saved on one device, with proper timestamps to easily match together all measurements and create unified data output. Sometimes, it is even possible to connect data-logging device to a telemetry channel and send online preview to the ground station. Access to online data is vital for in-flight decision-making: basing on the results, a decision to shorten or extend flight time ca be made.

Not disregarding the benefits of miniaturization, such as the aforementioned possibility to mount sensors on small drones, it is important to remember the drawbacks of smaller devices. For example, in situ aerosol equipment with closed measurement chamber which sucks air inside has smaller airflow in their miniaturized versions. Reduced airflow increases noise and causes worse overall performance in signal-to-noise ratio. This requires longer integration times, which in turn decreases spatial resolution of the data. Another factor affecting the results is less-controlled measurement conditions, especially air temperature and humidity. In full-scale devices, air could be dried and heated up to the defined values, whereas in their miniaturized versions, it is impossible, due to limited space and power. It is always important to verify the conditions (ambient or standardized) and parameters that we are measuring. Limitations of small sensors should be checked every time for selected measurement method and measured quantities.

## 2.2. Flight Range/Spatial Resolution

The most common concerns of small UAS users are flight time of the platform and its range of operation. The range depends on flight time, but there is also the second important factor: spatial resolution of data sampling. In vertical profiles, ascent/descent rate and sampling resolution of sensors determine spatial resolution of collected data. Small multi-rotors are usually able to ascend up to 8 m/s and descend up to 4.0 m/s (Pixhawk 2017). Vertical speed and mean flight time determine range, i.e., the maximum altitude of measured profiles. When sensors record data every second, spatial resolution is equal to the vertical speed, but in the case of time-averaging, spatial resolution is decreased. Due to high noise in miniature devices, most of the data need to be smoothed and averaged, and hence, to get better spatial resolutions, drones have to ascend or descend more slowly, which automatically reduces range of profiles. Differences in airflow around a drone during (Luo et al. 2016) ascent and descent require us to carry out multiple flights for different scenarios. To achieve the high spatial resolution and long vertical profile, while maintaining right safety margin, a scenario of fast ascent and slow descent can be chosen. In this scenario, drone ascends with sub-maximal speed until a defined battery level is reached, then it starts to descend slowly onto the ground level (the speed may be increased if the safety level of batteries allows that). To ensure the accuracy of our research, we also prepared a downward profile with higher spatial resolution and smaller airflow around a drone. Table 1 presents estimated vertical profile range, with different ascent/descent speed for 12-min flight (mean duration of aerosol atmospheric flight for flights in Laboratory of Atmospheric Physics on University of Warsaw).

Crucial stages of interesting aerosol events that are related to anthropogenic emissions take place in the planetary boundary layer, or even lower, at the short distance of a few hundred meters above the surface (Matthias et al. 2004; Nilsson et al. 2001). Consequently, the flight time can be used to improve spatial resolution rather than to collect high profiles. Another approach to this problem benefits from very fast battery changes and almost instant measurement of consequent profiles. Revisit times of around 3–5 min deliver valuable data in the case of very dynamic events. For flights up to 300 m, it is even possible to capture more than one profile on one battery set.

## 2.3. Control and Mission Planning

Most of small UAS are used for ground imaging purposes, so the most common way of controlling them is manual take-off/landing and pre-programmed flight, controlled by autopilot through waypoints defined by the ground control/mission planning software. This approach is the best when the point of interest is well defined spatially and the flight is conducted in stable conditions. For vertical sounding with sUAS, pre-programmed flight patterns can be used for low-altitude flights, when there is a safety margin for battery capacity. For higher flights, over a few hundred meters above the ground, wind conditions may be different than those on the ground due to

Table 1

*Vertical range in meters of sUAS for 12-min flight with different ascend/descend speed*

| | Descent (m/s) | | | | |
|---|---|---|---|---|---|
| | 4.5 | 4.0 | 3.0 | 2.0 | 1.0 |
| Ascend (m/s) | | | | | |
| 6.0 | 1851 | 1728 | 1440 | 1080 | 617 |
| 5.0 | 1705 | 1600 | 1350 | 1029 | 600 |
| 4.0 | 1525 | 1440 | 1234 | 960 | 576 |
| 3.0 | 1296 | 1234 | 1080 | 864 | 540 |
| 2.0 | 997 | 960 | 864 | 720 | 480 |
| 1.0 | 589 | 576 | 540 | 480 | 360 |

stronger winds and gusts. In such flights, constant careful observation of battery voltage and drone behaviour is a necessity. Autopilots can terminate pre-programmed mission if battery voltage threshold is reached; however, there are situations when the voltage drops only for a few seconds due to a wind gust. In such a case, autopilot will abort mission and collected profile is significantly shortened.

There is a very useful feature for simple vertical profiles collection, namely the stabilized mode. This mode allows the autopilot only to hold the drone on an altitude/latitude/longitude (based on pressure and GPS), but it is the operator who manually controls speed and flight direction. In stabilized manual flight, the operator and their ground team (typically, a team consists of three people: operator, operator assistant/ observer, and payload operator) make decisions about continuing or terminating the flight. When online data from sensors are available, it is even possible to change ascent/descent speed during the flight to focus on the measurements in the most interesting parts of the actual vertical profile.

## 3. Case Study

In the following section, an example of aerosol research with an aid of sUAS is presented. Institute of Geophysics, University of Warsaw, in cooperation with the Institute of Geophysics, Polish Academy of Science Geophysical Observatory at Swider, conducted an atmospheric aerosols experiment during the last quarter of 2014 and the first quarter of 2015. The goal of the experiment was to measure variability of black carbon concentration during the heating season in the center of Warsaw and in one of the surrounding towns. During winter, small towns suffer from high local emissions from household heating systems (Zawadzka et al. 2013). Central parts of Warsaw are heated by the central heating system, and hence, local emissions are mostly connected with transportation (Holnicki et al. 2017).

### 3.1. Location

Data used for the experiment were collected from various sites around the city of Warsaw (Poland) and

supported by radio-soundings from the third station. An overview map (Fig. 1) presents location of all of those sites. In Warsaw and Swider, in situ measurements with The Photoacoustic Extinctiometers (PAX) were conducted. In Swider Geophysical Observatory (52.11°N, 21.23°E, 94 m. asl), we had a Vaisala CL31 ceilometer operating and all of the described drone flights were made there. The station in Warsaw (52.21°N, 20.98°E, 112 m. asl) is located 4 km from the largest airport in Poland (Warsaw Chopin Airport ≈ 125,000 operations annually) and only 300 m from the Medical University of Warsaw Hospital's helipad. This is the reason why all of the drone operations were done in Swider. In Swider, the controlled traffic region (CTR) is located 600 m above ground level. To ensure safety during the experiment, special air zone was requested, ranging from the ground level up to 1000-m altitude, and 1-km diameter around the Swider station. To verify actual thermodynamical conditions and detect temperature inversion from meteorological soundings, we used the data coming from WMO #12374 Legionowo station (52.40°N, 20.95°E, 73 m. asl).

### 3.2. Instruments

Photoacoustic Extinctiometers 870 and 532 nm were used to measure scattering coefficient, absorption coefficient, single-scattering albedo, and BC concentration (Nakayama et al. 2015). The first device was placed in Swider Observatory, 2 m above ground level; the second operated in Warsaw, 18 m above the ground level. Data were integrated for over 60 s and then averaged with running mean and data window of 15 min. BC concentration was calculated with mass absorption cross section 7.75 $m^2/g$ for 532 nm and 4.74 $m^2/g$ for 870 nm. For analysis of BC concentration during low altitude, we used temperature inversion mean values averaged over 15 min from radiosonde start (twice per day, at noon and midnight UTC).

Vertical profiles of backscattering coefficient were collected at Swider station by Vaisala CL-31 ceilometer, which operates at 905 nm (Sokol et al. 2014). Vaisala ceilometers have internal overlap correction done almost to the ground level, which offers one of the best results among widely available

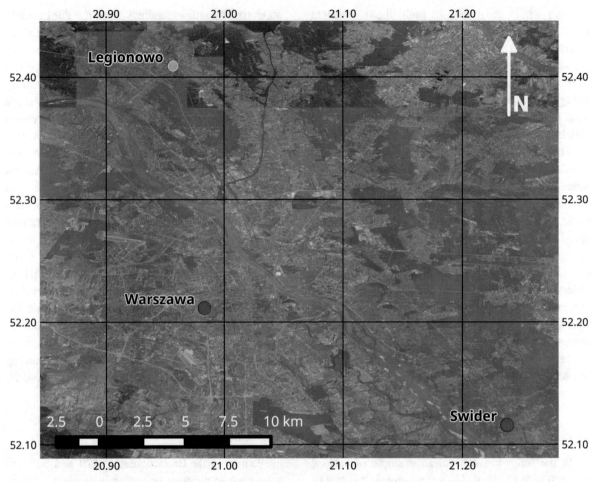

Figure 1
Overview map of measurement stations location (red dots) and WMO station in Legionowo (grey dot)

commercial ceilometers (Madonna et al. 2015). A ceilometer with overlap corrected as low as possible is the best for comparing drone profiles, registered instantly from the ground level. After verification, we assumed that usable profiles were from the range of 60–100 m above ground level. Presented profiles from the ceilometer are the range corrected signal (RCS).

BC concentration measurements were done with a micro-aethalometer AE-51, produced by AethLabs (Ferrero et al. 2014; Chilinski et al. 2016). AE-51 is fully autonomous, which has its own internal battery and memory, capable of operating for over 12 h. Measurements were done at 880 nm, with 1-s integration time and maximum airflow speed of 0.2l/min. Mass attenuation cross section was assumed

for 12.5 $m^2$/g and mass absorption cross section for 4.54$^2$/g. Total weight of the device is $\approx$ 280 g. The radiosonde used for acquiring thermodynamical profiles was well-known Vaisala RS92-SGP (Nash et al. 2010 with replaced original battery and widely used by WMO stations. Alkaline batteries were replaced with a lithium polymer 7.2V rechargeable battery. The battery replacement made the radiosonde lighter, now weighing $\approx$ 200 g.

sUAS utilized during the experiment was Versa X6sci hexacopter, manufactured by Versadrones from Ireland (Chilinski et al. 2016). The platform used six 15.5″ propellers, powered by 340 K/V brushless motors with total torque of 6.6 kg. The total mass of the system is 3.5 kg with 490 g of payload and the average flight time of the drone is around 12 min. During the experiment, flights were manually controlled by an operator with live data feed by 2.4-GHz data link (Fig. 2).

## 4. Results

In this section, we are going to discuss the outcomes of our experiment. Starting from data analysis, we will focus on vertical profiles of BC collected with the sUAS and background results from other instruments. First, we present differences in BC concentration, scattering coefficient, and absorption coefficient between station in Swider and Warsaw. Then, we proceed to the analysis of the influence of temperature vertical profile, especially low-temperature inversion, on BC concentrations. Next, we overview ceilometer performance in measuring aerosol events near to the ground level. Finally, we describe selected vertical profiles of BC concentration measured by sUAS compiled with corresponding profiles from ceilometer and ground results from PAX.

### 4.1. Ground Measurement Comparison

At both stations, data were collected for over 6 months, during the heating season of 2014/2015. The first step of the analysis was to determine if there are any significant differences between stations in daily mean values of measured by PAX coefficients and BC concentration. Table 2 presents mean values with 95% confidence intervals for both stations. Due to differences in wavelength of the devices in both stations, only BC concentration could be duly compared. For this reason, further analysis presents data results of monthly anomalies. The anomalies were calculated as percentage difference between mean value for the month and mean value for the entire experiment period (Table 2).

Differences between anomalies with 95% confidence intervals are presented on panels a, b, and c of Fig. 3. Panel d presents the absolute values of BC concentration. Our results show satisfactory correspondence between the stations during the experiment. Although measurements were done in the stations 20 km away from each other, situated in different environment (city center vs. suburban area), the behavior of daily mean anomalies follows the same trend. Discrepancies between the stations are almost completely covered by the uncertainty of results. When we compare the absolute values of BC concentration [Fig. 3, panel d shows slightly higher ($\approx 10\%$)] values for Warsaw, but this difference is below statistical significance. The preserved pattern of the monthly changes and difference of results between the stations below uncertainty level suggest that during the analysed period, any significant differences were not detected. This lack of discrepancies proves that despite our assumptions about differing emissions in those two places, higher local emission from heating in Swider is balanced by higher traffic pollution in Warsaw.

Figure 2
Hexacopter Versa X6sci with AE-51 micro-aethalometer and Vaisala RS92-SGP radiosonde

Table 2

*Mean values of coefficient and BC concentration from 10.2014 to 03.2015*

| Quantity | Swider (870 nm) | Warsaw (532 nm) |
|---|---|---|
| Scattering ($mM^{-1}$) | 91.6 ± 33.3 | 168.0 ± 60.7 |
| Absorption ($mM^{-1}$) | 15.6 ± 4.6 | 28.2 ± 6.7 |
| BC ($\mu g/m^3$) | 3.3 ± 0.9 | 3.7 ± 0.8 |

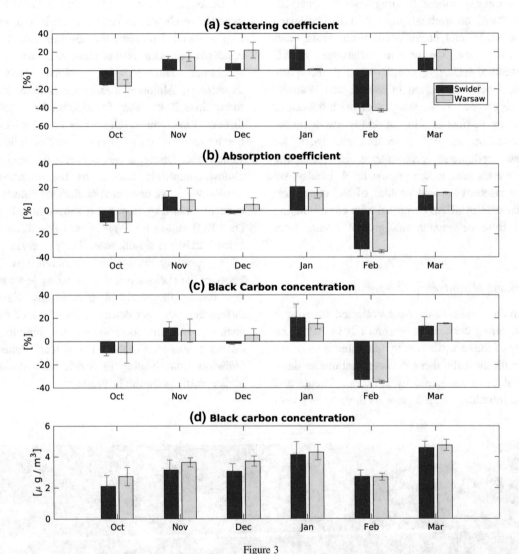

Figure 3
Monthly mean relative anomalies of scattering coefficient (**a**), absorption coefficient (**b**), BC concentration (**c**), and the absolute values of BC concentration (**d**) at the Swider and Warsaw stations. Blue bars for Swider and yellow for Warsaw. 95% confidence interval marked with red whiskers

## 4.2. Height of Temperature Inversion and BC Concentration

During the experiment, we examined the relation between smog conditions (days with high BC concentrations) and temperature inversion height. During low-altitude temperature inversion, conditions in the lowest parts of the atmosphere are thermodynamically stable. This prevents air from mixing and accumulates aerosols close to surface. Drones are especially useful for research of aerosols during such conditions when inversion is below its operational range and degree of aerosols accumulation can be investigated. To verify the hypothesis of temperature inversion influence on BC concentration, results from PAX were divided into two classes, basing on the height of temperature inversion. For this analysis, we selected only measurements made at noon and midnight UTC, when radiosonde was launched in the Legionowo station. Data from PAX were averaged for 15 min after radiosonde's launch time. Basing on the results of the radio-soundings, we first distinguished the class of low-altitude temperature inversion for inversion height of 600 m agl; other data points, with higher altitude inversion or no inversion, were categorized into the second class. This data set division is presented on Fig. 4. Mean values in Swider were: $4.69 \pm 0.74$ µg/m$^3$ with inversion below 600 m agl and $2.28 \pm 0.60$ µg/m$^3$ for higher altitude or no inversion. Corresponding results for Warsaw were: $4.54 \pm 0.47$ and $2.91 \pm 0.32$ µg/m$^3$. The difference between layers is significant, with almost 1.5–2 times higher BC concentration during days with low-altitude temperature inversion. This result confirms the potential of sUAS for measuring aerosols vertical distribution during smog conditions with low-altitude temperature inversion. Such conditions, where inversion is below operational range of drone, are especially interesting: greater load of aerosols is near the surface, and better signal-to-noise ratio and lower uncertainties make it easier to retrieve data from micro-aethalometers and particle counters (Chilinski et al. 2016).

## 4.3. Ceilometer and PAX Comparison

At this stage, the issue of detecting smog conditions at the surface during analysis of ceilometer signal was brought up. Due to technological limitations, tropospheric lidars (represented in our experiment by ceilometer Vaisala CL-31) suffer from insufficiently adjusted overlap, which means that devices are 'blind' to aerosols in the lowest layers of the atmosphere.

The overlap of lidar depending on its design could vary from ten up to hundreds of meters. Atmospheric aerosols scientists can benefit from synergy of lidar and drones profiles, but, on the other hand, drones can fill the gap for results made by overlap limitations. To verify the need for extending lidar measurements on the ground, we compared the differences between the results from the lowest bins of ceilometer with the ground results from PAX (Fig. 5). Simplification of not trivial task of comparing lidar measurements with ground measurements is not an easy task and simplification is necessary (Zieger et al. 2011; Welton et al. 2000). We decided to adopt a simple approach based on anomalies. It is important to mention that PAX measures scattering coefficient of samples in dry conditions in close measurement chamber, while ceilometer registers two-way attenuated backscatter coefficient in ambient conditions. Ceilometer RCS does not contain direct physical meaning; hence, we present only the figure with anomalies. Data from the Swider station were analysed for conditions where relative humidity was below 85%. The anomalies of daily means of scattering coefficient from PAX were compared with the anomalies of daily mean of range corrected signal at 5 lower most bins above overlap from Vaisala ceilometer. During data verification of data from ceilometer, we examined altitude bins between ground and 100-m agl. Detection of aerosols starts at 5th bin, which was around 50 m above the ground level. All bins between 5th and 10th (50–100 m) were examined and all showed the same dynamics, and hence, the average from all 5 bin was selected for the results. The presented figures contain uncertainties on 95% confidence intervals. The results show different level of analogies between measurements from both devices. In December, we have almost the same monthly mean, but in March, the difference is significant: the ceilometer shows strongly negative anomaly, while the PAX slightly positive or almost zero anomaly. In general, in the last quarter of 2014, results are more coherent in terms of value and sign

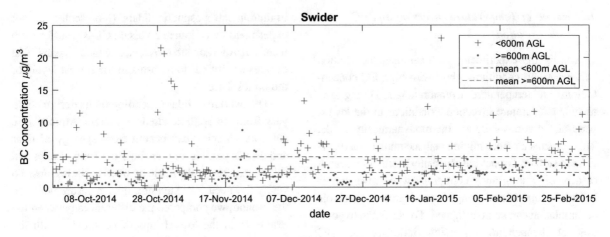

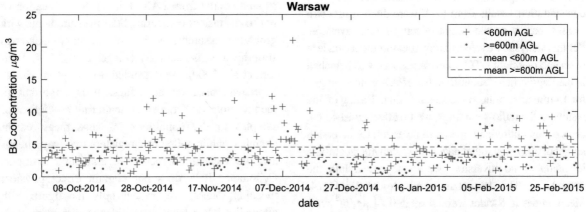

Figure 4

BC concentration in relation to altitude of temperature inversion (Swider on upper panel and Warsaw on lower panel). Blue '+' for measurements when temperature inversion was below 600 m above ground level and red dots for days with temperature inversion above 600 m above ground level. Mean values from all measurements for corresponding groups presented with dashed lines

than in the first quarter of 2015. The reasons for such discrepancies were not thoroughly investigated, but they suggest that there is a vertical variability of atmospheric aerosols distribution even between the lowest levels of Vaisala CL-31 ceilometer and the ground level. This hypothesis ought to be verified by vertical profiles acquired from drones.

### 4.4. Vertical Profiles of BC Measured by sUAS

During the experiment, vertical profiles of BC concentration with AE-51 aethalometer and RS92-SGP mounted below Versa X6 drone were collected when meteorological and technical conditions allowed. The range corrected signals from Vaisala CL-31 were plotted with BC concentration profiles as

a point for reference and comparison. The profiles from ceilometer are presented only in full overlap region (> 60 m agl). As shown in Sect. 4.2, temperature inversion is an important factor for high aerosols concentrations near to the surface. Thus, each flight was done with a radiosonde and corresponding thermodynamical profiles are presented together with aerosols profiles. In this section, five selected profiles are presented and divided into two flight group sessions: one in the end of October 2014 and the second in February 2015.

The first presented flight session had been conducted for over 24 h between noon of the 28th of October and noon of the 29th of October 2014 (Fig. 6). During those 2 days, a relatively clean airmass was transported from south east of Europe.

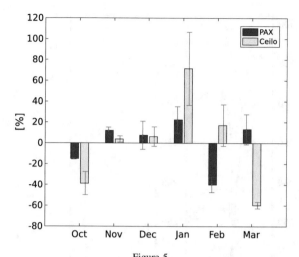

Figure 5

Scattering coefficient monthly mean anomaly from PAX@870 nm (blue bars) in comparison to Vaisala CL-31 lowest bins RCS@905 nm (yellow bars) monthly mean anomaly

AOD measured at the Swider station was only 0.078@500 nm and 0.027@870 nm with Ångstrom exponent 1.88. As we saw on RCS, plot boundary layer during that time was at altitudes below 650 m

agl. This gave us a chance to cover all of it during our flights with sUAS. The first flight up to 400 m agl (Fig. 7) was conducted at 11:23 UTC, when BC concentration reported by PAX was below monthly mean ($2.10 \pm 0.72$ μg/m$^3$). At that time, the temperature inversion of $\approx 2$ °C above 300 m agl was detected. In spite of that, BC profile was quite stable, with regional variations around 2.5 μg/m$^3$ and potential increase above the flight range in the area of deeper inversion. The ceilometer reported signal near to the background at lowest bins.

The second flight during that session was conducted at night, with very limited altitude, only up to 140 m (Fig. 8). The flight took place during heavy smog conditions, when BC reported by PAX was $\approx$ 54 μg/m$^3$, 25 times the monthly average. Temperature inversion was on the ground level with high-temperature gradient of 7.2 °C/100 m. The most interesting here is the profile of BC concentration: rapid decrease from $\approx 70$ μg/m$^3$ at the ground level to $\approx 4$ at 60 m agl, the change occurring between 40 m and 55 m agl. This resulted in very large local

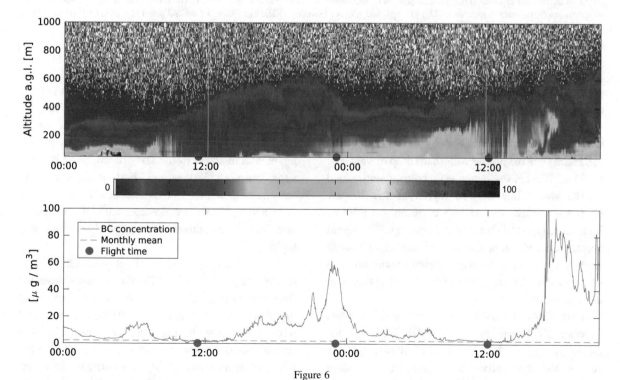

Figure 6

Measurements from 28/29.10.2014 at Swider station. RCS@905 nm from Vaisala CL-31 (upper panel). BC concentrations measured by PAX (lower panel)

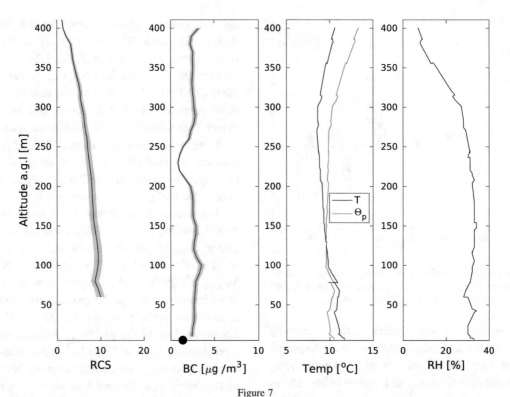

Figure 7

Flight on 28.10.2014 at 11:23 UTC. Vertical profiles of RCS@905 nm from Vaisala-CL31 (RCS), BC concentration from AE-51 (BC), air temperature ($T$), potential temperature ($\Theta_p$) (Temp), and relative humidity (RH). The black dot on BC panel represents BC concentration measured by PAX at the surface

gradient of BC concentration at the level of $\approx$ 380 $\mu g/m^3/100$ m. Figure 9 dedicated to BC concentration profile shows raw BC concentration output reported by AE-51. The concentration was so high that even without any smoothing or averaging, we can clearly see the profile with sharp drop of values at $\approx$ 50 m. What is significant is that almost entire load of BC was below overlap region of the Vaisala ceilometer, which has overlap corrected on a lower level than typical lidars and ceilometers. The signal presented in Fig. 6 at the time of this flight is only slightly higher than the one reported around noon. It confirms that the smog event was almost invisible to the ceilometer.

Last flight from the presented session from October 2014 was made at 12:03 UTC on the following day on 29th of October 2014 (Fig. 10). As on the day before, BC concentration values reported by PAX were close to the monthly mean. AOD on that day was slightly higher: 0.12@500 nm

and 0.041@870 nm with Ångstrom exponent 1.90. This time, the drone reached 480 m agl and entered zone of temperature inversion above 380 m agl. $\Theta_p$ shows well-mixed conditions (shallow convection). Both profiles of RCS and BC concentration ($\approx$ 2.7 $\mu g/m^3$) were stable to the level of inversion bottom. After reaching 380 m agl, both profiles start to constantly decrease. Correspondence between range corrected signal from ceilometer with BC concentration from aethalometer is easily visible during this flight.

The second flight session took place on the 14th of February 2015 (Fig. 11). The airmass advected from the south of Europe and AOD at the station at the time of the flights was 0.08@500 nm and 0.02@870 nm with Ångstrom exponent 1.96. We conducted two flights within the range of 90 min. Monthly mean value of BC concentration based on PAX measurements was $2.72 \pm 0.82$ $\mu g/m^3$, but during those flights, the measured values were almost

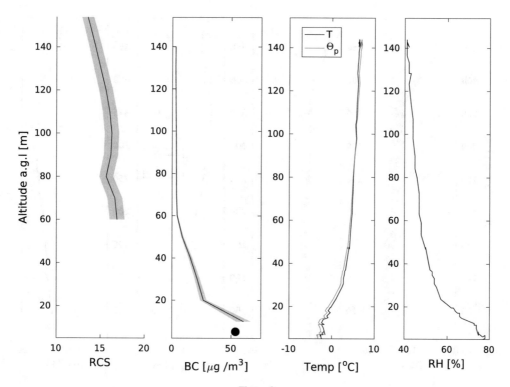

Figure 8
Flight on 28.10.2014 at 23:04 UTC. Panels description as in Fig. 7

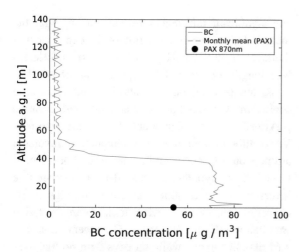

Figure 9
Flight on 28.10.2014 at 23:04 UTC. Vertical profile of BC concentration measured by AE-51@880 nm with monthly mean and result from PAX at surface

twice as high, $5.28 \pm 0.42$ µg/m³ (hourly mean). The flights (Figs. 12, 13) were done, respectively, at 08:19 UTC and 09:46 UTC. Such consecutive flights show drones' potential for tests with short revisit time and

measurement of dynamic events, such as rising temperature inversion. Elevation of well-mixed air layer extends rising temperature inversion. As in Fig. 12, variability of temperature and relative humidity allows for discerning three layers (0–250, 250–400, > 400 m). The same layers can be observed on the vertical profile of BC concentration and RCS. For the following flight (Fig. 13) shown thermodynamical profiles are simplified, but BC concentration and RCS still follow its pattern. Convection mixing moved the PBL higher, while temperature gradient increased. On 270 m agl, where lower boundary of temperature inversion was located, aerosols started to decrease significantly from 6.81 µg/m³ at 210 m agl to 1.62 µg/m³ at 310 m agl (gradient of 5.19 µg/m³/ 100 m). RCS profile followed the trend visible on the BC concentration vertical variability.

The flight session in February 2015 confirmed sUAS capability to measure dynamic changes of aerosol vertical profiles. What is clearly visible in the results is the significant correspondence between thermodynamical profile and profiles of aerosols.

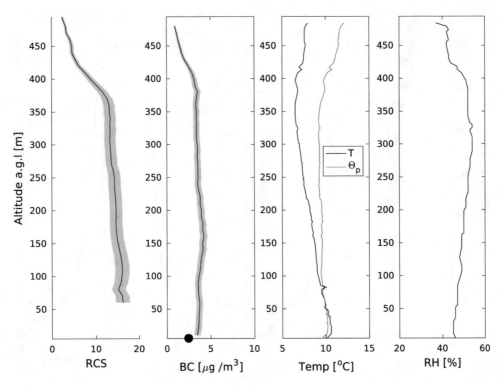

Figure 10
Flight on 29.10.2014 at 12:03 UTC.Panels description as in Fig. 7

This confirms that measuring aerosol optical properties' profiles should be done together with temperature and humidity profiles. The data obtained from meteorological radiosonde proved to be very useful for better understanding of aerosols vertical variability.

## 5. Summary

Small unmanned aerial systems are very useful for atmospheric aerosols research. First, we introduced the basics of how drones could be applied in atmospheric experiments what benefits they offer for the researchers, as exemplified by our measurement campaign held in Poland (metropolitan area of Warsaw) at the turn of 2014 and 2015. Two field stations were set up for the experiment, one in the center of Warsaw and the second in the suburbs, 20 km away in Swider.

The comparison of black carbon concentration monthly means between the stations did not reveal

any significant differences between them, with the mean value for the entire period of the experiment at the level of $3.48 \pm 0.91$ $\mu g/m^3$. Because of different wavelengths on which PAX operated at the two stations, the comparison of scattering and absorption coefficient was based on mean anomaly and, as in the previous case, no significant differences emerged. Verification of the influence of temperature inversion altitude on BC concentration at ground level confirmed our hypothesis that deeper temperature inversion at lower altitudes results in higher BC concentration. On days when temperature inversion was below 600 m agl, mean BC concentration was $4.61 \pm 0.43$ $\mu g/m^3$, while on days with no temperature inversion or with inversion on higher altitude, mean value was $2.60 \pm 0.34$ $\mu g/m^3$. Due to overlap problems and complex retrieval of lidar data, direct comparison of ground measurements with data from the lowest altitude bins is a difficult issue. Nevertheless, we attempt to do it during the experiment, using Vaisala CL-31 ceilometer, which has relatively well-corrected overlap at $\approx 50$ m agl. Although

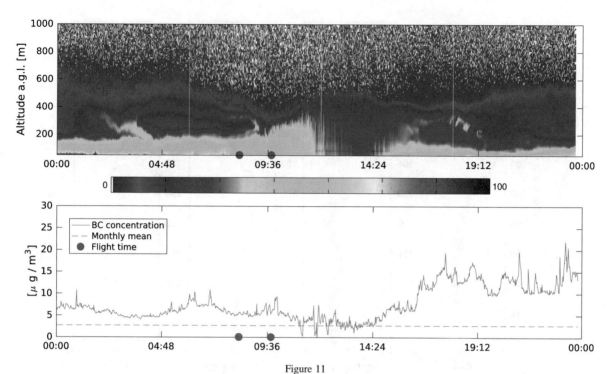

Figure 11
Measurements from 14.02.2015 at Swider. Range corrected signal from Vaisala CL-31 (upper). BC concentrations measured by PAX (lower)

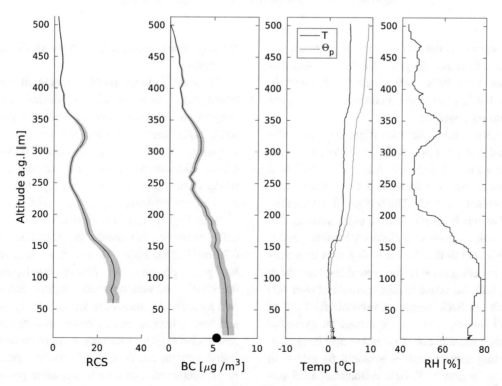

Figure 12
Flight on 14.02.2015 at 08:19 UTC. Panels description as in Fig. 7

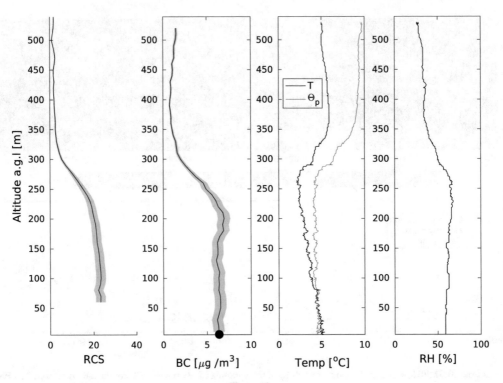

Figure 13
Flight on 14.02.2015 at 09:46 UTC. Panels description as in Fig. 7

initially promising, the performance of ceilometer in layers directly above the ground level revealed discrepancies with PAX, which are most probably related to the fact that low-altitude aerosol events are not visible on ceilometer.

The results from the two flight sessions with sUAS and the profiles which we collected in the range between 140 and 480 m agl also confirmed good performance of a drone as a support for atmospheric aerosols research. The five profiles presented above give much insight into aerosols vertical distribution near the ground in planetary boundary layer. Drone was able to deliver interesting data of a very strong smog event between the ground level and 60 m above it, with BC concentration gradient of over 400 $\mu g/m^3/100$ m. Such events are invisible for most of lidars/ceilometers, what makes drones an exclusive tool for their examination. Drones can also be applied in extending aerosol profiles acquired by lidars: in full overlap region of Vaisala ceilometer, RCS follows vertical variability of BC concentration, and cross-checking both profiles can serve to retrieve

more complex quantities, such as single-scattering properties.

Thermodynamical profiles measured by sUAS reveal dependence of aerosol vertical profile on temperature and humidity variability. Changes in stable layer height, elevation of inversion layer, and variability of humidity have visible influence on aerosol distribution. It supports our idea that measuring vertical profiles of aerosol properties should be checked against basic atmospheric profiles.

Sounding of aerosols with sUAS is a new concept, only recently presented in scientific journals (Chilinski et al. 2016). However, we believe that it has great potential for further development and introduction of new research schemes. Sensors carried by aerial platforms can be technically improved and thus deliver more and more valuable data. Current technological solutions already enable us to install particle counters, spectrometers, radiometers, or sunphotometers on drones. Sufficient potential can be found in solar radiation sensors on horizontal platforms and on sun-trackers, which deliver AOD profiles, as well as the profiles of heating rate. Moreover, improvement of battery technologies and increase in engines efficiency extend the potential of sUAS, both in terms of operational range and spatial resolution. The popularity of drones is booming and each year a wider variety of models is available on the global market. This offers new opportunities for scientists, and usage of sUAS for research, including atmosphere research, is certainly worth exploring.

## Acknowledgements

This research has been made within the Polish Grant no. 2012/05/E/-ST10/01578 of the National Science Center, coordinated by IGFUW. We acknowledge Brent Holben for preprocessing the AERONET data. CIMEL calibration was performed at the LOA AERONET-EUROPE calibration center, supported by ACTRIS-1 [(European Union Seventh Framework Program (FP7/2007e2013) under Grant agreement no. 262254].

## References

Arnott, W. P., Hamasha, K., Moosmuller, H., Sheridan, P. J., & Ogren, J. A. (2005). Towards aerosol light-absorption measurements with a 7-wavelength aethalometer: Evaluation with a photoacoustic instrument and 3-wavelength nephelometer. *Aerosol Science and Technology, 39*(1), 17–29.

Bond, T. C., Anderson, T. L., & Campbell, D. (1999). Calibration and intercomparison of filter-based measurements of visible light absorption by aerosols. *Aerosol Science and Technology, 30*(6), 582–600.

Bond, T. C., Doherty, S. J., Fahey, D. W., Forster, P. M., Berntsen, T., DeAngelo, B. J., et al. (2013). Bounding the role of black carbon in the climate system: A scientific assessment. *Journal of Geophysical Research-Atmospheres, 118*(11), 5380–5552.

Brady, J. M., Stokes, M. D., Bonnardel, J., & Bertram, T. H. (2016). Characterization of a quadrotor unmanned aircraft system for aerosol-particle-concentration measurements. *Environmental Science and Technology, 50*(3), 1376–1383.

Chan, C. Y., Xu, X. D., Li, Y. S., Wong, K. H., Ding, G. A., Chan, L. Y., et al. (2005). Characteristics of vertical profiles and

sources of pm2.5, pm10 and carbonaceous species in Beijing. *Atmospheric Environment, 39*(28), 5113–5124.

Chang, K., Rammos, P., Wilkerson, S. A., Bundy, M., & Gadsden, S. A. (2016). Lipo battery energy studies for improved flight performance of unmanned aerial systems. In *Conference on unmanned systems technology XVIII, volume 9837 of Proceedings of SPIE*, BELLINGHAM. Spie-Int Soc Optical Engineering.

Chilinski, M. T., Markowicz, K. M., & Markowicz, J. (2016). Observation of vertical variability of black carbon concentration in lower troposphere on campaigns in poland. *Atmospheric Environment, 137*, 155–170.

Cook, J., & Highwood, E. J. (2004). Climate response to tropospheric absorbing aerosols in an intermediate general-circulation model. *Quarterly Journal of the Royal Meteorological Society, 130*(596), 175–191.

Ferrero, L., Cappelletti, D., Busetto, M., Mazzola, M., Lupi, A., Lanconelli, C., Becagli, S., Traversi, R., Caiazzo, L., Giardi, F., Moroni, B., Crocchianti, S., Fierz, M., Močnik, G., Sangiorgi, G., Perrone, M. G., Maturilli, M., Vitale, V., Udisti, R., & Bolzacchini, E. (2016). Vertical profiles of aerosol and black carbon in the arctic: a seasonal phenomenology along 2 years (2011–2012) of field campaigns. *Atmospheric Chemistry and Physics,* 16(19):12601–12629. https://www.atmos-chem-phys.net/16/12601/2016/.

Ferrero, L., Castelli, M., Ferrini, B. S., Moscatelli, M., Perrone, M. G., Sangiorgi, G., et al. (2014). Impact of black carbon aerosol over italian basin valleys: High-resolution measurements along vertical profiles, radiative forcing and heating rate. *Atmospheric Chemistry and Physics, 14*(18), 9640–9663.

Hess, M., Koepke, P., & Schult, I. (1998). Optical properties of aerosols and clouds: The software package opac. *Bulletin of the American Meteorological Society, 79*(5), 831–844.

Holben, B. N., Eck, T. F., Slutsker, I., Tanre, D., Buis, J. P., Setzer, A., et al. (1998). Aeronet—a federated instrument network and data archive for aerosol characterization. *Remote Sensing of Environment, 66*(1), 1–16.

Holnicki, P., Kałuszko, A., Nahorski, Z., Stankiewicz, K., & Trapp, W. (2017). Air quality modeling for warsaw agglomeration. *Archives of Environmental Protection, 43*(1), 48–64.

Horvath, H. (1993). Atmospheric light-absorption—a review. *Atmospheric Environment Part A General Topics, 27*(3), 293–317.

IPCC (2013). *Climate change 2013: The physical science basis. Contribution of Working Group I to the Fifth Assessment Report of the Intergovernmental Panel on Climate Change.* Cambridge University Press, Cambridge. http://www.climatechange2013.org.

Koch, D., & Del Genio, A. D. (2010). Black carbon semi-direct effects on cloud cover: Review and synthesis. *Atmospheric Chemistry and Physics, 10*(16), 7685–7696.

Kupiszewski, P., Leck, C., Tjernström, M., Sjogren, S., Sedlar, J., Graus, M., Müller, M., Brooks, B., Swietlicki, E., Norris, S., & Hansel, A. (2013). Vertical profiling of aerosol particles and trace gases over the central arctic ocean during summer. *Atmospheric Chemistry and Physics,* 13(24):12405–12431. https://www.atmos-chem-phys.net/13/12405/2013/.

Loeb, N. G., Wielicki, B. A., Su, W. Y., Loukachine, K., Sun, W. B., Wong, T., et al. (2007). Multi-instrument comparison of top-of-atmosphere reflected solar radiation. *Journal of Climate, 20*(3), 575–591.

Luo, B., Meng, Q. H., Wang, J. Y., & Ma, S. G. (2016). Simulate the aerodynamic olfactory effects of gas-sensitive uavs: A numerical model and its parallel implementation. *Advances in Engineering Software, 102,* 123–133.

Lv, L. H., Zhang, T. S., Liu, C., Dong, Y. S., Chen, Z. Y., Fan, G. Q., Liu, Y., & Liu, W. Q. (2015). Atmospheric aerosols detection research with a dual field of view lidar. *Journal of Spectroscopy, 2015*(2015), 459460. https://doi.org/10.1155/2015/459460.

Madonna, F., Amato, F., Hey, J. V., & Pappalardo, G. (2015). Ceilometer aerosol profiling versus raman lidar in the frame of the interact campaign of actris. *Atmospheric Measurement Techniques, 8*(5), 2207–2223.

Markowicz, K., Ritter, C., Lisok, J., Makuch, P., Stachlewska, I., Cappelletti, D., Mazzola, M., & Chilinski, M. (2017). Vertical variability of aerosol single-scattering albedo and equivalent black carbon concentration based on in-situ and remote sensing techniques during the iarea campaigns in ny-Ålesund. *Atmospheric Environment, 164*(Supplement C):431–447. http://www.sciencedirect.com/science/article/pii/S1352231017303953.

Matthias, V., Balis, D., Bosenberg, J., Eixmann, R., Iarlori, M., Komguem, L., et al. (2004). Vertical aerosol distribution over europe: Statistical analysis of raman lidar data from 10 European aerosol research lidar network (earlinet) stations. *Journal of Geophysical Research Atmospheres, 109*(D18), 12.

McMurry, P. H. (2000). A review of atmospheric aerosol measurements. *Atmospheric Environment, 34*(12–14), 1959–1999.

Morawska, L., Thomas, S., Gilbert, D., Greenaway, C., & Rijnders, E. (1999). A study of the horizontal and vertical profile of sub-micrometer particles in relation to a busy road. *Atmospheric Environment, 33*(8), 1261–1274.

Myhre, G., & Samset, B. H. (2015). Standard climate models radiation codes underestimate black carbon radiative forcing. *Atmospheric Chemistry and Physics, 15*(5), 2883–2888.

Nakayama, T., Suzuki, H., Kagamitani, S., Ikeda, Y., Uchiyama, A., & Matsumi, Y. (2015). Characterization of a three wavelength photoacoustic soot spectrometer (pass-3) and a photoacoustic extinctiometer (pax). *Journal of the Meteorological Society of Japan, 93*(2), 285–308.

Nash, J., Oakley, T., Voemel, H., & Wei, L. (2010). World meteorological organization instruments and observing methods report no. 107. http://www.wmo.int/pages/prog/www/IMOP/publications/IOM-107_Yangjiang.pdf.

Nilsson, E. D., Rannik, U., Kulmala, M., Buzorius, G., & O'Dowd, C. D. (2001). Effects of continental boundary layer evolution, convection, turbulence and entrainment, on aerosol formation. *Tellus Series B Chemical and Physical Meteorology, 53*(4), 441–461.

Pixhawk (2017). Pixhawk autopilot firmware parameter reference. https://pixhawk.org/firmware/parameters.

Ramanathan, V., Crutzen, P. J., Lelieveld, J., Mitra, A. P., Althausen, D., Anderson, J., et al. (2001). Indian ocean experiment: An integrated analysis of the climate forcing and effects of the great indo-asian haze. *Journal of Geophysical Research Atmospheres, 106*(D22), 28371–28398.

Samset, B. H., & Myhre, G. (2011). Vertical dependence of black carbon, sulphate and biomass burning aerosol radiative forcing. *Geophysical Research Letters, 38,* 5.

Samset, B. H., Myhre, G., Schulz, M., Balkanski, Y., Bauer, S., Berntsen, T. K., et al. (2013). Black carbon vertical profiles strongly affect its radiative forcing uncertainty. *Atmospheric Chemistry and Physics, 13*(5), 2423–2434.

Schwarz, J. P., Spackman, J. R., Gao, R. S., Watts, L. A., Stier, P., Schulz, M., Davis, S. M., Wofsy, S. C., & Fahey, D. W. (2010). Global-scale black carbon profiles observed in the remote atmosphere and compared to models. *Geophysical Research Letters, 37*(18), L18812. http://dx.doi.org/10.1029/2010GL044372.

Sokol, P., Stachlewska, I. S., Ungureanu, I., & Stefan, S. (2014). Evaluation of the boundary layer morning transition using the cl-31 ceilometer signals. *Acta Geophysica, 62*(2), 367–380.

Spackman, J. R., Gao, R. S., Neff, W. D., Schwarz, J. P., Watts, L. A., Fahey, D. W., Holloway, J. S., Ryerson, T. B., Peischl, J., & Brock, C. A. (2010). Aircraft observations of enhancement and depletion of black carbon mass in the springtime arctic. *Atmospheric Chemistry and Physics, 10*(19):9667–9680. https://www.atmos-chem-phys.net/10/9667/2010/.

Vardoulakis, S., Fisher, B. E. A., Pericleous, K., & Gonzalez-Flesca, N. (2003). Modelling air quality in street canyons: A review. *Atmospheric Environment, 37*(2), 155–182.

Wandinger, U., & Ansmann, A. (2002). Experimental determination of the lidar overlap profile with raman lidar. *Applied Optics, 41*(3), 511–514.

Welton, E. J., Voss, K. J., Gordon, H. R., Maring, H., Smirnov, A., Holben, B., et al. (2000). Ground-based lidar measurements of aerosols during ace-2: Instrument description, results, and comparisons with other ground-based and airborne measurements. *Tellus Series B Chemical and Physical Meteorology, 52*(2), 636–651.

Welton, E. J., Voss, K. J., Quinn, P. K., Flatau, P. J., Markowicz, K., Campbell, J. R., et al. (2002). Measurements of aerosol vertical profiles and optical properties during indoex 1999 using micropulse lidars. *Journal of Geophysical Research Atmospheres, 107*(D19), 22.

Wiedensohler, A., Birmili, W., Nowak, A., Sonntag, A., Weinhold, K., Merkel, M., et al. (2012). Mobility particle size spectrometers: Harmonization of technical standards and data structure to facilitate high quality long-term observations of atmospheric particle number size distributions. *Atmospheric Measurement Techniques, 5*(3), 657–685.

Zarzycki, C. M., & Bond, T. C. (2010). How much can the vertical distribution of black carbon affect its global direct radiative forcing? *Geophysical Research Letters, 37,* 6.

Zawadzka, O., Markowicz, K. M., Pietruczuk, A., Zielinski, T., & Jaroslawski, J. (2013). Impact of urban pollution emitted in warsaw on aerosol properties. *Atmospheric Environment, 69,* 15–28.

Zieger, P., Weingartner, E., Henzing, J., Moerman, M., de Leeuw, G., Mikkila, J., et al. (2011). Comparison of ambient aerosol extinction coefficients obtained from in-situ, max-doas and lidar measurements at cabauw. *Atmospheric Chemistry and Physics, 11*(6), 2603–2624.

(Received March 31, 2017, revised December 18, 2017, accepted January 4, 2018, Published online January 25, 2018)

Pure Appl. Geophys. 175 (2018), 3343–3355
© 2018 The Author(s)
https://doi.org/10.1007/s00024-018-1931-9

▌Pure and Applied Geophysics

# Can Clouds Improve the Performance of Automated Human Detection in Aerial Images?

TOMASZ NIEDZIELSKI[1] and MIROSŁAWA JURECKA[1]

*Abstract*—The objective of this paper is to investigate the role of clouds in the effectiveness of automated human detection in aerial imagery acquired by unmanned aerial vehicles (UAVs). The automated processing is carried out with the nested k-means method applied to images taken in poor visibility caused by low-altitude clouds. Data were acquired during a field experiment carried out in the Izerskie Mountains (southwestern Poland). The fixed-wing UAV took RGB aerial photographs of terrain where persons simulated being lost in the wilderness. The UAV flights were conducted in the morning and around the noon, when clouds reduced clarity of aerial images. Subsequent UAV missions were performed in the afternoon and in the evening, when clouds had no impact on imagery. False hit rates $\geq 50\%$ correspond to clear imagery (8 of 9 non-cloudy cases). In contrast, images impacted by clouds reveal false hit rates $\leq 40\%$ (5 of 7 cloudy cases). Sensitivity analysis, carried out on a basis of artificially blurred imagery, confirms that reduced image clarity may improve automated human detection.

**Key words:** Unmanned aerial vehicle, wilderness search and rescue, image processing, geoinformatics, clouds.

## 1. Introduction

Automated detection of humans in aerial images is one of key elements in wilderness search and rescue (WiSAR) activities. According to Goodrich et al. (2009), there are two search and rescue (SAR) roles, namely information acquisition and information analysis. The use of unmanned aerial vehicles (UAVs) to collect aerial images along with their near-real-time processing to identify lost persons fits the two roles.

Handling Editor: Prof. Andrzej Icha.

[1] Department of Geoinformatics and Cartography, Faculty of Earth Sciences and Environmental Management, University of Wrocław, pl. Uniwersytecki 1, 50-137 Wrocław, Poland. E-mail: tomasz.niedzielski@uwr.edu.pl

The growing popularity of UAVs and the increasing number of UAV operations broaden their use in SAR applications (e.g. Van Tilburg 2017; Silvagni et al. 2017). They include scientific experiments (e.g. Doherty and Rudol 2007; Goodrich et al. 2008; Miller et al. 2008; Molina et al. 2012) as well as comprehensive systems or communities such as for instance: ALCEDO (http://www.alcedo.ethz.ch, access date: 24/09/2017), SHERPA (Marconi et al. 2013), SWARM, also known as SARDrones (http://sardrones.org, access date: 24/09/2017) and SARUAV (Niedzielski et al. 2017; Jurecka and Niedzielski 2017).

To quickly find a person who is lost in the wilderness, human-assisted searches (e.g. Goodrich et al. 2009; Molina et al. 2012) or automated algorithmic searches (e.g. Rudol and Doherty 2008; Flynn and Cameron 2013) are carried out. In the current state of the UAV-based SAR technologies, there is a general consensus that the automated procedures may support human-assisted searches, but cannot fully substitute an expert who operates a sensor (Molina et al. 2012; Niedzielski et al. 2018).

Cloudiness belongs to weather conditions influencing persons' needs for SAR team involvement, and they are often reported as the second group of environmental factors, after darkness, that contributes to the needs for SAR assistance (Gretchen 2004). Thus, SAR operations are likely to be carried out in cloudy weather. When visibility is poor it may constrain making appropriate decisions on where to go, in specific cases leading a person to be lost. In operational SAR missions, clouds lead to considerable limitations, in particular when airborne or satellite sensors are employed (O'Donnell 1999).

Clouds, fog or smoke may impact performance of different cameras, and therefore the presence of such barriers is treated as a disadvantage in remote sensing

(Oakley and Satherley 1998; Woodell et al. 2015; Gultepe et al. 2009; Djuricic and Jutzi 2013). In the context of UAV-assisted SAR work, bad weather not only influences image clarity but also limits the effectiveness of search through constraining flights and other field activities (Sumimoto et al. 2000; Niedzielski et al. 2018). However, the negative impact of deteriorated image clarity on the usability of such imagery can be reduced by data processing techniques (Steinvall et al. 1999; Woodell et al. 2015; Yuan et al. 2010). In this study, we approach the problem from the opposite perspective and verify the hypothesis that clouds form a natural filter which decreases false hit rates in the process of human detection. To verify the hypothesis, we apply the nested k-means algorithm to detect persons in aerial photographs taken by a UAV during the field experiment in the Izerskie Mountains (southwestern Poland).

## 2. Experimental Setup

The eBee UAV was controlled from a dedicated mobile UAV laboratory (Niedzielski et al. 2017, Fig. 5 therein). The stable altitude of 123 m above take-off location (ATO) was kept over the entire experiment conducted on 20/10/2016. Aerial images were taken using the visible light Canon S110 RGB camera. The ground resolution of imagery was approximately equal to 4.3 cm/px. The camera operated in the automatic mode set by the manufacturer. The wavelengths with maximum spectral responses are 660 (R), 520 (G), and 450 (B). Five persons were located in the field. Three individuals occupied fixed locations over the entire day and the other two persons were UAV operators who kept changing their sites. Figure 1 presents the study area of Polana Izerska and person locations (P1, P2, P3—persons at fixed locations who were subjects of detection; P45—UAV operators who were siting or standing very close to each other and their location varied over the day). Polana Izerska is a mountain meadow of size 250 × 170 m, with elevations ranging from 951 to 976 m a.s.l., located in the Izerskie Mountains in southwestern Poland.

Polana Izerska is specifically located, approximately 3 km eastward from Stóg Izerski which is the first orographic barrier for westerly winds that prevail in western Sudetes. As a consequence, low-altitude clouds and fog occur frequently in Polana Izerska. According to Błaś et al. (2002), the area between Stóg Izerski and Polana Izerska is very foggy, with the annual number of days with fog significantly exceeding 200. The UAV missions targeted at person detection are conduced at low altitudes above ground level to ensure high spatial resolution of imagery, with the recommended altitudes between 60 and 100 m (Goodrich et al. 2008). The weather on 20/10/2016 was typical for the study area. The variable cloud cover over the entire day enabled to collect both cloud-blurred and sharp imagery to allow the comparison.

For each period of day, a single UAV flight was performed to acquire a set of aerial images covering the entire study area. From each set of photographs—corresponding to morning, noon, afternoon or evening—we selected four images, the centres of which were located in the closest vicinity of persons P1, P2 and P3 (Fig. 2). Individual aerial images in the central projection, not orthophotos, were subject to further processing.

Over the entire exercise, air temperatures in Polana Izerska varied between 2.7 and 4.7 °C, while sea level pressure was stable and equal to 1014.7–1016.2 hPa (Table 1). The aerial images acquired in the morning and noon were highly impacted by clouds, while photographs taken in the afternoon and evening were clear due to better visibility. This enables the comparison of the performance of the nested k-means method for clean and cloud-blurred input images.

The analysis of a few meteorological characteristics of weather on 20/10/2016 in Liberec, i.e. the synoptic station no. 11603 (398 m a.s.l.) recognized by the World Meteorological Organization (WMO) and located close (approximately 25 km) to the study area, was carried out (Table 2). Stable weather conditions over the region, in which Polana Izerska and Liberec are located, were inferred from the comparison of Tables 1 and 2. Hence, the extrapolation of cloudiness characteristics from Liberec to Polana Izerska was done. The absolute heights of base of lowest clouds were of 698–997 m a.s.l. which, when extrapolated to Polana Izerska, confirms the low-level

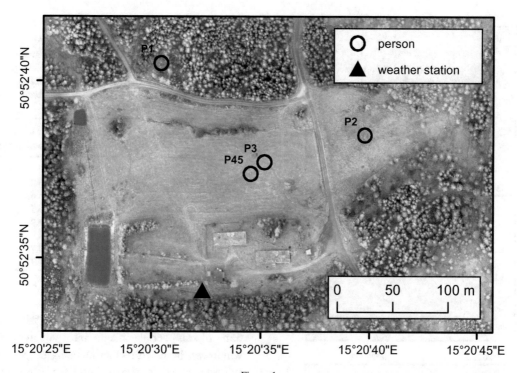

Figure 1
Orthophotomap of study area along with locations of persons who simulated to be lost (codes: P1, P2, P3) and tightly settled two UAV operators (joint code: P45)

cloudiness in the study area. Stratocumulus (Sc) prevailed in the morning and noon, with highly overcast sky reaching the maximum of 8 oktas in the noon. In the afternoon and evening Cumulus (Cu) co-occurred with Sc, reaching the minimum cloudiness of 3 oktas in the evening.

## 3. Methods

The following step-by-step procedure may be employed to investigate the role of clouds in the effectiveness of automated human detection in the UAV-acquired aerial imagery.

- A flight plan over a test area should be prepared so that a UAV operates at altitudes which are suitable for person detection in aerial images (Goodrich et al. 2008; Niedzielski et al. 2018) and flies above persons who pretend being lost in the wilderness.
- A specific day should be chosen so that changes in vegetation do not occur, but the amount of cloud cover varies over this day—ideally with and without cloud impact on visibility of terrain from the UAV.
- The UAV missions ought to be carried out in different cloud-related visibility conditions and JPG aerial photographs of terrain in which these persons are present should be taken.

  – (Cloud-blurred imagery) The images with deteriorated clarity should be processed by the nested k-means algorithm to highlight potential locations of persons (nested k-means + true clouds).
  – (Clear imagery) The clear images should be processed twofold:

    • by the nested k-means algorithm to highlight potential locations of persons (nested k-means + no clouds),
    • clear imagery should also be artificially blurred and processed by the nested k-means algorithm to highlight potential locations of persons (nested k-means + artificial clouds).

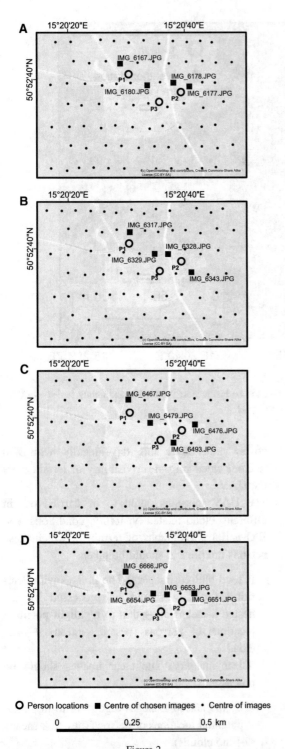

O Person locations    ■ Centre of chosen images    • Centre of images

0          0.25          0.5 km

Figure 2

Approximate locations of cameras mounted onboard the UAV along with locations of the selected images (IMG_) and settled persons (P1, P2, P3) in the morning (**a**), noon (**b**), afternoon (**c**) and evening (**d**)

Table 1

*Selected meteorological characteristics of weather on 20/10/2016 in Polana Izerska (969.8 m a.s.l.), based on the weather station of Świeradów Forest Inspectorate located within the study area, both at the moments of UAV image acquisition over Polana Izerska and at the closest round hours (see Table 3 for details). Source: Świeradów Forest Inspectorate*

| Time UTC | Temperature (°C) | Pressure (hPa) | Approximate sea level pressure (hPa) |
|---|---|---|---|
| 06:48:00 | 2.7 | 902.3 | 1016.1 |
| 07:00:00 | 2.8 | 902.4 | 1016.2 |
| 09:48:00 | 3.8 | 902.3 | 1015.6 |
| 10:00:00 | 4.6 | 902.3 | 1015.2 |
| 12:36:00 | 4.7 | 902.0 | 1014.9 |
| 13:00:00 | 4.7 | 902.0 | 1014.9 |
| 15:00:00 | 4.6 | 901.8 | 1014.7 |
| 15:12:00 | 4.6 | 901.8 | 1014.7 |

The nested k-means method assumes the segmentation of an aerial image into squares (moving windows), built with superimposing stripes to avoid splitting person's signal into two segments. The nesting is based on performing three consecutive tests in each segment. The first test uses the k-means method on a three-dimensional space of colours (RGB) to identify the most unique class, highly coherent in terms of colour composition. The second test uses the k-means method on a two-dimensional space of locations to check if the previously selected class is spatially coherent. The third test is applied to check if the size of the class fits the typical areas of standing or lying person. The details about the nested k-means method can be found in the paper by Niedzielski et al. (2017, Fig. 1 therein).

If the first test produces a flag 0, the overall output in a given segment is also equal to 0 (no person found). If the first test ends up with a flag 1 (person is likely to be present in a segment), the second test is run. If the second test produces a flag 0, the overall output in a segment is 0 (no person found). However, if the second test produces a flag 1 (person is likely to be present in a segment), the third test is run. If its application ends up with a flag 0, the overall procedure in a studied segment gives 0 (no person found). In contrast, if the third test produces a flag 1, the overall output in a given segment is equal to 1 (person found). The segment which gets 1 in the third test is hereinafter known as "find segment"

Table 2

*Selected meteorological characteristics of weather on 20/10/2016 in Liberec, based on the World Meteorological Organization (WMO) synoptic station no. 11603 (398 m a.s.l.) located in the vicinity of the study area, at the closest times to the moments of UAV image acquisition over Polana Izerska (see Tables 1, 3 for details). Source: http://www.ogimet.com (access date: 04/10/2017)*

| Time UTC | Height of base of lowest cloud (m) | Absolute height of base of lowest cloud (m a.s.l.) | Total cloud cover (oktas) | Total low cloud cover (oktas) | Horizontal visibility (km) | Low-level cloud type | Temperature (°C) | Pressure (hPa) | Sea level pressure (hPa) |
|---|---|---|---|---|---|---|---|---|---|
| 07:00:00 | 300–599 | 698–997 | 5 | 4 | 23 | Sc | 7.0 | 967.8 | 1016.0 |
| 10:00:00 | 300–599 | 698–997 | 8 | 8 | 15 | Sc | 8.1 | 967.7 | 1015.7 |
| 13:00:00 | 300–599 | 698–997 | 5 | 4 | 25 | Cu, Sc | 9.1 | 966.8 | 1014.6 |
| 15:00:00 | 300–599 | 698–997 | 3 | 3 | 27 | Sc, Cu | 8.0 | 966.4 | 1014.4 |

or "highlighted segment" (Niedzielski et al. 2017, Fig. 3 therein).

An aerial photograph processed using the nested k-means algorithm produces a new image which is a composition of the original photo and transparent squared segments. The intensity of segment greyness corresponds to a probability of finding a person (zero probability does not imply image colouring). There are three statistics associated with the analysis of a single aerial photograph (Fig. 3): (1) detection rate (the percentage of persons who were found, i.e.

100% × the number of correctly highlighted segments divided by a true number of persons within the entire image), (2) false hit rate (the percentage of highlighted segments which did not detect a person, i.e. 100% × the number of incorrect or simply overestimated highlighted segments divided by the total number of highlighted segments) and (3) percentage of segments suggested to be searched, abbreviated as POSSTBS (100% × the number of highlighted segments divided by all the possible segments within the entire image). The POSSTBS

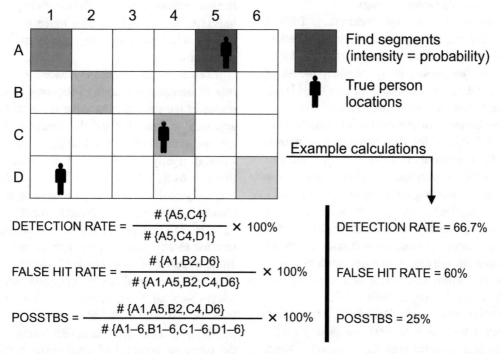

Figure 3
Statistical measures of person detection performance: a graphical presentation

statistics measures the fraction of terrain area which is advised to be searched by ground searchers.

## 4. Results

### 4.1. Detection Based on Cloud-Blurred Images

Figures 4 and 5 show the selected images, the locations of which are marked in Fig. 2, with superimposed highlighted segments. Figure 4 (left column) presents the performance of the nested k-means method for images targeted at the person P1, with no other human individuals present in the scene. Figure 4 (right column) offers the same presentation for the person P2 who also was the only one person in the scene. Figure 5 is targeted at the person P3, however, the other persons P1 or P2 and, in addition, the two UAV operators were also recorded in the images. The UAV operators were standing or sitting very close to each other, therefore they were treated as a single person P45. The low-level Sc clouds blurred meaningful parts of background images in first two rows of Figs. 4 and 5 (morning, noon), while Cu and Sc clouds did not deteriorate the clearness of background images in third and fourth rows of Figs. 4 and 5 (afternoon, evening).

For each aerial photograph under study, Table 3 juxtaposes information on image acquisition features (actual cloudiness which may constrain the view, code of targeted person, image number, period of day, detailed data collection times in CEST and UTC, number and codes of true persons in image) as well as on the performance of the nested k-means method (number of find segments, detection rate, false fit rate, POSSTBS). The records in Table 3 are grouped into four sets which correspond to observations of the similar scene containing the same person or persons (with one exception) in four periods of day covering cloudy and non-cloudy conditions. It should be noted that in all cloudy situations different levels of transparency through clouds were recorded, however, in the worst visibility cases the terrain view from the UAV was highly blurred but still visible.

The person P1 (images: 6167, 6317, 6467, 6666) was detected with rates of 100% for cloudy conditions, however, false hit rates were low (0%) when cloud cover was continuous and high (85.7%) when patchy clouds occurred. The latter statistics show

Figure 4

The UAV-acquired images with superimposed highlighted segments: left column—observations of the person P1 in four periods of day, right column—observations of the person P2 in four periods of day. Letters A–F mark characteristic points in geographical space to enable orientation

overestimation of find segments, probably due to the fact that cloud patches occupied minor parts of the image (the case is therefore similar to the non-cloudy situation). When clouds did not form a barrier between the terrain and the camera, the nested k-means method returned either perfect fit (100% detection rate, 0% false hit rate, 0.5% POSSTBS) or incorrect result (0% detection rate, 100% false fit rate). Overall, the analysis of four images targeted at the person P1 did not convey a clear message as to whether cloud cover improved the performance of person detection using the nested k-means method.

In contrast, the person P2 (images: 6177, 6343, 6476, 6651) was correctly detected with 100% detection rates in the four photographs, with and without cloud impact. However, the false hit rates were of 0% for cloudy conditions (patchy and continuous cover) and increased to 66.7% when images became clean after the clouds departed. That impacted the POSSTBS values which grew from 0.5 to 1.5% when switching from cloudy to non-cloudy conditions.

Even more pronounced evidence for the positive role of transparent clouds in improving the performance of human detection using the nested k-means approach is provided by the analysis of images targeted at the person P3 and, additionally, the person P1 and tightly settled two UAV operators P45 (images: 6180, 6329, 6479, 6654). The case covers two analyses with input images blurred by continuous clouds and the other two images without any cloud impacts. The number of highlighted segments was low for cloudy scenes, with high detection rate of 100% and false hit rates ranging between 0 and 40%. The values of statistics deteriorated when clean images were analysed. Indeed, detection rates varied between 100 and 66.7%, false hit rates were as high as 57.1 and 80%. The POSSTBS values confirmed the overrepresentation of highlighted segments for non-cloudy images (3.4 and 4.9%) in comparison to the cloudy scenes (1.5 and 2.5%).

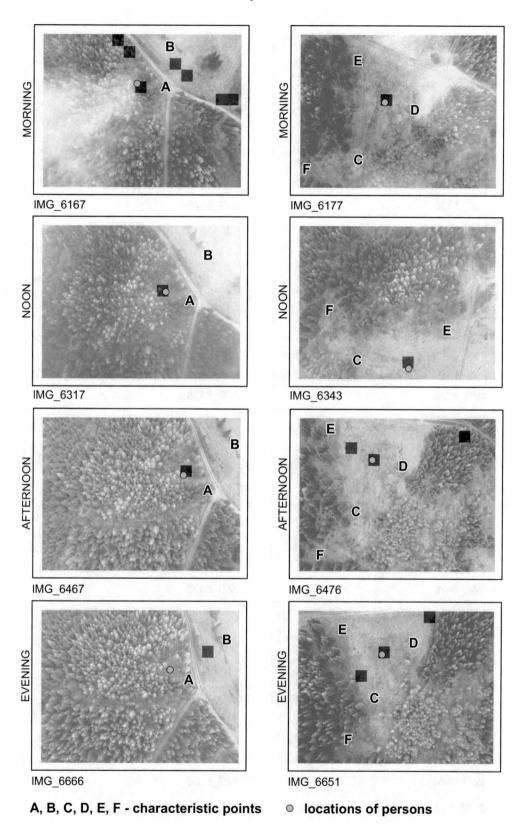

MORNING — IMG_6167

MORNING — IMG_6177

NOON — IMG_6317

NOON — IMG_6343

AFTERNOON — IMG_6467

AFTERNOON — IMG_6476

EVENING — IMG_6666

EVENING — IMG_6651

**A, B, C, D, E, F - characteristic points        locations of persons**

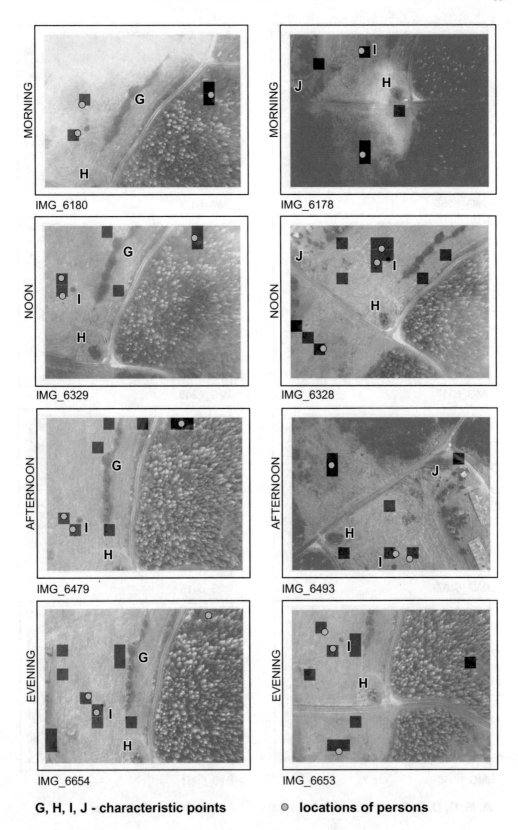

**G, H, I, J - characteristic points**   ● **locations of persons**

◄Figure 5
The UAV-acquired images with superimposed highlighted segments: left column—observations of the persons P3, P1 and P45 in four periods of day, right column—observations of the person P3, P2, P45 in four periods of day. Letters G–J mark characteristic points in geographical space to enable orientation

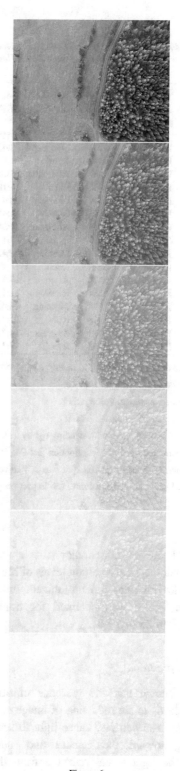

The other scene targeted at the person P3, also including the signals of P2 and P45 (images: 6178, 6328, 6493, 6653) does not enable the comparison between cloudy and non-cloudy cases. Patchy clouds were only present in the image no. 6178, but it did not captured P45 which constrains the thorough comparison. However, a lesson can be learnt from the analysis of non-cloudy situations. Although high detection rates of 100% were revealed, the image processing concurrently led to considerable false hit rates ranging from 50 to 66.7%.

The entire analysis shows that the increased false hits rates seem to characterize the analyses performed using clean images (8 of 9 non-cloudy cases reveal false hit rates $\geq 50\%$). In contrast, low false hit rates tend to correspond to the analyses carried out using cloud-blurred images (5 of 7 cloudy cases reveal false hit rates $\leq 40\%$). Therefore, the presence of transparent cloud cover is likely to reduce false hits and overrepresentation of highlighted segments.

## 4.2. Detection Based on Artificially Blurred Images

To confirm the above-mentioned findings obtained on a basis of cloud-blurred images, a numerical experiment was carried out. From the analyses presented in Figs. 4 and 5, clear images—those for which the nested k-means algorithm produced high false hit rates and potential locations were overrepresented (POSSRBS $\geq$ 2.9)—were selected (noon image: 6328; afternoon images: 6479, 6493; evening images: 6654, 6653). They were pre-processed by superimposing a new uniform white layer with different levels of transparency applied (25, 50, 75, 80 and 90%). As a result, the clarity of original image was artificially deteriorated to simulate the effect of cloud blurriness (Fig. 6). Such artificially blurred images became subject to further processing using the nested k-means approach. Table 4 juxtaposes performance statistics of the nested k-means algorithm for these five images and

Figure 6
Example of artificial blurring of image no. 6654 performed by superimposing a white layer with dissimilar transparencies (top to bottom): 0, 25, 50, 75, 80, 90%

Table 3

*Performance of the nested k-means method in detecting persons in UAV-taken aerial images acquired through continuous clouds (low visibility over the entire image) and patchy clouds (low visibility in parts of image) versus the skills of the same algorithm run using clean input photographs*

| Cloud cover in image | Vicinity of settled person | Image number | Period of day | Time CEST | Time UTC | True number of persons in image | Number of find segments | Detection rate (%) | False hit rate (%) | POSSTBS (%) |
|---|---|---|---|---|---|---|---|---|---|---|
| Patchy | P1 | 6167 | Morning | 08:41:39 | 06:41:39 | 1 (P1) | 7 | 100 | 85.7 | 3.4 |
| Continuous | P1 | 6317 | Noon | 11:44:23 | 09:44:23 | 1 (P1) | 1 | 100 | 0 | 0.5 |
| None | P1 | 6467 | Afternoon | 14:34:57 | 12:34:57 | 1 (P1) | 1 | 100 | 0 | 0.5 |
| None | P1 | 6666 | Evening | 17:12:55 | 15:12:55 | 1 (P1) | 1 | 0 | 100 | 0.5 |
| Patchy | P2 | 6177 | Morning | 08:42:29 | 06:42:29 | 1 (P2) | 1 | 100 | 0 | 0.5 |
| Continuous | P2 | 6343 | Noon | 11:46:28 | 09:46:28 | 1 (P2) | 1 | 100 | 0 | 0.5 |
| None | P2 | 6476 | Afternoon | 14:35:45 | 12:35:45 | 1 (P2) | 3 | 100 | 66.7 | 1.5 |
| None | P2 | 6651 | Evening | 17:11:50 | 15:11:50 | 1 (P2) | 3 | 100 | 66.7 | 1.5 |
| Continuous | P3 | 6180 | Morning | 08:42:39 | 06:42:39 | 3[a] (P3, P1, P45) | 3[b] | 100[ab] | 0[ab] | 1.5[b] |
| Continuous | P3 | 6329 | Noon | 11:45:23 | 09:45:23 | 3[a] (P3, P1, P45) | 5[b] | 100[ab] | 40.0[ab] | 2.5[b] |
| None | P3 | 6479 | Afternoon | 14:35:58 | 12:35:58 | 3[a] (P3, P1, P45) | 7[b] | 100[ab] | 57.1[ab] | 3.4[b] |
| None | P3 | 6654 | Evening | 17:12:01 | 15:12:01 | 3[a] (P3, P1, P45) | 10 | 66.7[a] | 80.0[a] | 4.9 |
| Patchy | P3 | 6178 | Morning | 08:42:33 | 06:42:33 | 2 (P3, P2) | 4[b] | 100[b] | 50.0[b] | 2.0[b] |
| None[c] | P3 | 6328 | Noon | 11:45:19 | 09:45:19 | 3[a] (P3, P2, P45) | 9[b] | 100[ab] | 66.7[ab] | 4.4[b] |
| None | P3 | 6493 | Afternoon | 14:36:59 | 12:36:59 | 3[a] (P3, P2, P45) | 6[b] | 100[ab] | 50.0[ab] | 2.9[b] |
| None | P3 | 6653 | Evening | 17:11:57 | 15:11:57 | 3[a] (P3, P2, P45) | 8[b] | 100[ab] | 62.5[ab] | 3.9[b] |

Statistics are defined in Sect. 3 and in Fig. 3

P1—settled person wearing blue jacket

P2—settled person wearing blue jacket

P3—settled person wearing red jacket

P45—two UAV operators sitting/standing tightly in bunch, wearing blue and red jackets (two people treated as one person)

[a] Statistics corrected due to the assumption that P45 is a single person,

[b] Statistics corrected due to person split into 2–4 adjacent segments (overlapping stripes—see Sect. 3),

[c] Small cloud patch near image corner (no impact on persons' view)

different levels of transparency. It is apparent from Table 4 that there exist certain levels of blurriness for which false hit rates are minimized. Indeed, these minimum values were attained for transparencies ranging from 25 to 75%.

## 4.3. Interpretation

It is known that the presence of atmospheric aerosols leads to the reduction of image contrast and colour. Aerosol particles cause light to scatter when travelling toward the camera and, additionally, weaken the light signal reflected from the ground (Oakley and Satherley 1998; Gultepe et al. 2009). Our analysis based on input images blurred by true clouds and the scrutiny based on the artificially blurred input photos revealed a decrease in false hit rates of human detection performed with the nested k-means method. Clouds and their artificial equivalents cause a significant reduction in contrast and colour of imagery, decreasing a number of details. Such details may be incorrectly classified as persons by the nested k-means method, contributing to false hits where no person is actually present. The clouds and their artificial equivalents work as filters which remove noise from imagery. As a result, the most evident signals remain visible. Hence, the nested k-means method seems to perform better because false hit rates are decreased due to reduced clarity of images. However, as shown above in the experiment with artificially blurred images, the effect of decreasing a number of false hits occurs to a certain level of

Table 4

*Performance of the nested k-means method in detecting persons in artificially blurred UAV-taken aerial images*

| Image number | Transparency (%) | True number of persons | Number of find segments | Detection rate (%) | False hit rate (%) |
|---|---|---|---|---|---|
| 6328* | 0 | 3[a] | 9[b] | 100[ab] | 66.7[ab] |
| 6328* | 25 | 3[a] | 5[b] | 100[ab] | 40.0[ab] |
| 6328* | 50 | 3[a] | 7[b] | 100[ab] | 57.1[ab] |
| 6328* | 75 | 3[a] | 8[b] | 100[ab] | 62.5[ab] |
| 6328* | 80 | 3[a] | 11[b] | 100[ab] | 72.7[ab] |
| 6328* | 90 | 3[a] | 10[b] | 100[ab] | 70.0[ab] |
| 6479 | 0 | 3[a] | 7[b] | 100[ab] | 57.1[ab] |
| 6479 | 25 | 3[a] | 7[b] | 66.7[ab] | 71.4[ab] |
| 6479 | 50 | 3[a] | 7[b] | 100[ab] | 57.1[ab] |
| 6479 | 75 | 3[a] | 5[b] | 66.7[ab] | 60.0[ab] |
| 6479 | 80 | 3[a] | 5[b] | 66.7[ab] | 60.0[ab] |
| 6479 | 90 | 3[a] | 8[b] | 66.7[ab] | 75.0[ab] |
| 6493 | 0 | 3[a] | 6[b] | 100[ab] | 50.0[ab] |
| 6493 | 25 | 3[a] | 5[b] | 100[ab] | 40.0[ab] |
| 6493 | 50 | 3[a] | 5[b] | 100[ab] | 40.0[ab] |
| 6493 | 75 | 3[a] | 5[b] | 100[ab] | 40.0[ab] |
| 6493 | 80 | 3[a] | 7[b] | 100[ab] | 57.1[ab] |
| 6493 | 90 | 3[a] | 10[b] | 100[ab] | 70.0[ab] |
| 6654 | 0 | 3[a] | 10 | 66.7[a] | 80.0[a] |
| 6654 | 25 | 3[a] | 9 | 66.7[a] | 77.7[a] |
| 6654 | 50 | 3[a] | 9 | 66.7[a] | 77.7[a] |
| 6654 | 75 | 3[a] | 7 | 66.7[a] | 71.4[a] |
| 6654 | 80 | 3[a] | 9 | 66.7[a] | 77.8[a] |
| 6654 | 90 | 3[a] | 11 | 66.7[a] | 81.8[a] |
| 6653 | 0 | 3[a] | 8[b] | 100[ab] | 62.5[ab] |
| 6653 | 25 | 3[a] | 7[b] | 100[ab] | 57.1[ab] |
| 6653 | 50 | 3[a] | 6[b] | 100[ab] | 50.0[ab] |
| 6653 | 75 | 3[a] | 6[b] | 100[ab] | 50.0[ab] |
| 6653 | 80 | 3[a] | 8[b] | 100[ab] | 62.5[ab] |
| 6653 | 90 | 3[a] | 10[b] | 100[ab] | 70.0[ab] |

*Small cloud patch near image corner (no impact on persons' view).

[a]Statistics corrected due to the assumption that P45 is a single person,

[b]Statistics corrected due to person split into 2–4 adjacent segments (overlapping stripes—see Sect. 3)

blurriness above which (when transparency becomes very low) a number of false hits increases again. This can be explained by too intensive reduction of clarity and a consequent removal of too many details, including persons. In the context of our exercise, such a noise reduction is particularly meaningful because persons who took part in the field experiment wore blue and red jackets, and the signal of their clothes remained strong.

## 5. Conclusions

This paper shows that the presence of transparent clouds in UAV-taken oblique aerial images may improve the performance of the automated human detection using the nested k-means method. Although no impact of cloudiness on person detection rates were reported, clouds were found to reduce false hit rates and, as a consequence, to reduce overrepresentation of sites suggested to be searched. Sensitivity analysis, in which the clarity of UAV-taken aerial images was artificially deteriorated by superimposing white layers with different transparencies to imitate the effect naturally caused by clouds, confirmed that the reduced clarity may improve the performance of the method. The decrease in the number of false hits was found for intermediate transparencies (25–75%), and the number of false hits tended to grow when clarity of images was highly reduced (superimposed white layers with 80–90% transparency).

The results may have a considerable practical potential—in a cloudy weather UAVs may be manoeuvered to fly at low altitudes so that cloud cover does not constrain the view but gives a transparent blurred picture of terrain. Such a flight planning for SAR purposes may, according to our preliminary study published in this paper, improve the performance of the automated human detection algorithms. Alternatively, artificial filtering to simulate the effect of transparent clouds may be considered. Usually the presence of clouds is considered to constrain the usefulness of different remote sensors (e.g. O'Donnell 1999; Woodell et al. 2015), but our results show that under specific circumstances the effect of clouds can be considered as a virtue. Although the findings are promising, further studies are required to investigate similar relationships when different person detection algorithms are employed. They may include either the investigation based on a bigger set of cloud-influenced images or the comparison of clean images with photographs artificially blurred by image filters. At present, the results may therefore be method-specific and may inherit certain limitations of the algorithms themselves. In the context of the nested k-means approach, discrepancies between snow-covered and snow-free terrain concerned overrepresentation, with high values of the

POSSTBS statistics recorded in winter (Niedzielski et al. 2017). Therefore, it is likely that transparent clouds may reduce differences between the performance of the method in various seasons of year.

## Acknowledgements

The research was financed by the Ministry of Science and Higher Education of Poland through the project no. IP2014 032773. The authors thank Joanna Remisz and Małgorzata Świerczyńska-Chlaściak for technical help in operating a UAV. The participants of the field exercise—Jacek Ślopek, Krzysztof Parzóch and Monika Fila—who simulated to be lost in the wilderness are also acknowledged. The authors are also indebted to Lubomir Leszczyński and Katarzyna Mecina, both representing Świeradów Forest Inspectorate, who supported our fieldwork and observational campaigns.

## REFERENCES

Błaś, M., Sobik, M., Quiel, F., & Netzel, P. (2002). Temporal and spatial variations of fog in the Western Sudety Mts, Poland. *Atmospheric Research, 64,* 19–28.

Djuricic, A., & Jutzi, B. (2013). Supporting UAVs in low visibility conditions by multiple-pulse laser scanning devices. *The International Archives of the Photogrammetry Remote Sensing and Spatial Information Sciences, XL-1/W1,* 93–98.

Doherty, P., & Rudol, P. (2007). A UAV search and rescue scenario with human body detection and geolocalization. In A. O. Mehmet & J. Thornton (Eds.), *AI 2007: Advances in artificial intelligence* (pp. 1–13). Berlin: Springer.

Flynn, H., & Cameron, S. (2013). Multi-modal people detection from aerial video. In R. Burduk, K. Jackowski, M. Kurzynski, M. Wozniak, & A. Zolnierek (Eds.), *Proceedings of the 8th international conference on computer recognition systems CORES 2013* (pp. 815–824). Berlin: Springer.

Goodrich, M. A., Morse, B. S., Engh, C., Cooper, J. L., & Adams, J. A. (2009). Towards using unmanned aerial vehicles (UAVs) in wilderness search and rescue: Lessons from field trials. *Interaction Studies, 10,* 453–478.

Goodrich, M. A., Morse, B. S., Gerhardt, D., & Cooper, J. L. (2008). Supporting wilderness search and rescue using a camera-equipped mini UAV. *Journal of Field Robotics, 25,* 89–110.

Gretchen, K. E. (2004). Epidemiology of wilderness search and rescue in New Hampshire, 1999–2001. *Wilderness and Environmental Medicine, 15,* 11–17.

Gultepe, I., Pearson, G., Milbrandt, J. A., Hansen, B., Platnick, S., Taylor, P., et al. (2009). The fog remote sensing and modeling field project. *Bulletin of the American Meteorological Society, 90,* 341–360.

Jurecka, M., & Niedzielski, T. (2017). A procedure for delineating a search region in the UAV-based SAR activities. *Geomatics Natural Hazards and Risk, 8,* 53–72.

Marconi, L., Leutenegger, S., Lynen, S., Burri, M., Naldi, R., & Melchiorri, C. (2013). Ground and aerial robots as an aid to alpine search and rescue: Initial SHERPA outcomes. In 2013 IEEE International symposium on safety, security, and rescue robotics (SSRR). *Linkoping, 2013,* 1–2. https://doi.org/10.1109/SSRR.2013.6719381.

Miller, A., Babenko, P., Hu, M., & Shah, M. (2008). Person tracking in UAV video. In R. Stiefelhagen, R. Bowers, & J. Fiscus (Eds.), *Multimodal technologies for perception of humans. Lecture notes in computer science 4625* (pp. 215–220). Heidelberg: Springer.

Molina, P., Colomina, I., Vitoria, T., Silva, P. F., Skaloud, J., Kornus, W., et al. (2012). Searching lost people with UAVS: The system and results of the close-search project. *International Archives of the Photogrammetry Remote Sensing and Spatial Information Sciences, 39*(B1), 441–446.

Niedzielski, T., Jurecka, M., Miziński, B., Remisz, J., Ślopek, J., Spallek, W., et al. (2018). A real-time field experiment on search and rescue operations assisted by unmanned aerial vehicles. *Journal of Field Robotics,.* https://doi.org/10.1002/rob.21784.

Niedzielski, T., Jurecka, M., Stec, M., Wieczorek, M., & Miziski, B. (2017). The nested k-means method: A new approach for detecting lost persons in aerial images acquired by unmanned aerial vehicles. *Journal of Field Robotics, 34,* 1395–1406.

Oakley, J. P., & Satherley, B. L. (1998). Improving image quality in poor visibility conditions using a physical model for contrast degradation. *IEEE Transactions on Image Processing, 7,* 167–179.

O'Donnell, J. E. D. (1999). Operational ocean search and rescue using AVHRR: Cloud limitations. *Journal of Atmospheric and Oceanic Technology, 16,* 388–393.

Rudol, P., & Doherty, P. (2008). Human body detection and geolocalization for UAV search and rescue missions using color and thermal imagery. In Aerospace conference (Ed.), 2008 (pp. 1–8). Big Sky: Institute of Electrical and Electronics Engineers.

Silvagni, M., Tonoli, A., Zenerino, E., & Chiaberge, M. (2017). Multipurpose UAV for search and rescue operations in mountain Avalanche events. *Geomatics Natural Hazards and Risk, 8,* 18–33.

Steinvall, O., Olsson, H., Bolander, G., Carlsson, C. & Letalick, D. (1999). Gated viewing for target detection and target recognition. In *Proceedings of SPIE* (vol. 3707, pp. 432–448).

Sumimoto, T., Kuramoto, K., Okada, S., Miyauchi, H., Imade, M., Yamamoto, & H., Arvelyna, Y. (2000). Detection of a particular object from environmental images under various conditions. In

2000 IEEE international symposium on industrial electronics, 2000 (col. 2, pp. 590–595), Cholula, Puebla.

Van Tilburg, V. (2017). First report of using portable unmanned aircraft systems (drones) for search and rescue. *Wilderness and Environmental Medicine, 28,* 116–118.

Woodell, G., Jobson, D.J., Rahman, Z., & Hines, G. (2015). Enhancement of imagery in poor visibility conditions. In *Proceedings of SPIE 5778, Sensors, and Command, Control, Communications, and Intelligence (C3I) Technologies for Homeland Security and Homeland Defense IV.* https://doi.org/10.1117/12.601965.

Yuan, H.-Z., Zhang, X.-Q., & Feng, Z.-L. (2010). Horizon detection in foggy aerial image. In International Conference on Image Analysis and Signal Processing (pp. 191–194). Zhejiang.

(Received  October 23, 2017, revised  June 15, 2018, accepted  June 20, 2018, Published online  July 6, 2018)

Pure Appl. Geophys. 175 (2018), 3357–3373
© 2018 Springer International Publishing AG, part of Springer Nature
https://doi.org/10.1007/s00024-018-1873-2

# Technical Report: Unmanned Helicopter Solution for Survey-Grade Lidar and Hyperspectral Mapping

Ján Kaňuk,[1] Michal Gallay,[1] ⓘ Christoph Eck,[2] Carlo Zgraggen,[2] and Eduard Dvorný[1]

*Abstract*—Recent development of light-weight unmanned airborne vehicles (UAV) and miniaturization of sensors provide new possibilities for remote sensing and high-resolution mapping. Mini-UAV platforms are emerging, but powerful UAV platforms of higher payload capacity are required to carry the sensors for survey-grade mapping. In this paper, we demonstrate a technological solution and application of two different payloads for highly accurate and detailed mapping. The unmanned airborne system (UAS) comprises a Scout B1-100 autonomously operating UAV helicopter powered by a gasoline two-stroke engine with maximum take-off weight of 75 kg. The UAV allows for integrating of up to 18 kg of a customized payload. Our technological solution comprises two types of payload completely independent of the platform. The first payload contains a VUX-1 laser scanner (Riegl, Austria) and a Sony A6000 E-Mount photo camera. The second payload integrates a hyperspectral pushbroom scanner AISA Kestrel 10 (Specim, Finland). The two payloads need to be alternated if mapping with both is required. Both payloads include an inertial navigation system xNAV550 (Oxford Technical Solutions Ltd., United Kingdom), a separate data link, and a power supply unit. Such a constellation allowed for achieving high accuracy of the flight line post-processing in two test missions. The standard deviation was 0.02 m (*XY*) and 0.025 m (*Z*), respectively. The intended application of the UAS was for high-resolution mapping and monitoring of landscape dynamics (landslides, erosion, flooding, or crops growth). The legal regulations for such UAV applications in Switzerland and Slovakia are also discussed.

**Key words:** Small UAV, 3D laser scanning, push-broom hyperspectral camera, legal regulations.

## 1. Introduction

Remote sensing with unmanned airborne systems (UAS) is becoming increasingly popular in a plethora of applications competing with manned aerial surveying. The advantage of the latter is in a greater payload capacity for the use of multiple sensors suitable for high-end surveying data quality. In case of the UAS, completions of the job in a single mission with multiple sensors require either sufficient payload capacity to carry the sensors or multiple missions need to be flown with different payloads. At the same time, without sufficient accuracy, the benefits of the UAS survey seem less appealing compared to conventional techniques despite the large amount of data collected. Nowadays, the recent development of light-weight remote-sensing instruments realizes new capabilities of UAS and both these problems are addressed (Colomina and Molina 2014; Mandlburger et al. 2015).

Recent studies by Wallace et al. (2014), Yang and Chen (2015), Mandlburger et al. (2016) and Eitel et al. (2016) showed how the reduction of the size and weight of laser scanner facilitated integration of lidar on unmanned aerial vehicle (UAV) platforms. In addition, hyperspectral imaging from UAVs has been deployed in mapping vegetation species and biomass (Aasen et al. 2015), precision agriculture (Zarco-Tejada et al. 2013; Bareth et al. 2015; Sima et al. 2016), forestry (Sankey et al. 2017; Wieser et al. 2017), or coastal studies (Jaud et al. 2018). Potential applications of hyperspectral imaging with UAS can be inferred from usage of manned systems (Ryan et al. 2014; Black et al. 2016; Shendryk et al. 2016). Currently, the number of UAV applications for high-resolution mapping rapidly grows and so does the development of new UAV platforms and sensors integrations (Colomina and Molina 2014; Nex and Remondino 2014; Toth and Józków 2016; Wieser et al. 2017; Sankey et al. 2017). However, not so many of them are capable of integrating high-precision measuring equipment such as a laser scanner and hyperspectral camera (Mandlburger et al. 2016).

[1] Institute of Geography, Faculty of Science, Pavol Jozef Šafárik University in Košice, Jesenná 5, 04001 Košice, Slovakia. E-mail: jan.kanuk@upjs.sk; michal.gallay@upjs.sk; eduard.dvorny@upjs.sk

[2] Aeroscout GmbH, Technikumstrasse 21, 6048 Horw, Switzerland. E-mail: eck@aeroscout.ch; zgraggen@aeroscout.ch

Although many European countries have undertaken mapping with airborne laser scanning on a national level, the density of point clouds available for users is in the order of several points per square metre and accuracy around a decimetre for flat open land. For example, to date, Slovakia or Hungary does not have lidar coverage for the entire state area. However, still, manned airborne missions are limited in terms of flying at low altitudes to achieve high spatial density of measurements, especially in challenging terrain. UAV coupled with survey-grade sensors such as laser scanner and hyperspectral camera provides means for very high-resolution mapping of sites, where the studied phenomenon or process operates on a smaller scale than the available data sets allow researching.

In 2013, there was a unique opportunity with the European Union funds for research infrastructure to develop a UAS which would be capable of carrying payloads up to 18 kg, enough to integrate lidar system or hyperspectral camera for survey-grade mapping. Therefore, the Institute of Geography of Pavol Jozef Šafárik University in Košice, Slovakia and Aeroscout GmbH, Switzerland collaborated on this project. While the academics of the Institute have experience with geospatial data processing and modelling (Hofierka and Knutová 2015; Kaňuk et al. 2015; Gallay et al. 2016a; Hofierka et al. 2017), Aeroscout developed its custom UAV solution the Scout B1-100 unmanned helicopter with integration of remote-sensing systems such as lidar or magnetometer (Eck and Imbach 2011; Morsdorf et al. 2017).

In this paper, the main aim is to demonstrate the performance, accuracy, precision, and level of detail achieved by the custom integration of sensors for laser scanning system, hyperspectral imaging, and digital photography. The integration comprises the state-of-the-art technologies within a UAV helicopter platform capable of high-resolution and high accuracy in both spatial and spectral domains. The UAS was briefly introduced in Gallay et al. (2016b). In this paper, we provide more details on the practical aspects of mission planning, data processing and also discussion of aspects on UAV flight legislation in Slovakia and Switzerland.

## 2. Technology and Sensors

### 2.1. UAV Platform

The presented Scout B1-100 unmanned helicopter (Aeroscout GmbH, Switzerland) is a vertical take-off and landing (VTOL) UAV according to Watts et al. (2012). The empty weight of the helicopter mechanics is about 42 kg and maximum take-off weight (MTOW) of 75 kg. Automatic landing of the UAV helicopter at a pre-set location is possible with submetric accuracy which is superior to manned helicopter operation. The UAV can fly a 90 min mission delivering spatial data with subcentimetric accuracy in real-time or after the post-processing phase. The advantage of the classical helicopter configuration is in higher homogeneity of data collection during continuous forward flight in comparison with a multi-rotor aircraft, in particular under changing wind conditions. The forward speed settings of a UAV mission can range typically from 0 up to 15 m/s. The benefit for high spatial resolution of mapping is in the possibility of slow speed flight which is not possible with fixed-wing aircrafts. The helicopter is water-resistant and it can be safely operated even under light rain or windy conditions up to 5.5 m s$^{-1}$ of wind speed. Technical details of the UAV platform are reported in Fig. 1 and Table 1. Figure 2 shows the UAV with two payloads. The flight altitude above ground is limited by the required sensor resolution and the sensor specification (e.g. maximal measurement range of the laser scanner, depending on measurement rate).

From the flight control point of view, the autopilot system is the most important component of the UAV as it controls and monitors the trajectory of the aircraft. The autopilot on-board is the flight control system wePilot3000 (WeControl 2017a). It comprises an inertial navigation system (INS) which includes a receiver of the signal from global navigation satellite systems (GNSS) and a temperature calibrated inertial measurement unit (IMU) with three gyros and three accelerometers. Other components comprise a magnetometer, an absolute and differential pressure sensor, radio control receiver, and a radio modem antenna. All on-board power supply converters are doubled for redundancy. The autopilot is controlled

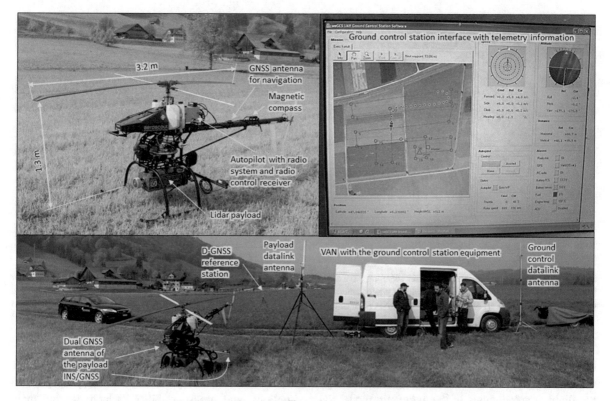

Figure 1
Scout B1-100 unmanned aerial system in the field

## Table 1

*Parameters of the UAV Scout B1-100*

| Item | Parameters |
| --- | --- |
| Length with rotor blades | 3.70 m |
| Height | 1.30 m |
| Width with balancer blades | 1.34 m |
| Main rotor diameter | 3.20 m |
| Length of tail | 2.26 m |
| Empty weight of system (without fuel, batteries, carbon fibre rotor blades, autopilot and payload) | 42.00 kg |
| Maximum take-of weight | 75 kg |
| Maximum payload capacity excluding fuel | 18 kg |
| Engine size | 100 cm² |
| Cooling | Air-cooled |
| Fuel | Max. 10 l, gasoline 95 mixed with motor oil (2.5% content) |
| Maximum flight duration with full fuel tanks | 90 min. |
| Operating temperature | − 7 to + 40 °C |

by special software which also records all the measured quantities for back-up.

The whole UAV system is designed to control flight missions in two operating modes, the manual mode and the automatic mode. In the manual mode, the pilot has full control over the helicopter with a radio controller via a 2.4 GHz channel. Optionally, the nominal value of revolutions per minute (RPM) of the main rotor can by regulated by autopilot (i.e. manual mode with governor), which helps the pilot operating the helicopter to focus on the helicopter manoeuvres. The pilot can switch the mode on the radio controller from manual to automatic mode anytime. However, the manual mode is designed mainly for safety reasons, because the pilot has more options to fully control the UAV in case of emergency. For normal operation and mapping purposes, the second, automatic mode is preferable. In automatic mode, the autopilot controls the helicopter and it is possible to perform various predefined procedures such as automatic take-off, performing

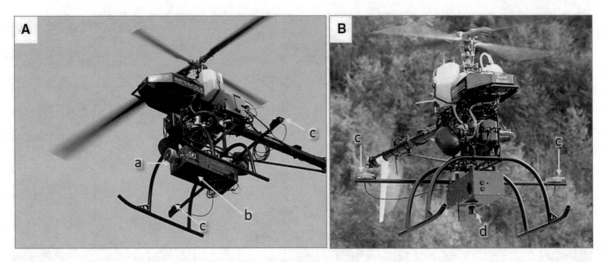

Figure 2
Scout B1-100 UAV with the integrated laser scanning payload (**A**) and hyperspectral payload (**B**). **A** comprises the VUX-1 laser scanner (Riegl, Austria) (a) and the Sony A6000 E-Mount photo camera (b). The position and orientation of the sensors is precisely monitored by dual GNSS antennas (c) and an embedded INS unit xNAV550 (Oxford Technical Solutions Ltd., United Kingdom) for the sensor attitude and position monitoring. **B** comprises the AISA Kestrel 10 hyperspectral camera (Specim, Finland) (d) and the INS (c)

the flight mission according to the set flight plan, return to home and automatic landing. Even though the flight is autonomous in the automatic mode, the pilot still can operate the UAV and control the UAV trajectory with the radio controller or with the ground control station (GCS). The use of GCS is a necessary condition in the automatic mode. The GCS includes a notebook (Panasonic Toughbook CF-53 in our case) with the weGCS software (WeControl 2017b), joystick (optional), radio modem antenna, and a power generator.

For comfortable mapping and safety reasons, two pilots are required for operating the helicopter. One of them operates the helicopter with the GCS according to the pre-set flight plan. The second operator is a back-up pilot and he or she uses the remote control in case of an emergency to switch the automatic mode to the manual mode.

GCS shows on screen telemetry information about position of the UAV, horizontal and vertical speed, inclination and direction of the UAV, temperature of engine, oxygen-to-fuel ratio in the engine (lambda sensor), percentage of the throttle opening, etc. (Fig. 1). Using the GCS the operator of GCS can manage and control a flight mission. The sensors controlling the payload are integrated within the payload box and completely separated

from the control of the UAV platform. The whole system is modular providing flexibility of payload options and transportability, which is the main advantage of this UAS. The UAV platform of Scout B1-100 was used in several studies for magnetometric mapping (Eck and Imbach 2011), fusion of lidar and image data (Yang and Chen 2015), or mapping forest structure (Morsdorf et al. 2017), but using different instruments than presented in this study.

### 2.2. Integration of Lidar and Hyperspectral Payloads Within the UAS

The presented UAS comprises two kinds of payload (Fig. 2). There can be only one kind of payload mounted on the helicopter at same time. In case the user requires using both lidar and hyperspectral sensors for mapping a particular site, the payloads must be alternated for such purpose and the flight mission must be performed twice. The reason is in the weight of both sensors and associated hardware which does not enable to integrate both sensors within a single payload. Both payloads comprise their own inertial navigation system (INS) for measuring the attitude and position of the mapping sensors. This navigation control is described in the next subsection followed by description of the laser scanning and hyperspectral payloads.

### 2.2.1 Navigation Control of the Payloads

Both payloads feature the inertial navigation system xNAV550 (Oxford Technical Solutions Ltd., United Kingdom). It is a computational system that integrates an inertial measurement unit (IMU) and two dual frequency (L1/L2) GNSS receivers (Fig. 2Ac, Bc) together with a data processing unit. Statistical filtering based on linear quadratic estimation proposed by Kalman (1960) is applied to the attitude data measured by the IMU (roll, pitch, and yaw) and geographic coordinates measured by the GNSS antennas. As a result, the best estimate of the payload position (moving platform for the mapping sensors) within the flight trajectory is calculated at a higher frequency than measured individually by the dual GNSS receiver or IMU. The INS solution of the Scout B1-100 payload provides up to 2 cm position accuracy of GNSS/IMU in the real-time kinematic (RTK) mode and constant heading performance of 0.1°, exceeding what can be achieved using magnetic sensors and without drift of position if measured only with GNSS. The differential corrections for RTK GNSS measurement are recorded by a dual frequency GNSS ground reference station (Fig. 1) allowing real-time transmission of the corrections to the payload sensors or post-processing of the flight trajectory. In addition to the flight trajectory monitoring, the position of the data acquired by the mapping sensor during that flight can be refined in the post-processing stage by recording the platform's (sensor's) trajectory during the mission. Combined forward–backwards (in time) post-processing of flight trajectory minimizes the position drift in cases when GNSS signal is not available. In this way, the flight trajectory can be significantly improved in comparison with only forward processing in real time if only the RTK method is used. The equipment of payload also includes vibration decoupling of sensitive electronics from the on-board sources of vibration such as fuel engine, main rotor, or tail rotor RPM.

Coupling the payload mapping sensors and other electronic devices with the UAV allows for on-board data storage as well as broadband data transmission to the payload ground control station and data visualization during the flight. The UAV was designed for monitoring various aspects of landscape dynamics in high spectral and spatial resolution, for example, landslides, soil erosion, flooding, or vegetation condition.

### 2.2.2 Laser Scanning Payload

The first payload integration comprises a pulse-based VUX-1 laser scanner (Riegl, Austria) coupled with a Sony A6000 E-Mount photo camera (Fig. 2a). The laser pulse repetition rate of the scanner can be set up to 550 kHz for high density of spatial coverage. The scanning system demonstrates a cutting edge topographic laser scanning technology of online waveform processing and multiple-time-around processing allowing for echo signal digitization of practically unlimited number of targets echoes (Pfennigbauer et al. 2014). Table 2 reports technical parameters of the payload and the laser scanner according its vendor's specifications by Riegl (2016). The 8.25 kg difference between the weight of the lidar payload and the weight of the laser scanner involves the digital camera, INS unit, a drive for data storage, cabling, metallic box and frame on which these parts are mounted. The lidar system is capable of measuring several thousands of points per square metre depending on the flight speed and altitude of the helicopter. This type of scanner has been successfully applied in other studies, e.g. in mapping forested flood plain habitats (Mandlburger et al. 2016; Wieser et al. 2017).

Table 2

*Parameters of the laser scanning system according*

| Items | Parameters |
| --- | --- |
| Weight of the laser scanner | 3.75 kg |
| Weight of the lidar payload | 12 kg |
| Laser wavelength | 1550 nm |
| Field of view | Up to 330° |
| Scanning method | Rotating mirror |
| Pulse repetition rate (PRR) | Up to 550 kHz |
| Scanning rate | 10–200 scan lines/s |
| Scanner power consumption | 60 W |
| Ranging accuracy | 10 mm |
| Ranging precision | 5 mm |
| Max. operating flight altitude above ground | 350 m at 50 kHz PRR, 55 m at 550 kHz PRR |
| Laser beam footprint diameter | 50 mm at 100 m range |

The collection of lidar data, its accuracy and time synchronization with the INS measurements is enabled and controlled by the Airborne Laser Scanning and Monitoring Integration (ALMI) technology (Aeroscout GmbH, Switzerland) developed for professional 3-D airborne lidar mapping with an UAV system. The development of the ALMI technology commenced within the EU funded research project "BACS" at ETH Zürich (2006–2010). The main benefit of the ALMI technology is to provide online feedback already during the flight of the UAV in relation with data quality, data recording, time synchronization, and network communication conditions. The lidar point density within a single scan line can be adjusted according to the requirements of a UAV mission and laser scanner parameters depending on the altitude above ground, forward speed, lidar measurement rate, accuracy requirements, etc. The performance of the ALMI technology has been successfully demonstrated on the Scout B1-100 UAV helicopter with different lidar scanners manufactured by Riegl, Austria, such as the VQ-480-U or LMS Q160. The ALMI technology can be applied on other UAS or even manned aircrafts. The point pattern generated by the scanner on a flat ground is regular with parallel scan lines providing uniform point density within a single lidar strip. The large field of view allows for clear scanning of vertical and subvertical sides of objects, such as trees, wires, antenna masts, buildings. The entire data set of an acquisition campaign is stored onto an on-board internal 240 GB SSD and is provided as real-time line scan data via the integrated LAN–TCP/IP interfaces.

The integrated digital camera acquires aerial imagery during the scanning according to the set interval between exposures with a 24 megapixel resolution. Using the imagery, the lidar point cloud can be coloured during the data processing stage and the imagery can be orthorectified for production of orthoimagery.

### 2.2.3 Hyperspectral Imaging Payload

The second payload was developed for mapping the landscape in high spectral resolution by passive hyperspectral scanning. The sensor is an AISA

Table 3

*Parameters of the hyperspectral camera*

| Items | Parameters |
|---|---|
| Spectral range | 400–1000 nm |
| Focal length | F/2.4 |
| Spectral sampling | 1.75/3.5/7 nm |
| Frame rate | Up to 130 Hz |
| Signal-to-noise ratio (peak) | 400–800 |
| Spatial resolution | 1312 or 2048 pixels |
| FOV | 40° |
| Total system power (camera, INS, DPU) | < 41 W |

KESTREL 10 camera by Specim (Finland) which is a new generation of low-weight hyperspectral cameras intended for UAVs and other platforms of limited payload size. The camera is a linear array (push-broom) imager recording the solar radiation reflected from land surface with a high light throughput and spatial resolution of 2048 pixels per scan line (Table 3). The spectral range of detectable electromagnetic radiation spans from 400 to 1000 nm. This range makes the payload system specifically suitable for mapping and monitoring vegetation as it contains visible, red-edge and near-infrared portion of the spectrum in which the plant condition and characteristics are most distinctly detectable. The data generated by the hyperspectral payload are radiometrically and spectrally stable with a high signal-to-noise ratio in variable real world remote-sensing conditions. The data acquisition is set up and controlled by the on-board Lumo Recorder software by Specim run on the internal computer system within the payload on-board the UAV (Fig. 3). This software is monitored and controlled by the ground control station via the Lumo GroundStation software by Specim. The camera was radiometrically calibrated by the vendor. The bore sight calibration of the camera with respect to the payload frame resulted from the test flight which hyperspectral data are presented in this study.

### 3. Performing the Mission and Data Acquisition

The UAV platform presented in this study can be categorized as a mid-sized UAV according to Watts et al. (2012) or as a small UAV according to Barnard

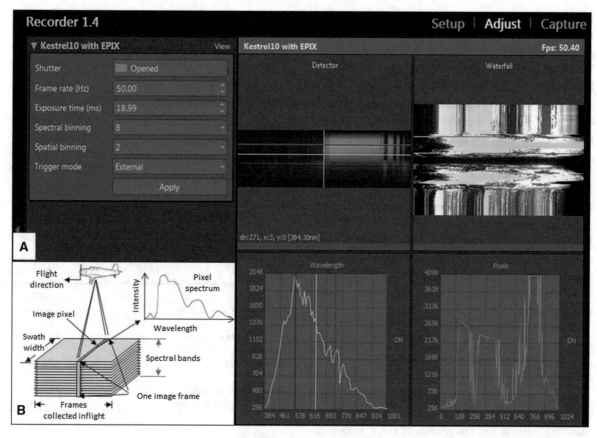

Figure 3
User interface of the Lumo Recorder interface by Specim developed for set up and control of the hyperspectral acquisition (**a**) which main principle is shown in (**b**)

(2007). However, its weight of 75 kg requires adhering to certain safety procedures for safe performance of the mission. In addition, the quality of the geospatial data to be collected by the mapping sensors is closely linked with good control and management of the flight mission. Careless handling of this UAV platform can eventually lead to dangerous situation, system damage or crash, potentially leading to injuries of the crew, observers, or damage to the third parties. The procedures for normal (standard) operation of the presented UAS are summarized in Table 4.

The procedures to be undertaken before the flight focus on the analysis of the mapped area and the related airspace, identification of possible risks of the mission failure, and preparation of scenarios in case of an emergency. Furthermore, the prestart actions involve planning the mission and preparation of the

Table 4

*List of procedures for normal operation of the Scout B1-100*

| Phase | Procedures |
|---|---|
| Prestart procedures | Meteo briefing |
| | Safety briefing |
| | Mission planning |
| | Pre-flight aircraft inspection |
| | Payload inspection |
| Flight mission | Ground control station start |
| | System start |
| | Take-off |
| | Flight mission |
| | Landing |
| | System shut down |
| Post-flight procedures | Data transmission |
| | Post-flight aircraft inspection |
| | Post-flight briefing |
| | Data processing |

helicopter and the payload for the flight. Usually, the UAV is flown in the automatic mode, i.e. autonomously, according to the set flight trajectory which has to be carefully planned. This involves setting a home point for automatic return of the helicopter and its landing. Position of a home point is defined by the operator of the ground control station (GCS). In case of data link failure, the home mode is activated after 30 s automatically or it can be activated anytime during the mission by the operator of the GCS. In the home mode, the autopilot flies the UAV in a straight line from actual position to the home waypoint; therefore, the relative height of this point to the ground or the altitude has to be sufficiently high with respect to the obstacles in the area. When the position of home point is reached, autopilot initializes landing, and after touch-down, the engine shuts down automatically.

Planning a flight mission is performed in the weGCS software which allows setting parameters such as position of waypoints and behaviour the UAV at the waypoints, speed, and altitude. The positions of waypoints are related to the shape of the mapped area and to the data quality requirements. In general, the planning mission for the presented UAV is indeed similar to that of a manned airborne vehicle which has been thoroughly developed over decades (Gandor et al. 2015).

Flight mission procedures involve starting the GCS, UAV, and payload. Good pre-flight preparation determines smooth running of a flight mission from take-off to landing. During the mission, the pilot and the UAV GCS operator monitor the behaviour of the UAV system and flight parameters via telemetry information and by visual contact with the UAV. The payload GCS operator monitors the data quality during the mission. Communication between the crew is required to inform each other about actual situation and mission progress.

After the landing and shutting-down the engine, the mapped data can be downloaded from the payload data storage. Post-flight briefing is first important step leading to the success of the following flight mission. The pilot, the operator of UAV GCS, and the operator of the payload GCS analyse and evaluate the flight mission and prepare technical report and recommendations for the future flights.

Table 5

*Summary of the laser scanning mission*

| Items | Parameters |
| --- | --- |
| Flight speed | 5 m/s |
| Flight height above ground | 30 m |
| Area size | 2 ha |
| Terrain elevation range | 6 m |
| Pulse repetition rate | 550 kHz |
| Point density | 1111 points/m$^2$ |
| Point spacing | 0.03 m |
| Exposure station baseline for digital photography | 8 m |
| Focal length of the digital camera | 0.02 m |

Table 6

*Summary of the hyperspectral scanning mission*

| Items | Parameters |
| --- | --- |
| Flight speed | 4 m/s |
| Flight height above ground | 140 m |
| Area size | 2 ha |
| Terrain elevation range | 3 m |
| Number of spectral bins | 92 |
| Integration time | 19 ms |
| Frame rate | 50 Hz |
| Width of spectral bands | 9 nm |
| Ground sampling distance | 0.1 m |

We present the results achieved in two individual test missions for two different areas near Lucerne, Switzerland. The first mission was dedicated to testing the lidar payload and it was conducted in April 2015 over two hectares of a meadow near Lucerne (47°02′36.0″N, 8°13′54.5″E). The sky was cloudy, the air temperature was 15 °C, and the wind speed was about 2 m/s. The hyperspectral payload integration was tested in October 2015 north of Lucerne (47°06′44.3″N, 8°15′35.4″E). The sky was overcast, the air temperature was 13 °C and the wind speed was about 1 m/s. Other technical parameters of the missions are summarized in Tables 5 and 6. GNSS reference station was set during both missions as it was described in Sect. 2.2.1.

## 4. Data Processing

The acquired data were processed using the software developed by the vendors of the sensors involved in mapping. In both missions, the flight trajectory was post-processed in the RT Post-process software utility (OXTS Ltd., UK) with a high GNSS weighting in combined forward and backward processing. By this approach, the benefit of measuring both GNSS and IMU data is exploited to minimize any drift error which accumulates along the trajectory. The resulting flight line was calculated in the World Geodetic System 1984 thus all data in the presented survey were derived in this coordinate system.

### 4.1. Lidar Data Processing

Before the mission start, the parameters for the lidar data acquisition were set in the RiACQUIRE software (Riegl, Austria) and time synchronization with INS measurements was performed with the ALMI software (Sect. 2.2.1). The lidar data were processed in the RiPROCESS software package (Riegl, Austria). After the flight, the raw laser data were downloaded from the on-board data storage unit and combined with the position and attitude data which resulted from the post-processing of the flight trajectory to get the 3D point cloud. The acquired point cloud was split into several flight lines (data strips) for their relative adjustment in the next step. Overlap of the multiple flight lines enabled improving the accuracy of their relative registration by least squares adjustment in RiPRO-CESS. The lidar points were filtered and classified into three main categories: ground, vegetation, and high noise. The high-noise data were removed from the data set as they were redundant. They originated along the flight lines due to laser reflection from the helicopter skids. The pictures acquired by the photo camera simultaneously with the lidar data were processed in RiPROCESS to assign the lidar points with natural colours as RGB values. The potential of the imagery is also in processing the photographs by image matching algorithms such as Structure from Motion given that the exposure centre geographic coordinates is calculated from the flight trajectory post-processing or using ground control points for bundle adjustment. However, this task was not performed in the presented study.

### 4.2. Hyperspectral Data Processing

The AISA Kestrel 10 camera is a push-broom scanner; therefore, the imagery originates line by line with considerable misalignment of the consecutive scan lines in the raw hyperspectral imagery. Therefore, the imagery has to be georectified and converted from DN numbers to radiances. These procedures were performed in the CaliGeoPRO pre-processing tools by Specim. The position and attitude data collected from the time synchronized on-board INS were used for georectifying and georeferencing of the AISA flight line images. The resulting image data from CaliGeo-PRO represented spectral radiance which was subject to atmospheric correction in the next step using the QUAC tools of the ENVI software version 5.3 (Harris Geospatial, USA). The results were smoothed using the THOR Spectral Smoothing tools of ENVI with the filter width settings of 4 and a second-order polynomial. Classification of the atmospherically corrected and smoothed hyperspectral data was performed by the Spectral Angle Mapper implementation of the ENVI software. Regions of interest had to be extracted from the imagery prior to the classification to account for the spectral variation in the scene.

### 5. Results

### 5.1. Results of UAV-Based 3D Laser Scanning

The lidar data were recorded within a 4 min mission resulting in 22 million of 3D points which were cut into five data strips (Fig. 4a). The data represent a flat meadow with adjacent electrified railway in the north displayed in detail by Fig. 4b, c. The main component of the absolute accuracy of the final point cloud is determined by the accuracy of the flight trajectory processing (Fig. 5) and adjustment of overlapping flight strips. The positional and vertical accuracy of the post-processed flight trajectory was 0.02 and 0.025 m, at $1\sigma$, respectively. The orientation accuracy of roll and pitch was 0.05° and 0.11° of yaw. The standard deviation of vertical differences between overlapping flight strips after adjustment was 4.6 cm. Despite there were no additional tests performed using independent GNSS reference points, the results demonstrate high overall quality of the

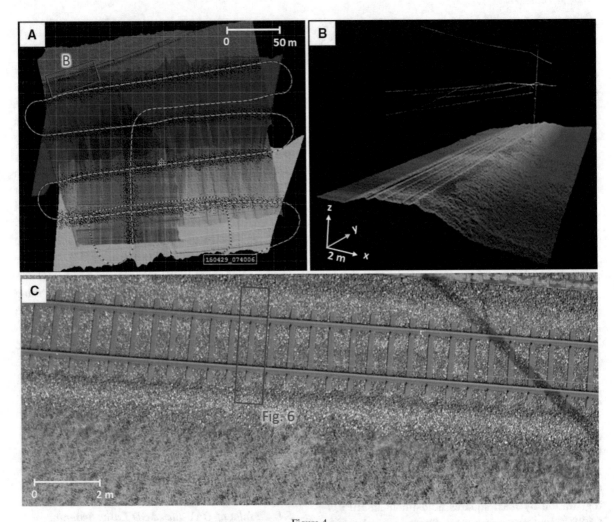

Figure 4
Lidar point cloud coloured according to flight strips with the flight trajectory (**a**), lidar point cloud of a railway and power lines coloured by elevation after lidar strip adjustment (**b**) located in **a** by the red rectangle, and as the orthophotoimage (**c**)

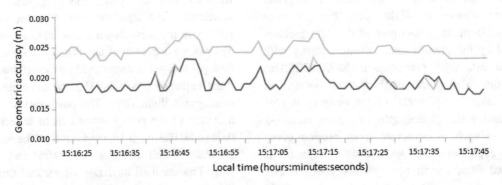

Figure 5
Geometric accuracy of a selected part of the flight trajectory along *x* (red), *y* (orange) and *z* (green) axes after combined backward and forward INS data post-processing in metres

final 3D lidar point cloud. The photographs acquired by the on-board digital camera were used to colourize the 3D lidar points in the RiPROCESS software which resulted in the orthoimage in Fig. 4c. Figure 6 visually conveys the achieved precision and the level of detail by showing the rails and the embankment as a 3D point cloud before the strip adjustment (Fig. 6a) and after the adjustment (Fig. 6b). Because of the accurate flight line processing, there were no ground control points needed for lidar data strip adjustment or bundle adjustment of imagery for generating the orthoimagery.

## 5.2. Results of UAV-Based Hyperspectral Scanning

The hyperspectral data collection over flat grassland with a road was performed in 3 min from 140 m above the ground. This altitude was relatively high in comparison with the lidar mission. It was chosen for the narrow field of view of the hyperspectral camera to achieve wider cross-track swath of sensing. The final unprocessed raw imagery (Fig. 7a) contained over 2500 scan lines each having 92 spectral bins of 9 nm spectral width (Table 6). The positional and vertical accuracy of the post-processed flight trajectory was 0.021 and 0.025 m, at $1\sigma$, respectively. The orientation accuracy of roll and pitch was 0.05° and 0.11° of yaw. Figure 7b demonstrates the scan line adjustment after applying the post-processed flight trajectory which is best perceived by the roads as straight cyan lines. Figures 7c and 8d zoom to orthorectified images of 10 cm spatial resolution in the area of the road crossing. Figure 7b, c portrays the mapped area as an RGB colour composite of two

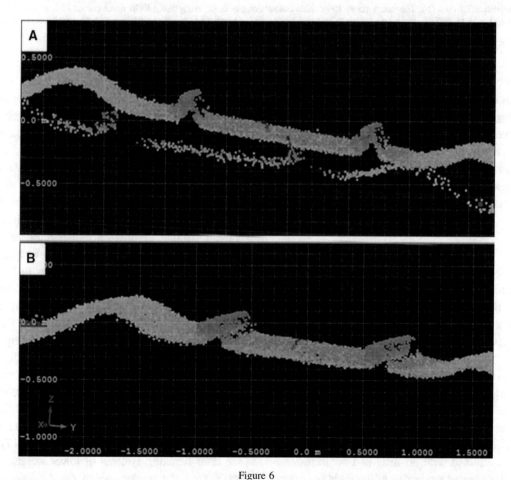

Figure 6
Perspective side view of a cross section of a railway located in Fig. 4c represented as 3D lidar point clouds of two overlapping data strips before strip adjustment (**a**) and after the adjustment (**b**), slight tilting of the point cloud is due to change of the viewing angle

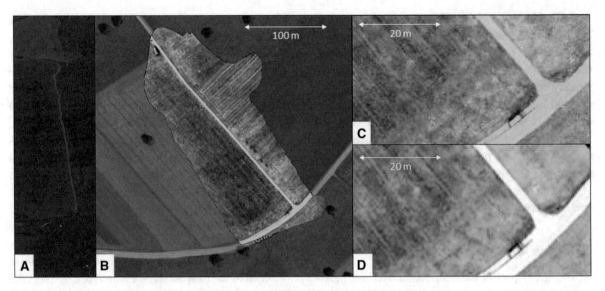

Figure 7
Hyperspectral imagery after acquisition as a non-georectified raster with DN numbers (**a**) and after time synchronization between the INS and the raw hyperspectral data (**b**). The detail views show false colour composite (**c**) comprising RGB combination of bands 713.79–720.68, 610.42–617.31, 521.11–527.96, and a true colour RGB combination (**d**) of bands 664.69–666.41, 559.69–561.40, 489.50, 491.21 nm. Image **b** contains the background orthoimagery by ©Google

bands from visible and one from near-infrared spectrum showing spectral differences in vegetation cover displayed in orange to red tones. The natural colour composite is displayed in Fig. 7d. These images visually demonstrate the high degree of the time synchronization between the on-board INS recordings and the data recorded by the hyperspectral.

## 6. Discussion

The results presented in this study clearly show that the UAS is capable of delivering survey-grade geospatial data in high accuracy and high resolution in both spatial and spectral domains. One of the key factors providing survey-grade quality data is in the highly accurate flight trajectory monitoring and post-processing in the order of centimeters. Therefore, no ground control points are required for adjustment of lidar scan swaths or bundle adjustment for photography. The INS solution is based on the GNSS dual-antenna combined with an IMU and combined forward and backward processing of the flight trajectory. Such an INS is essential when the accuracy of

heading is important and the vehicle has very low dynamics such as a helicopter has. Assessment of absolute accuracy with independent check points would provide a more robust support for the survey-grade potential of the presented UAS.

Applications of UAV lidar systems are rare to date but emerging. For example, Wieser et al. (2017) evaluated the accuracy of tree trunk diameter measured at breast height from lidar data acquired by the VUX-1 scanner on-board a lighter multi-rotor UAV than the Scout B1-100 unmanned helicopter. They achieved 19 mm standard deviation with respect to diameters from terrestrial laser scanning data after excluding four gross outliers. Their results confirm high geometric accuracy achievable with such as a lidar system. Similar values of vertical-strip adjustment were reported by Mandlburger et al. (2015) who also used that kind of UAS. Despite no ground reference data set was used to assess the absolute lidar data accuracy in our study, the results of the flight trajectory post-processing indicated similar relative accuracy as achieved by Wieser et al. (2017). There are laser scanning systems of lower weight than the VUX-1 on the market which can be integrated on mini-UAVs and provide promising results. Although

UAS of lower costs based on lower weight sensors and mini UAV platforms were developed (e.g. Jozkow et al. 2016; Sankey et al. 2017), the measurement accuracy of their sensors and achievable point density is lower in comparison with the presented UAS. Such lidar systems require further development to achieve survey-grade measurement accuracy and data quality comparable to the VUX-1 lidar system. Jozkow et al. (2016) and Sankey et al. (2017) reported geometric accuracy of UAV lidar data with respect to the GNSS reference points was in the order decimetres to metres.

Although our conducted missions concerned small sites, the presented UAS is capable of mapping relatively larger areas for having a gasoline power supply for up to 90 min. It will not outperform piloted airborne systems in terms of area coverage but the presented UAS is competitive to multi-rotor UAS powered by batteries which provide comparable flight speed and data quality but flight duration is limited to half an hour (e.g. Mandlburger et al. 2015).

Recent development in snapshot hyperspectral cameras (Aasen et al. 2015) has led to decreasing the size and weight of their integration on mini UAVs providing new perspectives for high-resolution landscape mapping. Results achieved with such snapshot cameras seem very competitive to push-broom hyperspectral sensors, such as was presented in this study. However, further development is needed to achieve survey-grade accuracy and data quality. Jaud et al. (2018) presented a custom solution of a lightweight low-cost UAS based on a similar type of hyperspectral sensor (push-broom) and IMU/GNSS constellation as we present. The more accurate IMU, wider baseline of the dual-antenna configuration and higher weight UAV platform in our case clearly provided more stable sensing conditions resulting in spatially consistent and geometrically accurate hyperspectral data.

Coupling active laser scanning and passive hyperspectral scanning technology for mapping with a high level of spatial detail has a very high potential for precision agriculture, forestry or geoscientific research. Unfortunately, our two test missions could not have been flown over the same area at that time to demonstrate these benefits as, for example in Shendryk et al. (2016) within a manned airborne mission.

The future perspective was demonstrated by Sankey et al. (2017) by the results achieved with a low-cost integration of a lidar and a passive push-broom hyperspectral camera on a multi-rotor UAV. Spatial resolution and spectral resolution of their hyperspectral data were comparable with data acquired with our UAS. The benefit of the highly accurate flight trajectory reconstruction and longer flight duration in our case provides considerable advantage which into certain extent, compensates much higher costs of the Scout B1-100 UAS in comparison with multi-rotor mini-UAS of lower weight.

## 7. Legal Regulations of UAV Mapping in Switzerland and Slovakia

Widespread application of UAVs much depends on conditions for their legal operation in terms of flying and mapping with UAVs. Furthermore, the rules of good practice in conducting the UAV missions are important. The growing market of UAV and their applications stimulate discussions on regulations for a safe and feasible use of UAVsQuery in many countries (Clarke and Moses 2014; Villa et al. 2016). In the European Union, individual member states control the use of UAVs themselves and the legal rules differ among the countries (Torresan et al. 2017). The presented UAS is based in Slovakia but the test missions were conducted in Switzerland.

From the Slovak perspective, the currently valid official regulations in the Slovak Republic present a barrier for a legal and wider spread use of UAVs. The reason is that the rules for manned aircrafts were adopted for regulation of the UAV operation. The Transport authority of the Slovak Republic, Division of Civil aviation (Transport Authority 2016), is the official authority responsible for adhering to international civil aviation regulations in the areas of technical, operational, and personal testing and certification. UAVs with the maximum take-off weight below 150 kg can perform flights in the airspace only under conditions respecting the safety regulations and these circumstances must be decided by the traffic authorities. The person holding the radio controller during a UAV flight mission is considered the UAV pilot. He or she is fully responsible for technical conditions of the UAV and for preparation and performing of the flight mission. The UAV needs to be insured against damage caused to third parties and it has to be registered by the Transport authority in Slovakia. Each UAV pilot must be registered and licensed. Certification and pilot license can be acquired after successfully passing a theoretical and practical examination. UAV certifications issued in other countries are not accepted in Slovakia to date. However, the international Private Pilot License is valid which makes its holder eligible also for flying UAVs in Slovakia. In either case, the UAV operator has to inform the Transport authority about flight mission in advance and he or she must be informed about the current flight traffic within the flight mission airspace. It is strictly prohibited to use UAVs in built-up areas, in areas with people, or in the controlled airspace around airports. Improper light conditions (i.e. night, dusk or dawn) are also considered a constraint for flying. The UAV must remain in direct sight of the pilot during the flight. Maximum distance from the UAV operator to the UAV is limited to 1000 m of a horizontal distance and 100 m of vertical distance. Currently, legal acquisition of data from airborne platforms in Slovakia, including UAS, is subject to authorization from the Ministry of Interior of the Slovak Republic according to the act on the protection of classified information. Our opinion is that meaningful reasons for this rule had been applicable decades ago; therefore this act requires revision to adapt to the contemporary trends in remote sensing.

In Switzerland, the Federal Office of Civil Aviation (FOCA 2016) is the official authority responsible for UAV flight regulations. The regulations are less complex than in Slovakia providing more flexibility to the UAV operators and to the data collection. This is important especially for scientific applications and research. According to the Swiss legislation, UAVs are considered to be aircraft models. They can be operated without a special permission up to the weight of 30 kg under the condition that the pilot maintains visual contact with the flying UAV at any time and the UAV must be in direct sight of the operator (i.e. the pilot). In certain exceptional cases, the Office may grant permission for operations without direct visual contact. In either case, the UAV

must hold a minimum distance of 100 m to inhabited areas (more than 10 inhabited houses) or to crowds with more than 23 people. The flight must stay below 150 m above ground in the air control traffic zones (CTR) and in the area up to 5 km away from such zones. The pilot must hold a minimum distance to the next airfield of 5 km. Take-off and landing can be performed with the approval of the land holder. Special approval of the local authority is needed for specific circumstances such as continuous noise emissions during the UAV flight. Operation of UAVs heavier than 30 kg requires submission of special application in which the emphasis is put on explaining how the safety of operation is secured and risks of damage and injuries to other people is considered.

## 8. Conclusion

This paper addressed current trends in the field of unmanned aerial remote sensing by presenting a custom solution of integrating high-end sensors within a mid-sized UAV platform of 75 kg of maximum take-off weight including 18 kg capacity for surveying and mapping sensors. The platform is the Scout B1-100 unmanned helicopter powered by a gasoline two-stroke engine and controlled by an autopilot system. The UAS is capable of performing the survey mission based on a flight plan autonomously or fully handled by a pilot. However, operating the UAS is not trivial and requires special training in flying the UAV and its maintenance.

The presented UAS solution comprises two types of payload dedicated for active and passive remote sensing in a very high resolution. The laser-scanning payload comprises a VUX-1 laser scanner and a Sony Alpha 6000 E-Mount digital camera. The scanner is capable of emitting up to 550,000 pulses per second and processing the laser echo waveform online (in real-time) for resolution of practically unlimited number of targets echo returns. The capability of flying at low speed of several metres per second and height above ground of tens of metres enables acquisition of thousands of lidar points per square metre and also a crisp digital photography for assigning the lidar points with natural colours or further processing by image matching techniques.

Hyperspectral imaging is enabled by the second kind of payload. The main sensor is the hyperspectral camera AISA Kestrel 10 which is a push-broom type of scanner for reflectance mapping in visible and near-infrared spectral domain in over 300 spectral bands. The accurate reconstruction of flight trajectory and sensors attitude in both payloads is assured by an INS with two dual frequency GNSS antennas and IMU.

Such a technological solution allowed for achieving high accuracy of the flight line post-processing of less than 3 cm in two test missions near Lucerne, Switzerland. The laser scanning test mission over a flat terrain with railway embankment resulted in a 3D point cloud capturing the terrain shape and railway features in ultra-high-resolution. With the average point density of more than 1100 points/m$^2$, a vertical lidar strip adjustment of less than 5 cm the experiment demonstrated the high potential of UAS-borne laser scanning for geosciences and applications.

Different legislation on national level is a serious barrier to using UAV systems widely. Some countries have very strict legislation because procedures of manned aircraft operation are adapted to the UAS. However, a good example of a more flexible legislation on national level is Switzerland. Requirements to operate UAV systems are different from demands of the manned aircraft crew. Despite the differing legislation among countries the standards for normal UAV operation are identical. We argue that one of the reasons for missing common worldwide legislation which better adapts to the use of UAS is in the missing standards of procedures for operation of UAV. We addressed these aspects based on the experience with the presented UAS and the two test flights.

The future development in UAV-based mapping will likely lead to even greater miniaturization of sensors which will enable their integration within one payload on the same UAV platform or even mini UAVs. Another approach is in integrating new types of sensors which are capable of mapping 3D geometry and spectral properties at once such as hyperspectral laser scanning systems. In either case, it will decrease costs and increase efficiency of UAV mapping. The challenge remains in real-time transmission of the measured data from the UAS to the GCS and real-time processing of such massive data. The potential of survey-grade accuracy mapping with UAS with multiple kinds of sensors is very promising in applications requiring high resolution in spatial, spectral, and temporal domain.

*Acknowledgements*

Production of the presented UAV and development of the lidar and hyperspectral payloads was financed within two projects of the University Science Park TECHNICOM for innovative applications supported by knowledge technologies—phase 1 and phase 2 (ITMS 26220220182, ITMS2014 + 313011D232). The projects were co-funded by the European Union Structural Funds and the Ministry of Education, Science, Research and Sport of the Slovak Republic, the executive authority for the Operational Programme Research and Development. The research presented in this paper was funded by the Slovak Research and Development Agency within the scientific project APVV-15-0054 and financial support was also provided by the Slovak Research Grant Agency within the projects VEGA 1/0474/16 and VEGA 1/0963/17. We would like to thank Benedikt Imbach and Christoph Fallegger for assistance in preparing the flight missions. We also want to thank the Specim team for the support with the hyperspectral data processing. Our special thanks go to the reviewers and the editor for their comments which helped us to better convey the conducted research to the reader.

References

Aasen, H., Burkart, A., Bolten, A., & Bareth, G. (2015). Generating 3D hyperspectral information with lightweight UAV snapshot cameras for vegetation monitoring: From camera calibration to quality assurance. *ISPRS Journal of Photogrammetry and Remote Sensing, 108*(10), 245–259.

Bareth, G., Aasen, H., Bendig, J., Gnyp, M. L., Bolten, A., Jung, A., et al. (2015). Hyperspectral full-frame cameras for monitoring crops: Spectral comparison with portable spectroradiometer measurements. *Photogrammetrie Fernerkundung Geoinformation, 1*, 69–79.

Barnard, J. (2007). Small UAV (< 150 kg TOW) *Command, control and communication issues*. Technical Report. Institution of Engineering and Technology.

Black, M., Riley, T. R., Ferrier, G., Fleming, T. R., & Fretwell, P. T. (2016). Automated lithological mapping using airborne hyperspectral thermal infrared data: A case study from Anchorage Island, Antarctica. *Remote Sensing of Environment, 176,* 225–241.

Clarke, R., & Moses, L. B. (2014). The regulation of civilian drones' impacts on public safety. *Computer Law and Security Review, 30*(3), 263–285.

Colomina, I., & Molina, P. (2014). Unmanned aerial systems for photogrammetry and remote sensing: A review. *ISPRS Journal of Photogrammetry and Remote Sensing, 92,* 79–97.

Eck, C. & Imbach, B. (2011). Aerial magnetic sensing with an UAV helicopter. *ISPRS International Archives of the Photogrammetry, Remote Sensing and Spatial Information Sciences, Zürich, Switzerland, Vol. XXXVIII-1/C22(1).*

Eitel, J. U. H., Höfle, B., Vierling, L. A., Abellán, A., Asner, G. P., Deems, J. S., et al. (2016). Beyond 3-D: The new spectrum of lidar applications for earth and ecological sciences. *Remote Sensing of Environment, 186,* 372–392.

Federal Office of Civil Aviation (FOCA). (2016). *Drones and aircraft models.* https://www.bazl.admin.ch/bazl/en/home/good-to-know/drones-and-aircraft-models.html. Accessed 09 Dec 2016.

Gallay, M., Eck, C., Zgraggen, C., Kaňuk, J., & Dvorný, E. (2016a). High resolution airborne laser scanning and hyperspectral imaging with a small UAV platform. *International Archives of the Photogrammetry Remote Sensing and Spatial Information Sciences, XLI-B1,* 823–827. https://doi.org/10.5194/isprs-archives-xli-b1-823-2016.

Gallay, M., Hochmuth, Z., Kaňuk, J., & Hofierka, J. (2016b). Geomorphometric analysis of cave ceiling channels mapped with 3D terrestrial laser scanning. *Hydrology and Earth System Sciences, 20,* 1827–1849.

Gandor, F., Rehak, M., & Skaload, J. (2015). Photogrammetric mission planner for RPAS. *The International Archives of the Photogrammetry Remote Sensing and Spatial Information Sciences, XL-1/W4,* 61–65.

Hofierka, J., & Knutová, M. (2015). Simulating spatial aspects of a flash flood using the Monte Carlo method and GRASS GIS: A case study of the Malá Svinka Basin (Slovakia). *Open Geosciences, 7*(1), 118–125.

Hofierka, J., Lacko, M., & Zubal, S. (2017). Parallelization of interpolation, solar radiation and water flow simulation modules in GRASS GIS using OpenMP. *Computers and Geosciences, 107,* 20–27.

Jaud, M., Le Dantec, N., Ammann, J., Grandjean, P., Constantin, D., Akhtman, Y., et al. (2018). Direct georeferencing of a pushbroom, lightweight hyperspectral system for mini-UAV applications. *Remote Sensing, 10*(2), 204. https://doi.org/10.3390/rs10020204.

Jozkow, G., Toth, C., & Grejner-Brzezinska, D. (2016). UAS topographic mapping with Velodyne lidar sensor. *ISPRS Annals of the Photogrammetry, Remote Sensing and Spatial Information Sciences, III-1,* 201–208.

Kalman, R. E. (1960). A new approach to linear filtering and prediction problems. *Journal of Basic Engineering, 82*(1), 35–45.

Kaňuk, J., Gallay, M., & Hofierka, J. (2015). Generating time series of virtual 3-D city models using a retrospective approach. *Landscape and Urban Planning, 139,* 40–53.

Mandlburger, G., Glira, P., & Pfeifer, N. (2015). UAS-borne lidar for mapping complex terrain and vegetation structure. *GIM International, 29*(7), 30–33.

Mandlburger, G., Pfennigbauer, M., Wieser, M., Riegl, U., & Pfeifer, N. (2016). Evaluation of a novel UAV-borne topo-bathymetric laser profiler. *International Archives of the Photogrammetry, Remote Sensing and Spatial Information Sciences ISPRS Archives, 2016-January,* 933–939.

Morsdorf, F., Eck, C., Zgraggen, C., Imbach, B., Schneider, F. D., & Kükenbrink, D. (2017). UAV-based LiDAR acquisition for the derivation of high-resolution forest and ground information. *Leading Edge, 36*(7), 566–570.

Nex, F., & Remondino, F. (2014). UAV for 3D mapping applications: A review. *Applied Geomatics, 6*(1), 1–15.

Pfennigbauer, M., Wolf, C., Weinkopf, J., & Ullrich, A. (2014). Online waveform processing for demanding target situations. *Proceedings of SPIE The International Society for Optical Engineering.* https://doi.org/10.1117/12.2052994.

Riegl. (2016). *RIEGL laser measurement systems GmbH. RIEGL VUX-1 Data Sheet.* http://www.riegl.com/. Accessed 20 Dec 2017.

Ryan, J. P., Davis, C. O., Tufillaro, N. B., Kudela, R. M., & Gao, B.-C. (2014). Application of the hyperspectral imager for the coastal ocean to phytoplankton ecology studies in Monterey Bay, CA, USA. *Remote Sensing, 6,* 1007–1025.

Sankey, T., Donager, J., McVay, J., & Sankey, J. B. (2017). UAV lidar and hyperspectral fusion for forest monitoring in the southwestern USA. *Remote Sensing of Environment, 195,* 30–43.

Shendryk, I., Tulbure, M., Broich, M., McGrath, A., Alexandrov, S., & Keith, D. (2016). Mapping tree health using airborne laser scans and hyperspectral imagery: a case study for a floodplain eucalypt forest. *Geophysical Research Abstracts, 18,* EGU2016-355. Available online: http://meetingorganizer.copernicus.org/EGU2016/EGU2016-355.pdf. Accessed 09 Dec 2016.

Sima, A., Baeck, P.-J., Delalieux, S., Livens, S., Blommaert, J., Delauré, B., & Boonen, M. (2016). A new COmpact hyperSpectral Imaging system (COSI) for UAS. *Geophysical Research Abstracts, 18,* EGU2016-5504. http://meetingorganizer.copernicus.org/EGU2016/EGU2016-5504.pdf. Accessed 09 Dec 2016.

Torresan, C., Berton, A., Carotenuto, F., Di Gennaro, S. F., Gioli, B., Matese, A., et al. (2017). Unmanned aerial vehicles for environmental applications. *International Journal of Remote Sensing, 38*(8–10), 2029–2036.

Toth, C., & Jóźków, G. (2016). Remote sensing platforms and sensors: A survey. *ISPRS Journal of Photogrammetry and Remote Sensing, 115,* 22–36.

Transport Authority. (2016). Podmienky vykonania letu UAV (drony). http://letectvo.nsat.sk/2015/08/25/podmienky-vykonania-letu-uav/. Accessed 09 Dec 2017.

Villa, T. F., Gonzalez, F., Miljievic, B., Ristovski, Z. D., & Morawska, L. (2016). An over view of small unmanned aerial vehicles for air quality measurements: Present applications and future prospectives. *Sensors, 16*(7), 1072.

Wallace, L., Watson, C., & Lucieer, A. (2014). Detecting pruning of individual stems using airborne laser scanning data captured from an unmanned aerial vehicle. *International Journal of Applied Earth Observation and Geoinformation, 30,* 76–85.

Watts, A. C., Ambrosia, V. G., & Hinkley, E. A. (2012). Unmanned aircraft systems in remote sensing and scientific research:

Classification and considerations of use. *Remote Sensing, 4,* 1671–1692.

WeControl. (2017a). Autopilot wePilot300. http://www.wecontrol. ch/products/wePilot3000/flyer_wePilot3000.pdf. Accessed 19 Dec 2017.

WeControl. (2017b). Ground control station weGCS http://www. wecontrol.ch/products/weGCS/flyer_weGCS.pdf. Accessed 19 Dec 2017.

Wieser, M., Mandlburger, G., Hollaus, M., Otepka, J., Glira, P., & Pfeifer, N. (2017). A case study of UAS borne laser scanning for measurement of tree stem diameter. *Remote Sensing, 9*(11), art. no. 1154.

Yang, B., & Chen, C. (2015). Automatic registration of UAV-borne sequent images and LiDAR data. *ISPRS Journal of Photogrammetry and Remote Sensing, 101,* 262–274.

Zarco-Tejada, P. J., Guillén-Climent, M. L., Hernández-Clemente, R., Catalina, A., González, M. R., & Martín, P. (2013). Estimating leaf carotenoid content in vineyards using high resolution hyperspectral imagery acquired from an unmanned aerial vehicle (UAV). *Agricultural and Forest Meteorology, 171–172*(4), 281–294.

(Received  October 7, 2017, revised  March 23, 2018, accepted  April 19, 2018, Published online  May 8, 2018)

Pure Appl. Geophys. 175 (2018), 3375–3390
© 2018 Springer International Publishing AG, part of Springer Nature
https://doi.org/10.1007/s00024-018-1807-z

**∎ Pure and Applied Geophysics**

CrossMark

# Technical Report: The Development and Experience with UAV Research Applications in Former Czechoslovakia (1960s–1990s)

JAROMÍR KOLEJKA[1] and LADISLAV PLÁNKA[2]

*Abstract*—The use of unmanned aerial vehicles in a number of fields of human activity represents the second wave of interest in the development and application of automated flying remotely controlled machines to collect aerial data. The former Czechoslovakia was one of the world's leading countries in the 1960s–1990s in terms of an unprecedented boom of development and applications of flying machines for imaging the Earth's surface. The reasons for their use were the same as today. Since the mid-1960s, radio-controlled (RC) models of aircraft carrying various types of photographic cameras have been developed. In spite of many administrative constraints, kite helicopters, fixed-wing aircrafts, and rogallo-wing aircrafts gradually began to be used in research. The photographic cameras for 1, 2, 4, and 6 bands carried by RC-aircraft models were developed in cooperation with leading Czech companies. These cameras used colour and black-and-white films, positive and negative films, and panchromatic, spectrozonal, and multispectral films. The general methodology and the RC-aircraft model application rules were both developed. The dominant processing method was the visual image interpretation, with and without the assistance of instruments. Optical and digital image mixers were used in Czechoslovakia, so it was possible to use natural and unnatural colour composites to highlight the studied phenomenon. A number of examples of the techniques and the scientific applications are presented in the article.

**Key words:** RC-aircraft model, mono- and multi-band aerial cameras, methods, rules, experience, applications.

## 1. Introduction

Remote sensing of the Earth by low-flying carriers has been experiencing an intensive renaissance in the last decade (Miřijovský 2013). In addition to the development of remote sensing and transmission technology at the background of modern progress, there has been an innovation in the field of unmanned flying sensor carriers (Aber et al. 2010). Drones of various types have become widely available thanks to the development of mass production for both the professional community and the various spheres of application, entertainment, documentaries, and even high-risk activities. These are the achievements of twenty first century technology for a number of experts and lay people. The truth is, however, that the present use of unmanned aerial vehicle (UAV) applications has been experiencing more of a recurring wave of professional and public interest, mainly due to better accessibility to this technology and substantial liberalisation of its operating rules. It is a relatively little known fact that unmanned flying means for remote sensing of the Earth from low heights (up to 500 m above the surface) began to be used for civilian research, technical, and other purposes already in the post-World War II period (Przybilla and Wester-Ebbinghaus 1979). After 1965, Czechoslovakia was one of the world's leading countries in the development and use of this technology (Stehlík 1967).

Motivation for the development and use of this technology in the research on the Earth was the same as today:

1. lower equipment costs than the traditional manned aviation technology,
2. lower cost of equipment than manned and automatic satellite technology,
3. lower costs of staff training,
4. closer and more intense observation of objects on the Earth's surface,

[1] Institute of Geonics, Czech Academy of Sciences, Studentská 1768, 70809 Ostrava, Czech Republic. E-mail: jkolejka@centrum.cz

[2] Faculty of Mining and Geology, Institute of Geodesy and Mine Survey, VŠB-Technical University of Ostrava, 17. listopadu 15/2172, 70833 Ostrava, Poruba, Czech Republic. E-mail: ladislav.planka@vsb.cz

5. higher resolution of scanning techniques flying relatively close to the studied surface,

6. high operability and repeatability, and short-term action applicability,

7. moderate dependence on atmospheric parameters and the weather, especially clouds,

8. low demands on ground equipment, take-off, and landing areas,

9. low demands on maintenance and repairs,

10. quick achievement of the operating level for imaging or scanning,

11. possible slow to stationary flight as required,

12. easy to move techniques to the work site and visual inspection of the areas and objects,

13. easy and repeatable route guidance to the target area in the event of a previous unsuccessful flight, and

14. organizational easier flight preparation and land sensing under valid legislation, etc.

For a long time, aerial imagery on the territory of the socialist Czechoslovakia could only be carried out by state-driven and strictly controlled organisations only. Since the second half of the 1970s, it has been possible, by reducing the impact of the Act on Classified Information on non-metric imagery for civilian purposes, to perform such surveys by other organisations, mainly for research and design purposes with radio-controlled (RC) aircraft models of limited weight and load capacity. The civil use of such technology was already known abroad at that time (e.g., in Japan, Kawamura 1984). The military usage of UAVs started by the US and Israeli armies in early 1950s (Newcome 2004).

The fundamental liberalisation of air traffic following the political and economic changes in Czechoslovakia after 1989 briefly pushed the deployment of small manned aircrafts and RC-aircraft models to the background. A significant revival took place in the new millennium, coupled with the use of modern unmanned aerial vehicles—UAVs (drones) and Global Navigation Satellite System (GNSS) methods.

Because the beginnings of RC-model aircraft use in Czechoslovakia fall into the "pre-Internet" era, and most of the publications about them were written in Russian, the expert community abroad is little informed about the early beginnings of the remote sensing of the Earth from low altitudes using radio-controlled aircraft models. The present paper intends to rectify this shortcoming.

## 2. Development of unmanned flying camera carriers

Permits for remote sensing of the Earth's surface in the territory of Czechoslovakia were issued by the power ministries. According to the first historical document (found in the archive of the Institute of Geography of the Czechoslovak Academy of Sciences in Brno) dated in 1965, the Ministry of National Defence (MNO) authorised aerial photography on the condition that only "a captive hovercraft and not a helicopter or aircraft" could be used and the Ministry would be informed about the sites in which the imaging took place. In that year, the project for the documentation and the monitoring of soil erosion on agricultural land began. It was necessary to take photographic images from a height of about 100 m for this purpose. At that time, it was possible to bring the camera to such a height only with a tied balloon. However, guiding it to the desired site with anchor ropes over a complex terrain of large areas of arable land with developed crops was very difficult or even impossible. In an attempt to maximally comply with the official opinion of the MNO, the Brno aircraft model maker Mr Josef Vymazal was asked to develop a tethered helicopter. He created a two-lane helicopter model powered by 2.5 $cm^3$ MVVS engines. The load capacity of the helicopter was very limited and its deployment was not satisfactorily operational. That is why initially aircraft models with a hard fixed wing were developed and used for taking photos (Fig. 1).

During the subsequent experimentation, the concept of an RC-aircraft model was adopted due to the requirement of having a low-speed flight. This model used the sail (parachute) wing (commonly used on "rogalls"). The very low performance of the model engine limited its use to ideal no-wind conditions only. However, the first-quality photographic images were taken with the use of this machine, and the usability of low-altitude Earth's surface imaging method using RC-aircraft models (carriers) was

Figure 1
RC-camera carrier Ekolet 01 with a hard fixed wing was constructed by the Svazarm Modelklub Chomutov for the Institute of Landscape Ecology of the Czechoslovak Academy of Sciences, the most branch (Source: the archive of L. Plánka)

demonstrated. In 1967, Mr. Otakar Stehlík published a description and guidance on the use of this RC-aircraft model for research purposes at an international level (Stehlík 1967). In 1969, the aircraft model maker, Mr. František Štěrbák, constructed a new, bigger, and better RC-aircraft model with a parachute wing capable of carrying diverse photographic devices. The hull of the aircraft was made of laminated glass fabric and equipped with a w-shaped wing made of polystyrene. The model had traditional tail parts and was equipped with a wheeled chassis. In this variation, a lot of operational deployments were achieved and their results were highly appreciated by the professional community.

In the first half of the 1970s, the limiting factors of the design of RC-aircraft models were shown primarily in RC units, propulsion units, and engine propellers. Nevertheless, RC-aircraft models operated in Czechoslovakia could reliably capture images of the Earth's surface from a height of 1 km using simple and lightweight cameras. Such success was

also internationally accepted (Nagiyev et al. 1990). For the Academy of Sciences of the USSR, two aircraft models of a classic wing were delivered from the beginning, followed by another ten units of new carriers with complete operator training at the point of delivery. These new carriers were in the form of a two-engine helicopter each with a 3 m rotor diameter. The helicopter was driven by two separate MVVS motors of 2.5 $cm^3$ located directly on the rotor blades.

The leading RC-aircraft model maker Mr Jiří Trnka constructed a more advanced model with a triangular wing for the Institute of Geography in Brno. The camera was installed in a detachable container in the chassis. The tail rods were T-shaped, and the single V-shaped wing was reinforced with beams and coated with paper (Fig. 2).

The RC-aircraft models for the remote sensing were operated on an "inappropriate" terrain (e.g., forest clearings, quarries, developed agricultural crops, peat bogs, etc.) from the viewpoint of the

Figure 2
RC-aircraft camera carrier "Rogallo 1976" on the preparation table before flight (Source: the archive of L. Plánka)

synchronous movement of the ground pilot under a flying aircraft model on foot or at longer distances in a moving car. Loss of the RC-aircraft model from the view of the pilot or from the reach of the radio transmitting unit was unlikely. Throughout the history of this type of RC-aircraft camera carrier being operated, this has happened only once in a dozen flights per year, in a difficult weather situation.

The first carrier labelled "Rogallo 76" was innovated very rapidly in 1977 on the basis of the previous practical experience. The new "Rogallo 1977" model was originally designed as a measuring probe carrier for remote measurement of the air temperature and humidity at low altitudes above the ground. However, after practical tests, it was very soon adapted for imaging the Earth's surface with various cameras weighing from 350 to 2500 g. The model was equipped with an OS MAX motor, and later with an Enya engine. It had small dimensions (length 90 cm, wingspan 180 cm, wing area 1.27 m$^2$) and a low total weight of 5.5 kg (including the camera). It was possible to launch the model simply by throwing it by hand, namely without any demands on the surface quality of the starting site (Plánka 1984). The model was able to reach a height of up to 600 m in a very short time, so it took about 30–45 min to capture 12–36 shots for a relatively quiet flight (Fig. 3).

At the beginning of the 1980s, a more robust modification of the RC-aircraft model with a parachute wing called "Rogallo 1981" was constructed (Fig. 4). This carrier was primarily designed for the transport of the Flexart M-6 six-band multispectral camera. It had a paraglider wing with an area of 3.24 m$^2$ with a span of 280 cm, a hull of 100 cm long, and a total weight of 13 kg, later 16.5 kg. Because of its weight, it was no longer possible to launch it by hand, even though the original 15ccm Webra engine was soon replaced by the significantly more powerful 35 cm$^3$ Webra Bully engine. To protect the camera, the runway of about 5 m long was provided with plastic plates and the acceleration of the take-off of the carrier to the operating height was accelerated with a rubber catapult. The same benefits of piloting and landing which controlled the flight of the weaker and lighter "Rogallo 1977" model were applied to the "Rogallo 1981", as well (Fig. 4).

needs of aircraft modelling. As it was commonly assumed that flights would be carried out over or around the inhabited areas, "absolute" operational reliability was the basic requirement for their design. Therefore, the RC-aircraft model of a geometric-shaped paraglider ("rogallo") made from translucent polyethylene film was preferred to those with a standard hard fixed wing (Plánka 1984). J. Trnka then constructed other RC-aircraft models in several modifications from 1976 and they were successfully operated until the early 1990s. A great advantage of this wing layout was the relatively low speed of the model flight, which could be so low that under favourable wind conditions and with limited fuel supply to the engine, the carrier could be "fixed" above one point. This option was very effectively used when landing in a limited space and/or against a strong wind without the risk of damage to the carrier and connected cameras. Relatively low speed allowed

Figure 3
Launch of the RC-aircraft model "Rogallo 1977" with a one-channel camera "by hand" (Source: the archive of L. Plánka)

The flight control all of the above-mentioned models was provided by ROBBE, or by Multiplex RC equipment to a distance of 800–1000 m from the flight control site. This included the radio control of the motor speed, the direction and the height of the flight, the opening and closing of protective objective doors, and the rewind and the exposition of films.

Until the political changes in 1990, land imaging using RC models was carried out only to the maximum authorised height of 300 m in Czechoslovakia, although exceptions were easily permitted.

### 3. Applied Imaging Systems

Cameras working with a narrow, easily accessible, and inexpensive 3 cm-wide film (Fig. 5) were used initially. Although these positive and negative films originated from the production of significant brands, their geometric resolution was limited (e.g.,

Figure 4
RC-six-band camera carrier "Rogallo 1981" before taking off from a runway made up of plastic plates with the help of a rubber catapult
(Source: the archive of L. Plánka)

Figure 5
Most frequently used single-band cameras (from left to right and to the front): Leica, Prakti II, Polaroid, and Flexaret (Source: the archive of L. Plánka)

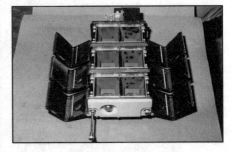

Figure 6
Two-band, four-band, and six-band arrangements of the Flexaret cameras constructed for Czechoslovak RC-aircraft models (Source: the archive of L. Plánka)

by the quality of photographic films). Later on, for the purpose of increasing the resolution, the Flexaret cameras produced by the Czech company Meopta Přerov in various designs, working with 6 cm-wide roll films, were used exclusively (Fig. 6). While the single and two-band cameras were used without filters (two-band cameras exposed synchronously both a panchromatic black-and-white and a colour film, positive and/or negative), filters from the six-band cameras in various combinations were inserted into a four-band camera (Table 1). The permanent vertical

Table 1

*Basic features of the Flexaret M-6 camera filters*

| Band | Wave length (nm) | Transmittance (%) |
|------|------------------|-------------------|
| 1 | 477.5–522.5 | 29 |
| 2 | 513.5–554.5* | 30 |
| 3 | 530.7–557.3 | 52 |
| 4 | 591.9–632.1* | 60 |
| 5 | 671.2–734.8* | 62 |
| 6 | 785.7–846.3* | 36 |

*Filters commonly used in four-band camera

axis of the camera was secured with the camera cabinet placed in an aluminium frame and protected against vibrations by effective silent blocks.

The need to take photos in the infrared (IR) band led to the construction of Flexaret multiband cameras. The relevant part of the multiband camera used to expose the black-and-white IR material had to allow sharpening other than "infinity" shots, as was possible with the conventional panchromatic materials. This part also allowed the placing of the appropriate "colour" additive filter in front of the lens. In the end, the six-band Flexaret M-6 multispectral photographic camera consisted of three pairs of coupled Flexaret cameras. This arrangement in pairs of cameras allowed synchronous shots to be captured through two different filters side by side on one film strip, moving automatically after two frames. Under these circumstances, it was possible to use three different types of film material simultaneously for the respective pair of the multispectral camera. The film shift was provided by a servomechanism with intervals controlled by a switch coupled with the camera exposure. The camera exposure was provided by another remotely controlled servomechanism operating the exposure of the camera. The six-band camera was equipped with special additive filters (Table 1), whose optical properties were very close to the properties of the additive filters used in the professional MKF 6 M photographic cameras. MSK 4 was stationed on Soviet satellites, carrying out the Interkosmos Earth Remote Sensing Programme. These filters were provided by the Astronomical Institute of the Czechoslovak Academy of Sciences. The Flexaret cameras were equipped with Belar 3.5/ 80 lenses and with a Pentacon-Prestor shutter (Kolejka et al. 2001). All Flexaret camera modifications were performed in cooperation with the Czech company Meopta Brno (Fig. 6).

The black-and-white panchromatic negative films ORWO NP 20 or 400 Fortepan and colour slide film AGFA CT 18 or Kodak Ektachrome, and the infrared colour films Kodak Aerochrome 2443 Infrared or the infrared black-and-white films IR 840 and ORWO NP 22 were used for imaging from Czechoslovak RC-aircraft models.

The photos were mostly taken for "scientific purposes" for various institutions that needed reliable geospatial information of the required quality for their research activities (e.g., monitoring the aerial forest stands, the search for potential archaeological sites, the survey of badly accessible areas, and crop monitoring on large areas of arable land, the water areas, and wetlands survey, etc.). This information had to be cheap and operationally available. In addition, the participating institutions did not require the acquisition of rigorously accurate aerial metric imagery (Hanzl 1986). However, the metric properties of the images were tested in special experiments (Marek et al. 1982). It was found that under poor flight conditions at a scale of 1: 3 500, the position errors were about 2–3 cm and the elevation errors were about 20 cm.

The aerial imaging method from small altitudes using RC-aircraft models became an integral part of the so-called "sub-satellite" experiments of the Interkosmos' Remote Sensing Working Group during the period 1981–1985. The relevant methodology was developed mainly by the Institute of Geography of the Czechoslovak Academy of Sciences in Brno in close collaboration with the Zentralinstitut für Physik der Erde (ZIPE) in Potsdam (German Democratic Republic—DDR) and by the substantial contribution of the Institute of Landscape Ecology of the Czechoslovak Academy of Sciences in Most. The final version was published in 1984 (Plánka 1984).

### 4. Imaging Experiments and Data Acquisition

The technology of acquiring and processing aerial imagery from remotely controlled RC-aircraft models has been adopted by experts in a number of fields of science, research, design, economy, and state administration (Stehlík 1981). For their needs, aerial imaging campaigns have been applied, based on a well-established methodology, but always adapted to the task specifics.

1. In agriculture, attention has been focused, in addition to soil erosion studies, also on the identification of agricultural crops, their phenological phases or damages, and on waterlogging of soils or their toxic contamination (Stehlík 1967).

2. In forestry, as well as identification of the species composition of stands, monitoring of the level of air pollutant damage to trees, and documentation of forest calamities caused by wind, frost, or insects was carried out (Nagiyev et al. 1986; Plánka 1990).

3. For water management, aerial photographs were taken to document water bodies, siltation, and overgrowing them with plants, in some cases also to monitor water pollution and shoreline development, and wetlands were also investigated (Kolejka and Petch 1989).

4. A number of aerial imaging campaigns served to document and to monitor difficult-to-reach landscape types such as wetlands, river mouth areas in natural and artificial lakes, important from the viewpoint of nature protection (Žaloudík 1990).

5. Aerial imagery taken with RC-aircraft models was also used to study the development and distribution of snow cover.

6. Some attempts have also been made to verify the applicability of the photos taken by RC-aircraft models to determine the soil humus and soil moisture content (Feranec et al. 1985).

7. The aerial imagery from the RC-aircraft models also served very well for archaeological research (Gojda 2000).

The most frequent demand for aerial photographs of inaccessible and inconspicuous localities involved peat bogs and forest stands.

The methodology of multispectral aerial photography from low altitudes included a sequence of phases (Table 2). The formulation of tasks was carried out on the basis of the requirements of the client of the aerial imaging campaign. This included setting the experiment goals, defining and describing the imaging territory, and determining the type and characteristics of the areas and objects planned for imaging. It also included the formulation of legal relations between the ordering party, the user, and the maker/provider of the photographic material, the determination of the authority granting the exemption from the ban on aerial photography, and the determination of the degree of confidentiality of the acquired material. At the same time, the method of reimbursement of costs associated with the

experiment was determined. The responsibility of the client and the provider was given in the sense of ensuring the technical, material, personnel, and organizational needs, which will be part of the campaign preparation, the field work, and the laboratory image processing (Plánka 1987).

After the imaging campaign, the aerial imagery obtained was passed on to the user for thematic processing.

## 5. Technology and procedures for image processing obtained by RC-aircraft models

The visual interpretation of the image material obtained by RC-aircraft models for scientific purposes in Czechoslovakia was the main processing method used in the 1960s and 1990s. This was done by (1) the direct visual interpretation of the printed image in the required magnification or (2) using instrument support.

The direct visual (thematic) interpretation was logically the cheapest and most accessible way. The interpretation took place based on individual black-and-white or colour photographs (photos or slides), as well as photographs glued to photo-plans on a solid base. The results of the interpretation were simultaneously or subsequently plotted on the basic topographic map. According to the plot, a special thematic map was usually compiled. There was a strong influence of the processor's subject on the result depending, both on the correctness of the thematic evaluation of the image and on the positional accuracy of the result (Tollinger 1988).

The level of instrument support was related to the financial possibilities of the processor and to the state-of-the-art technological progress. Special instruments for analogue processing of the remotely sensed data were often based on photogrammetric devices. The easiest tools were represented by stereoscopes (for processing of image stereo pairs). A more complicated device was the stereoscope supplemented with a stereometer to determine elevation differences (the instrument INTERPRETOSKOP was commonly used in Czechoslovakia). A certain standardization in the thematic image processing was provided by a re-drawer, especially of a more modern

Table 2

*Phases of preparation and imaging by RC-aircraft models in Czechoslovakia*

| Phase | Title | Content | Result |
|---|---|---|---|
| 1 | Personal group composition | Head of working group<br>Ground pilot of RC-aircraft model<br>Machinist<br>Photographer<br>Navigator<br>Archivist | Working group |
| 2 | Technical equipment | Camera carrier equipped with radio signal receiver<br>Ground radio control station<br>Pointers for flight navigation<br>Photographic camera<br>Standard photographic laboratory<br>Drawing machine for aerial imagery<br>Special motor car for working group and material transportation | Imaging and transportation equipment |
| 3 | Planning | Imaging area field reconnaissance<br>Inventory of critical objects<br>Inventory of sites suitable for camera carrier take-off and landing<br>Inventory of sites important for ground pilotage and navigation<br>Inventory of natural and artificial barriers influencing carrier control, navigation and manoeuvring | Flight schedule |
| 4 | Preparation | Determining the type of ground navigation and flight height<br>Determination of the side and altitude angles of the position of the navigated carrier in flight space<br>Determination of the flight path according to the intensity of sunlight, air temperature and humidity, soil temperature and humidity, wind direction and force<br>Checking the absence of possible disturbing radio signals in the imaging territory<br>Checking for the absence of quarries or other plants electrically igniting explosives<br>Mounting the kite of the carrier<br>Preparation of the runway and the starting catapult<br>Engine and carrier control function check<br>Preparing and testing the camera | RC-aircraft model and camera ready to flight |
| 5 | Supporting ground works | Ground spectrometric measurements<br>Meteorological measurements<br>Soil and vegetation observation | Documentation of ground measurements and observations |
| 6 | Fieldwork | Launch<br>Guiding the carrier to imaging tracks<br>Film exposures<br>Landing<br>Flight documentation | Flight |
| 7 | End of imaging | Disconnection of the carrier and camera<br>Checking proper camera operation and film exposure<br>Performing an "Aerial Imaging Report"<br>Film designation by the number of the flight and the number of the respective spectral band (if applicable) | Raw imagery material |
| 8 | Laboratory image pre-processing | Processing of ground observations records<br>Invoking exposed films<br>Acquisition of indicative photographic positives<br>Design of image mosaic of study area | Imaging documentation |
| 9 | Archiving materials | Forwarding to the archive<br>Designing the confidentiality degree<br>Numbering images<br>Image clustering according to experiments and their types | Archive of image data |

type, into which a limited range of the colour composite was implanted, or a light correction image into sides (e.g., RECTIMAT C). Many so-called "mixing projectors" were used to process photographic multispectral images taken from 2- to 6-channel cameras carried by RC-aircraft models. These devices were able to project several (mono) spectral (zonal) images through an optional set of colour and grey correction filters on the common screen, where a colour composite was generated from 3 to 4 zonal images from the same territory. The mutual registration of the individual bands (zonal images) was performed mechanically based on the crossing points. The resulting colour composite (in natural and non-natural colours could either be photographed from the screen, or copied to the colour-inverse paper placed in the photographic cassette. By carefully choosing the filters in the mixer, it was possible to achieve a higher colour contrast, which was mainly used to highlight the studied phenomenon with other phenomena in the area of interest. A typical instrument example was the MSP-4C Mixing Projector (produced by the company Zeiss Jena, GDR) originally designed to process the professional multispectral images from the MKF-6 M or MSK-4 cameras used in the framework of the international INTERKOSMOS Remote Sensing

Figure 7
False colour composite of a multispectral image of the Czechoslovak agricultural area taken from a height of about 100 m compiled on the screen of the MSP-4 instrument; the approximate ground area size is 90 × 90 m (Source: the archive of J. Kolejka)

Program (Fig. 7). This equipment operated in several institutions in Czechoslovakia and has been used for a wide range of research tasks (from soil erosion damages, agricultural crop, and forest stand diseases) through the nature conservation inventory and documentation to an archaeological objects survey and search for potential sites.

At the beginning of the 1980s, there was another upgrade of technological re-equipment in selected institutions in the Czechoslovak Socialist Republic. The electronic mixer and image analyser Multicolor Data System model 4200F (produced by Japanese company NAC Inc., Tokyo, Japan) was used for territorial research. The mixer was used to process multispectral images of all commonly used sizes (on a transparent material—film) simultaneously from four bands of the electromagnetic spectrum (Koželuh 1985). The device was simply controlled using a button control and it performed interactive imaging operations. It was used as a preliminary analyser for computer image processing, i.e., since 1983 at the Institute Geography of the Czechoslovak Academy of Sciences in Brno. The device was equipped with an analogue input and an output on the one hand, and with a digital video record and processing on the other hand. Individual zonal images were recorded with a black-and-white camera (digital zoom from 16 to 160 mm), allowing the scanning of areas from 30 × 30 mm to 250 × 250 mm. The analogue video signal from the television camera was digitized and recorded in a fast 8-bit converter and stored in a maximum of four image memories, each with a capacity of 512 × 512 pixels. The video data went on a colour TV monitor (Fig. 8) with a size of 280 × 280 mm (512 × 512 pixels) depending on the selected function (Koželuh 1988).

The exact positional registration of zonal images was realised by mechanically moving the illuminating table and checking the output on the colour monitor. The correction of errors caused by image non-homogeneity and by optical lens defects was also possible. The image processing was supported with a black-and-white display, colour composite, image magnification (ZOOM), single-band density slice (SLS), multiband density slice (MLS), the share of different area calculations (AREA), the histogram calculation (HIST), and subtraction of two video

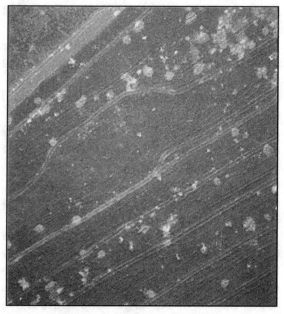

**Figure 8**
False colour composite assembled by the NAC Multicolor Data System model 4200F from a multispectral image of an alfalfa field attacked by a cocoon. The image was taken by an RC-aircraft model from a height of about 100 m and the approximate ground area size is 110 × 114 m (Source: the archive of J. Kolejka)

memories (SUB). The instrument was proved to produce colour composites, to run density-metric image analyses (point HIST), as well as to compile thematic cartograms (MLS).

## 6. Selected Examples of Study of the Physical Environment by Means of RC-Aircraft Camera Carriers

The thematic processing of aerial imagery taken by RC-aircraft models for the soil erosion studies was initiated by O. Stehlík in the 1960s. The imaging usually took place after torrential rains on affected areas of arable land. In this way, areas of varying intensity of damages were determined, and their extent was measured directly on the image. These findings were then followed by the measurement of the height of the soil out-flow directly in the field and the areas with different erosion were finally specified (Fig. 9).

Numerous imaging experiments concerned the agricultural land. Synchronous multilevel imaging

**Figure 9**
Aerial image from the RC-aircraft model of the field affected by the soil erosion after heavy rain in 1978, scale of negative 1: 1611, flight height 64.5 m, and the approximate ground area size is 223 × 146 m (Source: the archive of L. Plánka)

and measurement campaigns were conducted (using a ground spectrometer, RC-aircraft model taking photos, professional aerial imaging, and satellite imaging from the US—Landsat, USSR—KATE). The aim of these experiments was to find changes in the spectral characteristics of a number of agricultural crops in various phenological phases, depending on the height of the imaging (Fig. 10). Studies of soil erosion on agricultural land were repeated (Kolejka and Petch 1989, 1991).

The weed damages to agricultural crops were well traceable and, if necessary, various methods of precise farming were applied for crop protection after an appropriate interpretation of multispectral imagery. Using the MSP-4 optical mixer, the affected areas were highlighted in false colour composites, and their interpretation results were redrawn to topographic maps and delivered to the farm (Fig. 11).

The authorities of nature conservation in Czechoslovakia needed images of difficult-to-reach or clumsy sites for their work. Aerial imaging took place over water objects, wetlands, forest units, and extremely dissected terrain or over large, homogeneous areas without appropriate natural landmarks. Typical

examples were surveys and inventories of wetland status (Fig. 12).

A number of imaging projects were related to the mouth zones of rivers, both natural and artificial water reservoirs, for the purpose of study and monitoring processes of material sedimentation, coastal development, and wetland vegetation development. Attention was paid to the area of Nové Mlýny water reservoirs especially. The interpretation of colour slides (in the original 6 × 6 cm) arranged in various magnifications was carried out both by simple visual interpretation (Fig. 13) and by the NAC Multicolor Data System model 4200F tools (Fig. 14).

The NAC device was able to perform more demanding operations with the multispectral, spectrozonal, and panchromatic colour and black-and-white image material acquired by RC-aircraft model camera carriers. By selecting the appropriate analytical tools and imaging techniques, the effects of some relief forming processes in the river bed were visualised (Fig. 14). In this way, it was possible to identify the formation of sedimentation benches, to highlight the presence of sediments in the water at the junction of rivers (Kolejka and Petch 1989).

Figure 10
Agricultural applications of aerial images from RC-aircraft camera carriers (from left to right): the quality of growth and density of the sugar beet culture—area size 230 × 340 m, the research area of the Institute of Plant Agriculture Prague-Ruzyně near Střítež u Třebíče—area size 340 × 340 m, the Dendrological garden in Průhonice, 1988—area size 340 × 340 m, all figures are based on black-and-white negatives 6 × 6 cm (Source: the archive of L. Plánka)

Figure 11
Redrawn weed damage (in pink) over a properly created false colour composite of a four band of multispectral image in MSP-4; the approximate ground area size is 1120 × 790 m (Source: the archive of J. Kolejka)

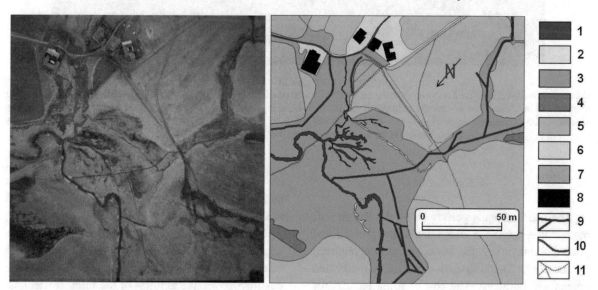

Figure 12
Aerial image of a wetland provided by an RC-aircraft model carrier designed for nature conservation purposes (exact delimitation of wetlands, inventory of natural water bodies, and identification of former drainage channels, both natural and artificial). Key: 1—natural water bodies, 2—dry natural drainage channels, 3 waterlogged land, 4—forest, 5—agricultural meadows, 6—urban lawns, 7—fields, 8—built-up areas, 9—human-made drainage channels, 10—paved roads, and 11—ground roads, approximate location of the image centre: N 49.620298, E 16.115360 (Source: the archive of J. Kolejka)

Figure 13
Colour slide in the original size 6 × 6 cm, covering the Svratka river mouth to the Middle Novomlýnská nádrž reservoir taken from a height of about 150 m. Readers can observe the turbidity of the polluted waters registered in the upper part of the image, the formation of two generations of sedimentary aggradation dykes—bottom left and the bush succession on dykes. The approximate ground area size is 115 × 115 m, approximate location of the image centre: N 48.906926, E 16.611668 (Source: the archive of J. Kolejka)

## 7. Conclusion

The aerial imaging performed by the RC-aircraft models was a suitable alternative to the professional piloted aerial survey, the use of which was hindered by numerous administrative, organizational, and operational problems. The cost of using RC-aircraft models was incomparably lower and provided the benefits already mentioned. The use of RC-aircraft models with a parachute wing as carriers of one to six-band cameras taking pictures up to 6 × 6 cm in the visible and close IR-bands was successfully tested in Czechoslovakia.

However, the use of RC-aircraft models faced the relatively difficult navigation of the aircraft camera carrier, which was based almost exclusively on the qualified subjective abilities of the pilot. It was possible to take up to 12 analogue images in one flight (in the case of the most frequently used single-band Flexaret camera), and only 6 images in the case of the

Figure 14

Thematic processing of the colour panchromatic images taken by the RC-aircraft model from a height of about 100 m above the confluence of two rivers (since 1990, this area has been flooded with the Lower Novomlýnská nádrž reservoir). From left to right: the highlighted image, its digitized form, the masking of the area outside the rivers, and classification of pixels in the riverbed, approximate location of the image centre: N 48.887714, E 16.654717 (Source: the archive of J. Kolejka)

flight with the Flexaret M-6 multispectral camera, but it was also possible to repeat flights a few times per day at short time intervals. The clear application advantage of the above-mentioned technical means was the operability of their deployment, which meant the acquisition of the image material immediately after the need for documenting the phenomenon.

The aerial imaging techniques using RC-aircraft models from small elevations, developed and operated in Czechoslovakia, have no significant presence in the contemporary literature and on the Internet, although this subject has been sufficiently written on in the past. Many of the procedures that have been discovered in connection with the use of present UAV carriers were already known in the past, and the latest technological progress brings to this technology legitimate popularity and versatility in a wide range of human activities. A big milestone in the current phase of UAV use is the fact that they do not face many different restrictions as it was in the past. The liberalisation of the human society and the technological progress together represent another impetus to the development of new practices that unfortunately did not happen in the past. There is no doubt that the use of UAVs is an important step in the democratisation of the access to aerial information.

## REFERENCES

Aber, J. S., Marzolff, I., & Ries, J. B. (2010). *Small-format aerial photography: principles, techniques and geoscience applications* (p. 268). Amsterdam: Elsevier.

Feranec, J., Kolář, J., Kudela, K. Sabol, T. (1985). Zisťovanie povrchového zamokrenia pôd pomocou aerokozmických snímok. In Sborník přednášek 2. konference o dálkovém průzkumu Země, Ústí nad Labem, Dům techniky ČSVTS, pp. 97–113.

Gojda, M. (2000). *Archeologie krajiny* (p. 238). Praha: Academia.

Hanzl, V. (1986). Přímá lineární transformace snímkových souřadnic s eliminací radiálního zkreslení objektivu. *Geodetický a kartografický obzor, 74*, 113–117.

Kawamura, N. (1984). *Japan's technology farm* (p. 148). Sofia: Selskostopanska akademiya.

Kolejka, J., & Petch, J. (1989). Geografické vyhodnocení digitalizovaných leteckých snímků vodních objektů. *Geografie, 94*, 241–248.

Kolejka, J., & Petch, J. (1991). Hodnocení zpracovatelských metod dálkového průzkumu Země pro zjišťování eroze půdy na území Středomoravských Karpat. *Zprávy Geografického ústavu ČSAV, 28*, 55–74.

Kolejka, J., Plánka, L., & Trnka, J. (2001). Rádiem řízené modely snímkují naši krajinu. *GEOinfo, 7*, 41–45.

Koželuh, M. (1985). Interpretace družicových snímků pomocí přístroje NAC MCDS 4200F. *Zprávy Geografického ústavu ČSAV, 22*, 29–39.

Koželuh, M. (1988). NAC Multicolor Data System 4200F. Informační zpravodaj čs. pracovní skupiny pro dálkový průzkum Země, 2, pp. 33–35.

Marek, K.-H., Söllner, R., Weichelt, H., Röser, S., & Marek, G. (1982). Erste Erfahrungen beim Einsatz von RC-Bildflugzeugen fur Grundlagenforschungen zur Fernerkundung. *Vermessungstechnik, 30*, 329–331.

Miřijovský, J. (2013). *Bezpilotní systémy* (p. 170). Sběr dat a využití ve fotogrammetrii: Olomouc, Palackého univerzita v Olomouci.

Nagiyev, P., Guseynov, K., Dzhafarov, E., Mamedov, R., Kolář, J., Kříž, B., et al. (1986). Avtomatizirovannaya klassifikaciya selskokhozyaystvennykh obyektov po materialam skanernoy aerosyomki. *Issledovaniye Zemli iz Kosmosa, 2*, 96–102.

Nagiyev, P., Guseynov, K., Dzhafarov, E., Mamedov, R., Kolář, J., Kříž, B., et al. (1990). Automated classification of agricultural crops on the basis of aerially scanned images. *Soviet Journal of Remote Sensing, 6*, 279–289.

Newcome, L. R. (2004): Unmanned Aviation: A Brief History of Unmanned Aerial vehicles. Reston, American Institute of Aeronautic and Astronautic, p. 166.

Plánka, L. (1984). Metoda leteckého snímkování z malých výšek. *Zprávy Geografického ústavu ČSAV, 21,* 3–12.

Plánka, L. (1987). The Use of Radio-controlled Aeromodels for Photography with the View of Remote Sensing of the Earth. In United Nations Training Course "Remote Sensing Applications to Geological Sciences", October 5–24, 1987, Dresden, Veröffentlichungen des Zentralinstituts für Physik der Erde, Potsdam, ZIPE, pp. 58–69.

Plánka, L. (1990). Spektrální charakteristiky zemědělských plodin a jejich využití k objektivizaci interpretace leteckých multi-spektrálních snímků. *Zprávy Geografického ústavu ČSAV, 27,* 39–49.

Przybilla, H.-J., & Wester-Ebbinghaus, W. (1979). Bildung mit ferngelenktem Kleinflugzeug. *Bildmessung und Luftbildwesen, Zeitschrift für Photogrammetrie und Fernerkundung, 47,* 137–142.

Stehlík, O. (1967). Contribution aux méthodes de l'ínvestigation de l'erosion du soil. In Travaux du symposium international de géomorphologie appliquée, Mai 1967, Bucureşti, pp. 69–75.

Stehlík, O. (1981). Dálkový průzkum Země z malých výšek. In Sborník přednášek „Konference o dálkovém snímání a dálkovém průzkumu Země, Brno, 14.–16. září 1981, Brno, Dům techniky ČSVTS, pp. 46–50.

Tollinger, V. (1988). Některé problémy sběru a využívání dat DPZ získaných z malých výšek. In Sborník referátů porady řešitelů Cílového projektu základního výzkumu č. 602 „Dálkový průzkum Země"konané ve dnech 24.–25.2.1988 v Březůvkách, 2 volumes, Praha/Březůvky, Ústav teorie informace a automatizace ČSAV/JZD Mír, pp. 23–31.

Žaloudík, J. (1990). Úloha dálkového průzkumu Země v Česko-slovenské koncepci výzkumu vodní složky krajiny. *Využití dálkového průzkumu Země ve vodním hospodářství* (pp. 27–44). Dům techniky ČSVTS: Praha.

(Received August 24, 2017, revised January 16, 2018, accepted February 16, 2018, Published online February 28, 2018)

Printed in the United States
By Bookmasters